T0345060

HANDBOOK OF KNOWLEDGE MANAGEMENT FOR SUSTAINABLE WATER SYSTEMS

Challenges in Water Management Series

Editor:

Justin Taberham

Independent Consultant and Environmental Advisor, London, UK

Other titles in the series:

URBAN WATER SECURITY
ROBERT C. BREARS
2017
ISBN: 9781119131724

WATER RESOURCES: A NEW WATER ARCHITECTURE
ALEXANDER LANE, MICHAEL NORTON AND SANDRA RYAN
2017
ISBN: 9781118793909

INDUSTRIAL WATER RESOURCE MANAGEMENT
PRADIP K. SENGUPTA
2017
9781119272502

HANDBOOK OF KNOWLEDGE MANAGEMENT FOR SUSTAINABLE WATER SYSTEMS

Edited by

MEIR RUSS

University of Wisconsin-Green Bay
Green Bay, WI, USA

This edition first published 2018

Registered Office(s)
John Wiley & Sons, Inc., 111 River Street, Hoboken, NJ 07030, USA
John Wiley & Sons Ltd, The Atrium, Southern Gate, Chichester, West Sussex, PO19 8SQ, UK

Editorial Office
The Atrium, Southern Gate, Chichester, West Sussex, PO19 8SQ, UK

For details of our global editorial offices, customer services, and more information about Wiley products visit us at www.wiley.com.

Wiley also publishes its books in a variety of electronic formats and by print-on-demand. Some content that appears in standard print versions of this book may not be available in other formats.

Library of Congress Cataloging-in-Publication Data:

Names: Russ, Meir, 1968- editor.
Title: Handbook of knowledge management for sustainable water systems /edited by Meir Russ.
Description: Hoboken : Wiley-Blackwell, 2018. | Series: Challenges in water management series | Includes bibliographical references and index. | Identifiers: LCCN 2017044560 (print) | LCCN 2017053007 (ebook) | ISBN 9781119271666 (pdf) | ISBN 9781119271673 (epub) | ISBN 9781119271635 (hardback)
Subjects: LCSH: Water-supply–Management. | Water resources development. | Sustainable development. | BISAC: TECHNOLOGY & ENGINEERING / Environmental / Water Supply.
Classification: LCC HD1691 (ebook) | LCC HD1691 .R877 2018 (print) | DDC 363.6/10684–dc23
LC record available at https://lccn.loc.gov/2017044560

Cover Design: Wiley
Cover Image: © RomoloTavani/Gettyimages

Set in 10/12pt MeliorLTStd by SPi Global, Chennai, India

Printed in Singapore by C.O.S. Printers Pte Ltd

10 9 8 7 6 5 4 3 2 1

"I have acquired insight from all my teachers"
(Psalms 119:99)
"My teaching shall drop as the rain ... (Deuteronomy 32:2)"

This book is dedicated to all the teachers from all walks of life and all over the world I have encountered and to those who will continue my teaching, for their wisdom, integrity, strong will, dedication and patience. Wisdom and water are more precious when shared among people and between generations.
May this be their legacy.

Contents

List of contributors

Laura Albareda School of Business and Management, Lappeenranta University of Technology, Finland; Department of Strategy, Deusto Business School, Deusto University, Avenidad de las Universidades, Bilbao, Spain

Stephen Atkins Otago Polytechnic of New Zealand, Dunedin, New Zealand

Chiara Bartolacci Department of Economics and Law, Università degli Studi di Macerata, Macerata, Italy

Isabelle Bourdon University of Montpelier, 34090 Montpellier, France

Jose Antonio Campos Department of Marketing, Deusto Business School, Department of Industrial Technologies, Faculty of Engineering, Deusto University, Bilbao, Spain

Vallari Chandna University of Wisconsin–Green Bay, Green Bay, Wisconsin, USA

Bin Chen State Key Laboratory of Water Environment Simulation, School of Environment, Beijing Normal University, Beijing 100875, China

Cristina Cristalli Department of Research for Innovation, Loccioni Group, Ancona, Italy

Delin Fang State Key Laboratory of Water Environment Simulation, School of Environment, Beijing Normal University, Beijing 100875, China

Lesley Gill Otago Polytechnic of New Zealand, Dunedin, New Zealand

Ali Guna School of Environment and Natural Resources, Renmin University of China, Beijing 100872, China

Janet G. Hering Eawag, Swiss Federal Institute for Aquatic Science and Technology, CH-8600 Dübendorf, Switzerland; Swiss Federal Institute of Technology (ETH) Zürich, IBP, CH-8092 Zürich, Switzerland; Swiss Federal Institute of Technology Lausanne (EPFL), ENAC, CH-1015 Lausanne, Switzerland

Eduard Hochbichler Institute of Silviculture, Department of Forest and Soil Sciences, University of Natural Resources and Life Sciences, Vienna, 1190 – Vienna, Austria

Daniela Isidori Department of Research for Innovation, Loccioni Group, Ancona, Italy

Ana Iusco University of Wisconsin–Green Bay, Green Bay, Wisconsin, USA

Chris Kimble KEDGE Business School, 13009 Marseille, France

Roland Koeck Institute of Silviculture, Department of Forest and Soil Sciences, University of Natural Resources and Life Sciences, Vienna, 1190 – Vienna, Austria

Kay Lion Otago Polytechnic of New Zealand, Dunedin, New Zealand

Shuk-Ching Li University College Utrecht, Maupertuusplein 1–320, Utrecht, The Netherlands

Fabien Martinez EM Normandie, Métis Lab, Dublin Campus, 19–21 Aston Quay, Dublin 2, Ireland

Federico Niccolini Department of Economics and Management, Università degli Studi di Pisa, Pisa, Italy

Lothar Nunnenmacher Eawag, Swiss Federal Institute for Aquatic Science and Technology, CH-8600 Dübendorf, Switzerland

Breanne Parr University of Wisconsin–Green Bay, Green Bay, Wisconsin, USA

Meir Russ University of Wisconsin–Green Bay, Green Bay, Wisconsin, USA

Marje Schaddelee Otago Polytechnic of New Zealand, Dunedin, New Zealand

Dajun Shen School of Environment and Natural Resources, Renmin University of China, Beijing 100872, China

Tonny Tonny Otago Polytechnic of New Zealand, Dunedin, New Zealand

Bilgehan Uzunca Utrecht University School of Economics, Kriekenpitplein 21–22, 3584 EC Utrecht, The Netherlands

Harald Vacik Institute of Silviculture, Department of Forest and Soil Sciences, University of Natural Resources and Life Sciences, Vienna, 1190 – Vienna, Austria

Harald von Waldow Eawag, Swiss Federal Institute for Aquatic Science and Technology, CH-8600 Dübendorf, Switzerland

Xuedong Yu School of Environment and Natural Resources, Renmin University of China, Beijing 100872, China

Series Editor Foreword – Challenges in Water Management

The World Bank in 2014 noted:

'Water is one of the most basic human needs. With impacts on agriculture, education, energy, health, gender equity, and livelihood, water management underlies the most basic development challenges. Water is under unprecedented pressures as growing populations and economies demand more of it. Practically every development challenge of the 21st century – food security, managing rapid urbanization, energy security, environmental protection, adapting to climate change – requires urgent attention to water resources management.

Yet already, groundwater is being depleted faster than it is being replenished and worsening water quality degrades the environment and adds to costs. The pressures on water resources are expected to worsen because of climate change. There is ample evidence that climate change will increase hydrologic variability, resulting in extreme weather events such as droughts floods, and major storms. It will continue to have a profound impact on economies, health, lives, and livelihoods. The poorest people will suffer most.'

It is clear there are numerous challenges in water management in the 21st Century. In the 20th Century, most elements of water management had their own distinct set of organisations, skill sets, preferred approaches and professionals. The overlying issue of industrial pollution of water resources was managed from a 'point source' perspective.

However, it has become accepted that water management has to be seen from a holistic viewpoint and managed in an integrated manner. Our current key challenges include:

- The impact of climate change on water management, its many facets and challenges – extreme weather, developing resilience, storm-water management, future development and risks to infrastructure
- Implementing river basin/watershed/catchment management in a way that is effective and deliverable
- Water management and food and energy security
- The policy, legislation and regulatory framework that is required to rise to these challenges
- Social aspects of water management – equitable use and allocation of water resources, the potential for 'water wars', stakeholder engagement, valuing water and the ecosystems that depend upon it

This series highlights cutting-edge material in the global water management sector from a practitioner as well as an academic viewpoint. The issues covered in this series are of critical interest to advanced level undergraduates and Masters Students as well as industry, investors and the media.

Justin Taberham, CEnv
Series Editor
www.justintaberham.com

Preface

The amount of water on earth is fixed. It does not change. The amount of fresh water that is available for human use is less than 3% of all available water on earth and even this amount is continuously contaminated by human acts. Also, the population in the world is growing rapidly. Just a few years ago the world population reached 8 billion people and in 20–30 years it will reach 9 billion. It means that the demand for potable water and water needed for agriculture, industry, and energy is growing faster than the natural water can be supplied and thus the gap between the demand for fresh water and the available natural resources is growing. Adding to that is the increasing need for cleaning the water and waste water collection and treatment in the world, which makes this book very interesting and beneficial for the generations to come.

To be able to face these important challenges, a holistic approach regarding the use of fresh water must be taken by decision-makers all over the globe, in developed, developing and undeveloped countries, to face the problems and overcome them in a sustainable way. To do so, such a holistic approach must contain the following activities in parallel:

- **Master planning for water and waste water** – the master plan should consider the future for at least the next 40 years, taking into consideration climate change, population growth, standard of living improvements, energy and food.
- **Saving water** – to include education, promotion, fighting against water loss in the piping system, advanced irrigation systems, regulation and water pricing.
- **Water reuse** – includes reasonable sanitation systems, waste water collection and treatment, and using the effluent for agriculture, industry and gardening.
- **Production of new resources of water** – such as cleaning the rivers, building water treatment plants, and building desalination brackish and sea water desalination plants.

This book addresses these steps and more. It will be of great help to those who will make intelligent decisions to stand up to these enormous challenges our generation, and generations to come, are and will be facing.

It will also assist to solve problems of transboundary waters and reduce the danger of wars over water and become a strong sustainable bridge towards peace between nations.

Abraham Tenne

Former head of the Desalination Division in the Israeli Water Authority and chairman of the WDA (water desalination administration) of the government of Israel

Introduction and a theoretical framework for Knowledge Management for Sustainable Water Systems

Meir Russ

University of Wisconsin–Green Bay, Green Bay, Wisconsin, USA

According to the World Health Organization (WHO), in 2009, about one fifth of the world's population lived in countries that did not have enough water for their use. By 2025, 1.8 billion people will experience absolute water scarcity, and by 2030, almost half the world will live under conditions of high water stress. Yet, only recently has the science of coupled human-water system been initiated (Partelow, 2016; Sivapalan & Blösch, 2015) and transdisciplinarity research utilized for societal sustainability problem-solving (Polk, 2014). But, the understanding of needs for data and knowledge transfer bridging organizational boundaries, and technological aspects that challenge the praxis of policy making and planning are paradoxically increasing or even worse, lacking (Cash *et al.*, 2003; Hinkel, Bots, & Schlüter, 2014; Polk, 2014; Thomson, El-Haram, Walton, Hardcastle, & Sutherland, 2007). For example, in a recent Water JPI (2014) paper presenting eight major water topics for Europe (Horizon 2020), while identifying the gaps and game changers, Knowledge Management was listed directly and indirectly in ALL of them. These real life issues and academic research gaps are the motivators for this handbook. Managing knowledge more effectively and efficiently might be a solution to many of the critical water issues that humans in the 21st century are, and will be, facing. Knowledge commons (e.g. Brewer, 2014) and virtual, digital spaces of learning (Niccolini *et al.*, 2018) provide some unexpected and surprising rays of hope, of what might happen when knowledge is created and managed well. The Israeli experience (Jacobsen, 2016; Siegel, 2015) is an illustration of a water miracle (not phantoms) that can happen in the desert, and by extension, everywhere, where and when people will set their minds to it.

Handbook of Knowledge Management for Sustainable Water Systems, First Edition. Edited by Meir Russ.

As the new knowledge-driven economy continues to evolve, knowledge is being recognized as a key asset and a crucial component of organizational, inter-organizational and national strategy. The ability to manage knowledge, therefore, is quickly becoming vital for securing and maintaining survival and success. As a result, organizations at all levels are investing heavily in Information Systems (IS) and/or Knowledge-Based Systems (KBS) technologies. Unfortunately, such investments frequently do not meet expected outcomes and/or returns. For the purpose of this handbook we will recognize Knowledge Management (KM) as a socio-technical phenomenon in which the basic social constituents such as person, team and organization require interaction with IS/KBS applications to support a strategy and add value to the organization (Russ, 2010) while improving the sustainability of a water system. Many organizations and their executives recognize that the critical source of sustainable competitive advantage is not only having the most ingenious product design, the most brilliant marketing strategy, or the most state-of-the-art production technology, but also having the ability to attract, retain, develop and manage its most valuable human assets (talent) and their knowledge and innovation. Furthermore, such an interaction of talent, processes and systems is what enables organizations to develop and manage knowledge for success.

Sustainability has been defined as economic development that meets the needs of the present generation without conceding the ability of future generations to meet their own needs (e.g., Russ, 2014b.) With growing pressure from customers and regulators toward environmental and social issues, organizations and governments at all levels are increasingly expected to shoulder greater responsibility for making sustainable development a reality. Recent droughts and water shortages worldwide and the advanced scientific understanding and documentation of the impact of demographic and economic forces on water footprint and embedded water make the need for sustainable development and management of water systems only more acute. This requires policy makers, planners and executives to balance economic, business, social and environmental concerns and outcomes. For that to happen, leaders need to quantify the relationships of all those aspects across different time horizons and link their organizational knowledge-base to strategy and outcomes so that they can consider the tradeoffs of different alternatives for their long-term success.

This book is envisioned as a manuscript that will provide a robust scientific foundation for an interdisciplinary, multi-perspective theory and practice of Knowledge Management in the context of, and for the advancement of, sustainable water systems. The book goes beyond the current literature by providing a platform for a broad scope of discussion regarding KM4SWS, and, more importantly, by encouraging an interdisciplinary/transdisciplinary fusion between diverse disciplines. Specifically, the call for proposals for this book solicited chapter proposals from a multidisciplinary array of scholars to discuss socio-hydrology sustainable systems within the present political (legislative), economic and technological context from a number of disciplines/perspectives, including: Economic Development, Financial,

Systems-Networks, IT/IS Data/Analytics, Behavioral, Social, Water Systems, Governance Systems and Related Ecosystems. Multi-level and multi-discipline chapters that synthesize diverse bodies of knowledge were strongly encouraged. When appropriate, plurality of empirical methods from diverse disciplines that can enhance the building of a holistic theory of Knowledge Management for Sustainable Water Systems were also encouraged.

While preparing for, and editing this book, a number of alternative theoretical frameworks were considered (e.g. Elliot, 2011). The multi-level framework that was adopted (described briefly below) is an amalgamation of a number of models reviewed (some are listed in the bibliography below) with the addition of models I developed regarding Knowledge Management over the last 20 years of teaching and studying the subject.

The first building block (see Figure 1) is the model of co-evolution of the Human Systems (political, economic, technological, social; see discussions and indicators in, for example: Partelow, 2016, Vogt, Epstein, Mincey, Fischer, & McCord, 2015; mostly based on Ostrom, 2009) and the Sustainable (in our case) Natural and Engineered Water Systems (see for example Sivapalan & Blöschl, 2015). Such co-evolution results, of course, from the impact human activities have (mediated by technology) on the systems and the responses and outcomes of the water systems to these activities. The co-evolutionary model (e.g. Sivapalan & Blöschl, 2015) was modified and enhanced by adding on the human system side: the complexity of the different potential units of analysis involved on the human systems side, starting with an individual, teams, organizations and then going up in complexity to inter-organization, national, regional and global units and scales. Each unit has its own learning complexity and more complex units, issues, and boundary management aspects (see excellent discussions of the importance of this complex management in Cash *et al.*, 2003). On the sustainable water system side, the different levels of the systems were added (e.g. household, city, river basins, etc.). Each one of them is connected to the framework by models that are used and/or understood by the human actors (see the interesting discussion in Sivapalan & Blöschl, 2015, about two models: stylized and comprehensive). Finally, Knowledge Management (KM) was added at the heart of the Human Systems' section and the Co-evolution's section (the two KMs are of course related and intertwined).

The second level of the model is the construct of Knowledge Management (see Figure 2). Here, the model developed by Russ, Fineman, and Jones, (2010) was used, with focus on the actors, (or talent), the process, or specifically the learning and decision-making, and the systems, or in this case the knowledge based systems. The majority of the chapters in this book touch on all three factors and illustrate different aspects (e.g. content and process) of KM.

In the third level of the model, each one of the three constructs used as building blocks for KM (listed above) was broken down into its specific models (see Figure 3.a–e). For example, learning might focus on tacit knowledge using the Kolb active learning model (1976), or on codified learning using the

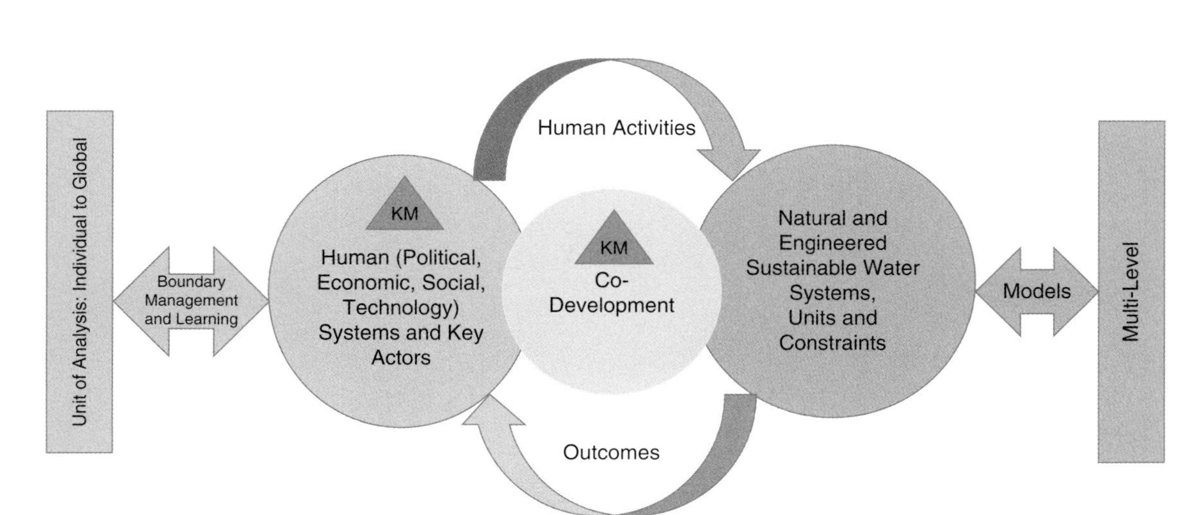

Figure 1 The coevolution of human and water systems and Knowledge Management.

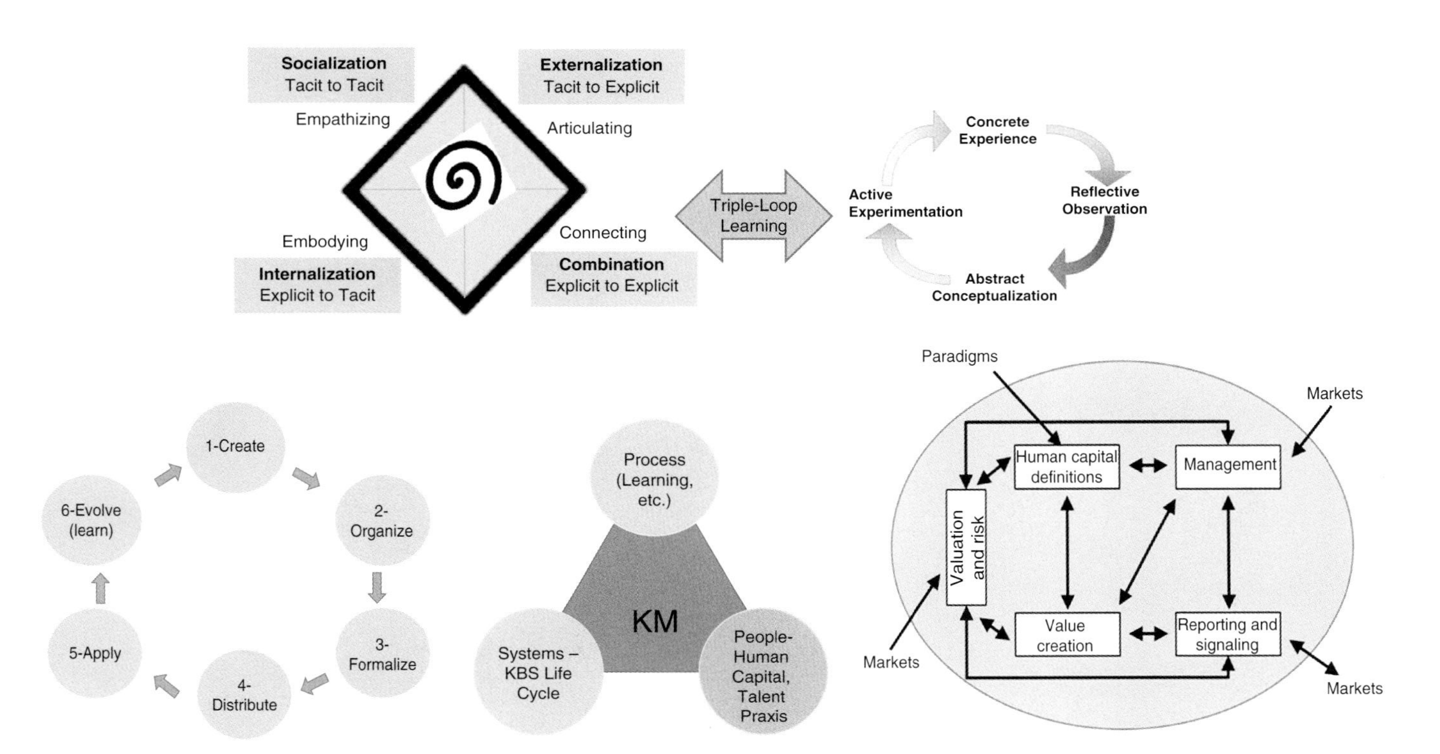

Figure 2 Knowledge Management in Sustainable Water Systems.

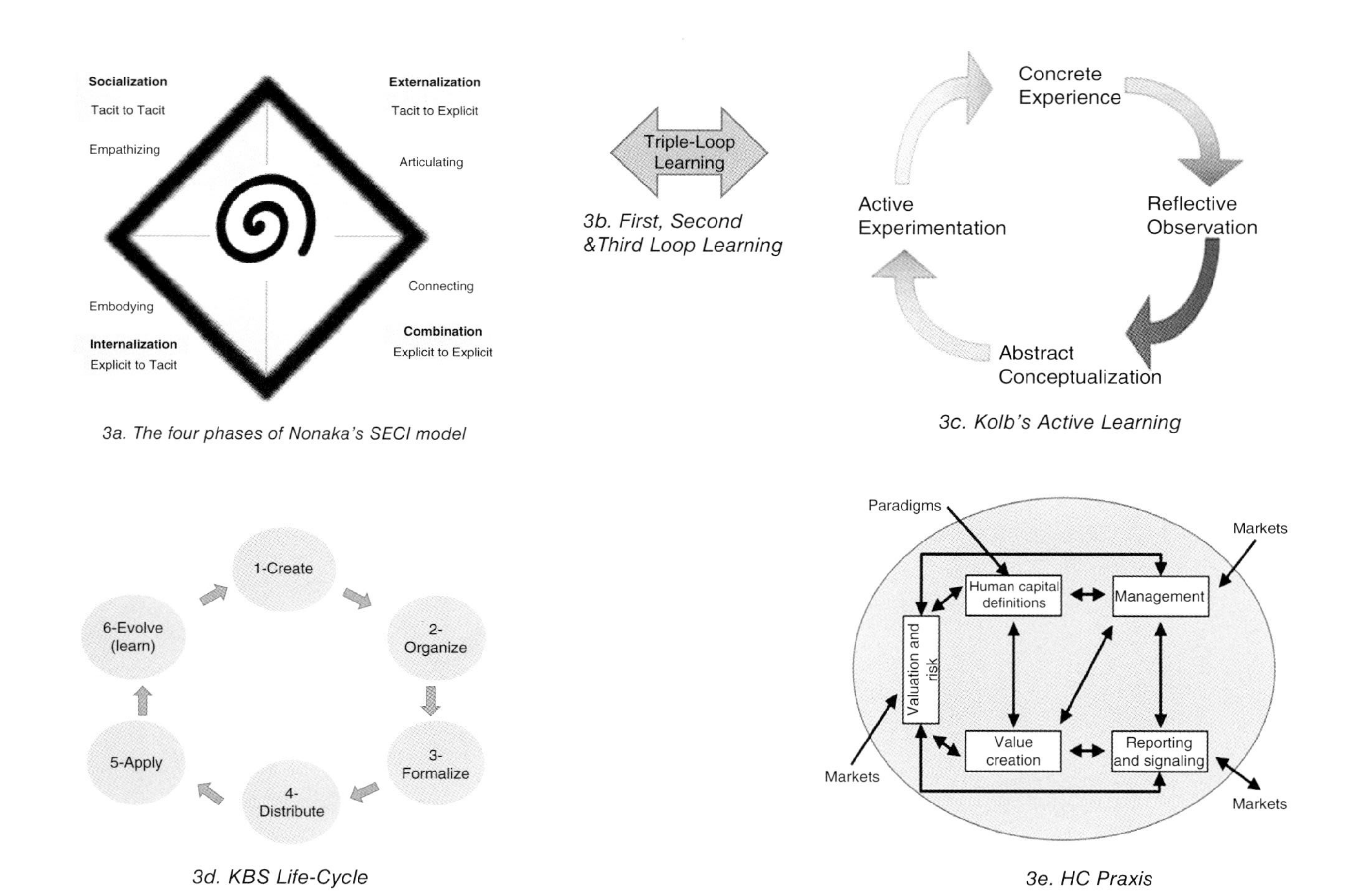

Figure 3 The three elements of Knowledge Management in Sustainable Water Systems (detailed).

virtual Ba model illustrated by Niccolini *et al.* (See Chapter 11), or any mix of the two; or others as appropriate for the case; all (potentially) using up to the three feedback loops of learning (e.g., Argyris' double-loop learning, 2002; or the review in Tosey, Visser, & Saunders, 2012 of triple loop learning). The human actors', talent was modeled using the HC praxis model (Russ, 2014a) and the Knowledge-Based-Systems (KBS) using the six life cycle stages of KBS (e.g., Russ, Jones, & Jones, 2008), including the sustainability aspect of the KBS as well as consideration (Elliot, 2011).

The complexity of the reality of KM in SWS are overwhelming, as they are illustrated in the chapters in this book. Such complexity is a result of the nature of Knowledge Management which could cut across ALL levels of the model, as well as across the unit of analysis in the water systems. Add to that the complexity of the diverse scientific areas and the diverse styles of learning and decision-making of the different actors, and you can see a complex networked system at its best.

But the truth of the matter is that one aspect is still missing from this analysis and must be added to the proposed framework, thus adding the fourth level, time (see Figure 4). Again and again, while teaching my KM classes, consulting with clients, researching and reading other practitioner and academic research, I was perplexed by the failures of all the constituencies to understand the importance of time, its complexities for strategic decision-making, managing knowledge or human capital, among many other aspects. Time is one of the hidden assumptions (dimensions) we have and use continually without giving it a second thought.

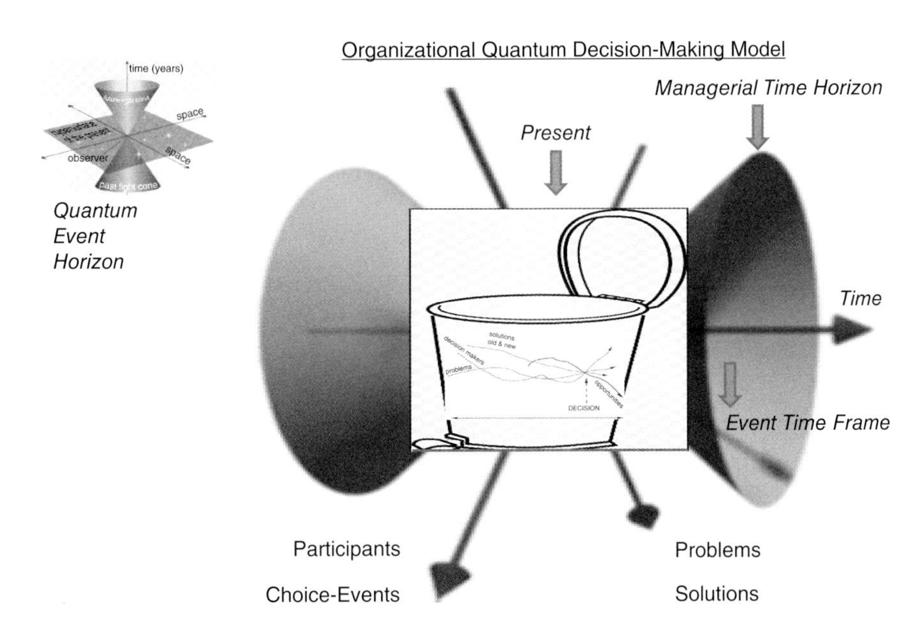

Figure 4 The time aspect of the Organizational Quantum Decision-Making Model.

In practice, the misalignment of time horizons (present and future) and time frames of events, makes an enormous difference, but is rarely explicitly identified as an issue and/or studied (see a rare very recent exception in Myllykoski, 2017). In our context of Knowledge Management for Sustainable Water Systems, the key players who take actions or make decisions, not only might have a diverse set of expertise, understating and knowledge of the subject at hand, but they also operate in a different time "space". Politicians' time horizons of the relevant future for their decision-making is different from that of the farmer, the hydrologist and the weather scientist. Their understanding of an event (and its time frame) and the implication the event might have within a complex system, including the impact of time-lags and complex feedback loops, could be startlingly different and diverse (from others) within the realm of their intentions, goals and their time horizon. This factor (of time) can explain by itself why knowledge is not used effectively regarding important aspects of the sustainability of water systems. Corralling all the different actors into a single space of knowledge and coherent time for the purpose of advancing Sustainable Water Systems can happen rarely if at all. One case in which all of this might potentially happen is in a major disaster. Recent history should tell us that even in the vast majority of the cases of disaster this is not sufficient to create such a coherent space. To illustrate the praxis of such situations, an updated model of the "garbage can model of decision-making" (Cohen, March, & Olsen, 1972) is proposed here, where time is a multi-dimensional construct having a synchronized (or not) time horizon, time frame, and event-time, or what I would define as the "quantum model of organizational time". This model advances the model of time described by Myllykoski (2017) (building on Hernes, 2014) in which she described the past and the future as a stream of events, that get their "true" meaning at the present time, resulting from a stream of events, creating or enabling a decision to be made, seeing time as agentic (p. 23). Events, or the bits of information perceived by the observer that are remarked by the actor as events, are seen as collapsing at the time of the decision. As such, it enables us to freeze an understanding of the past, from the present perspective and the planning for the future, confirming a present rational for the decision regarding the future (Tsoukas, 2016). Viewing time as agentic, and seeing the stream of bits of information coming from the past and going toward the future, collapsing at a time of a decision into one interpretation, brings Schrödinger's cat from the quantum realm into individual decision-making, and results in what I would call, the "individual quantum model of decision-making". Adding the complexity of multiple key actors with different time frames, etc. results in an "Organizational Quantum Model of Decision-Making", which is illustrated in Figure 4. Finally, one plausible explanation for the success of role playing (Sivapalan & Blöschl, 2015) and simulation-based learning (Deegan *et al.*, 2014) in a complex context as described here, is because it allows for the time horizons of the participating individuals and the time frame of the event they engage with to become coherent, enabling an improved process of decision-making.

The chapters in this book not only illustrate the framework proposed above, and suggest new venues for improving our water systems; more than that, they illuminate opportunities to eliminate the risks of a thirsty world.

Acknowledgment

The call for chapters for this book stimulated the authors to synthesize multi-level and multi-discipline diverse bodies of knowledge by encouraging an interdisciplinary fusion between diverse disciplines. The authors were invited to contribute chapters to the book based on proposals approved by the editor. Each complete chapter received external, blind review in addition to the editor review. The editor thanks Kelly Anklam for her assistance in editing this introduction and a number of chapters. He wishes to thank also Dr. Vallari Chandna for editing a number of chapters, as well as the reviewers for their in-depth reviews. The editor would also like to thank Prof. Knut Ingar Westeren for his help with the indexing of this book, as well as Dr. Robert Wenger and Dr. Jack Day for sharing their professional network and making this volume a richer resource. Finally the editor wishes to thank the Philip J. and Elizabeth Hendrickson Professorship in Business at UW–Green Bay for partial financial support. As always, all mistakes are his.

Bibliography

Argyris, C. (2002). Double-loop learning, teaching, and research. *Academy of Management Learning & Education, 1*(2), 206–218.

Baldassarre, G. D., Kooy, M., Kemerink, J. S., & Brandimarte, L. (2013). Towards understanding the dynamic behaviour of floodplains as human-water systems. *Hydrology and Earth System Sciences, 17*(8), 3235–3244.

Baldassarre, G. D., Viglione, A., Carr, G., Kuil, L., Salinas, J. L., & Blöschl, G. (2013). Socio-hydrology: Conceptualising human-flood interactions. *Hydrology and Earth System Sciences, 17*(8), 3295–3303.

Blackburn, W. (2007). *The sustainability handbook: The complete management guide to achieving social, economic and environmental responsibility.* Environmental Law Institute. ISBN: 978-1585761029

Bogardi, J. J., & Kundzewicz, Z. W. (Eds) (2002). *Risk, reliability, uncertainty, and robustness of water resource systems.* Cambridge, UK: Cambridge University Press.

Brewer, J. (2014). Harvesting a knowledge commons: Collective action, transparency, and innovation at the Portland Fish Exchange. *International Journal of the Commons, 8*(1), 155–178.

Cash, D. W., Clark, W. C., Alcock, F., Dickson, N. M., Eckley, N., Guston, D. H., & Mitchell, R. B. (2003). Knowledge systems for sustainable development. PNAS *100*:8086–8091. Downloaded August 19, 2013 from http://www.pnas.org/content/100/14/8086.full.pdf+htm

Cohen, M. D., March, J. G., & Olsen, J. P. (1972). A garbage can model of organizational choice. *Administrative Science Quarterly, 17*(1), 1–25.

Deegan, M., Stave, K., MacDonald, R., Andersen, D., Ku, M., & Rich, E. (2014). Simulation-based learning environments to teach complexity: The missing link in teaching sustainable public management. *Systems, 2*(2), 217–236.

Eccles, R. G., & Serafeim, G. (2013). The performance frontier. Innovating for a sustainable strategy. *Harvard Business Review*, May, 50–60.

Elliot, S. (2011). Transdisciplinary perspectives on environmental sustainability: A resource base and framework for IT-enabled business transformation. *MIS Quarterly, 35*(1), 197–236.

EPA. (2012, October). *Resource guide to effective utility management and lean.* Available at http://water.epa.gov/infrastructure/sustain/upload/EUM-and-Lean-Resource-Guide.pdf

EPA. (2014, April*). Moving toward sustainability: Sustainable and effective practices for creating your water utility roadmap.* Available at http://water.epa.gov/infrastructure/sustain/upload/Sustainable-Utilities-Roadmap-12-10-14_508.pdf

FEWresources.org. (xxxn.d.). *Water scarcity.* Available at http://www.fewresources.org/water-scarcity-issues-were-running-out-of-water.html. (Updated 06/15/15; downloaded Oct 5, 2015).

Hernes, T. (2014). *A process theory of organization.* Oxford, UK: Oxford University Press.

Hernes, T. (2017). In Langley, A. & Tsoukas, H. (Eds). *The SAGE handbook of process organization studies* (pp. 601–606). London: Sage Publications.

Hinkel, J., Bots, P. W., & Schlüter, M. (2014). Enhancing the Ostrom social-ecological system framework through formalization. *Ecology and Society, 19*(3). Article 51.

Hodgson, A. (2010, September 13). Global water shortages will pose major challenges. *Euromonitor International.* Available at http://blog.euromonitor.com/2010/09/special-report-global-water-shortages-will-pose-major-challenges.html

Hoekstra, A.Y. (2015). The water footprint of industry. In J. J. Klemes (Ed.) *Assessing and measuring environmental impact and sustainability* (pp. 221–254). Oxford, UK: Butterworth-Heinemann (Elsevier).

Kasemir, B., Jäger, J., Jaeger, C. C., Gardner, M. T. (2003). *Public participation in sustainability science, a handbook.* Cambridge, UK: Cambridge, University Press.

Khatri, K., & Vairavamoorthy, K. (2011). *A new approach of risk analysis for complex infrastructure systems under future uncertainties: A case of urban water systems.* In *Reston, VA: ASCE copyright Proceedings of the First International Symposium on Uncertainty Modeling and Analysis and Management (ICVRAM 2011), and the Fifth International Symposium on Uncertainty Modeling and Analysis (ISUMA); Hyattsville, Maryland, April 11 to 13, 2011, d 20110000.* American Society of Civil Engineers.

Khatri, K. B. (2013). *Risk and uncertainty analysis for sustainable urban water systems.* UNESCO-IHE, Institute for Water Education.

Kolb, D. A. (1976). Management and the learning process. *California Management Review, 18*(3), 21–31.

Lago, P., Koçak, S. A., Crnkovic, I., & Penzenstadler, B. (2015). Framing sustainability as a property of software quality. *Communications of the ACM*, *58*(10), 70–78.

Mounce, S. R., Brewster, C., Ashley, R. M., & Hurley, L. (2010). Knowledge Management for more sustainable water systems. *Journal of Information Technology in Construction, 15*, 140–148.

Mukheibir, P., Howe, C., & Gallet, D. (2015). Institutional issues for integrated "one water" management. *Water Intelligence Online, 14*, 9781780407258. Available at http://www.waterrf.org/PostingReportLibrary/4487a.pdf

Myllykoski. J. (2017). *Strategic change emerging in time*. Dissertation. University of Oulu. G91.

Newell, S., Robertson, M., Scarbrough, H., & Swan, J. (2009). *Managing knowledge work and innovation*. New York, NY: Palgrave Macmillan.

Newig, J., Pahl-Wostl, C., & Sigel, K. (2005). The role of public participation in managing uncertainty in the implementation of the Water Framework Directive. *European Environment, 15*(6), 333–343.

Niccolini, F., Bartolacci, C., Cristalli, C., & Isidori, D. (2018) Virtual and inter-organizational processes of knowledge creation and Ba for sustainable management of rivers. In M. Russ, (Ed). *Handbook of Knowledge Management for Sustainable Water Systems* (pp. 261–285). Hoboken, NJ: John Wiley & Sons Ltd.

Ostrom, E. (2009). A general framework for analyzing sustainability of social-ecological systems. *Science, 325*, 419–422.

Pahl-Wostl, C. (2002). Towards sustainability in the water sector – The importance of human actors and processes of social learning. *Aquatic Sciences, 64*(4), 394–411.

Pahl-Wostl, C., Holtz, G., Kastens, B., & Knieper, C. (2010). Analyzing complex water governance regimes: The management and transition framework. *Environmental Science & Policy, 13*(7), 571–581.

Pahl-Wostl, C., Mostert, E., & Tàbara, D. (2008). The growing importance of social learning in water resources management and sustainability science. *Ecology and Society, 13*(1), Article 24.

Pahl-Wostl, C., Sendzimir, J., Jeffrey, P., Aerts, J., Berkamp, G., & Cross, K. (2007). Managing change toward adaptive water management through social learning. *Ecology and Society, 12*(2), Article 30.

Partelow, S. (2016). Coevolving Ostrom's social–ecological systems (SES) framework and sustainability science: Four key co-benefits. *Sustainability Science, 11*(3), 399–410.

Polk, M. (2014). Achieving the promise of transdisciplinarity: A critical exploration of the relationship between transdisciplinary research and societal problem solving. *Sustainability Science, 9*(4), 439–451.

Raine, S. R., Meyer, W. S., Rassam, D. W., Hutson, J. L., & Cook, F. J. (2007). Soil–water and solute movement under precision irrigation: Knowledge gaps for managing sustainable root zones. *Irrigation Science, 26*(1), 91–100.

Rockström, J., Karlberg, L., Wani, S. P., Barron, J., Hatibu, N., Oweis, T., & Qiang, Z. (2010). Managing water in rainfed agriculture – The need for a paradigm shift. *Agricultural Water Management, 97*(4), 543–550.

Russ, M. (Ed.) (2010). *Knowledge Management strategies for business development*. Hershey, PA: Business Science Reference.

Russ, M. (2014a). What kind of an asset is human capital, how should it be measured, and in what markets? In M. Russ (Ed.) *Management, valuation, and risk for human capital and human assets: Building the foundation for a multi-disciplinary, multi-level theory* (pp. 1–33). New York, NY: Palgrave-Macmillan.

Russ, M. (2014b). *Homo Sustainabiliticus* and the "new gold". In M. Russ (Ed.) *Value creation, reporting, and signaling for human capital and human assets: Building the foundation for a multi-disciplinary, multi-level theory* (pp. 1–6). New York, NY: Palgrave-Macmillan.

Russ, M., Fineman, R., & Jones, J. K. (2010). Conceptual theory: What do you know? In M. Russ, (Ed.) *Knowledge Management strategies for business development* (pp. 1–22). Hershey, PA: Business Science Reference.

Russ, M., Jones, J. G., & Jones, J. K. (2008). Knowledge-based strategies and systems: A systematic review. In M. Lytras, M. Russ, R. Maier & A. Naeve (Eds). *Knowledge Management strategies: A handbook of applied technologies* (pp. 1–62). Hershey, PA: IGI Publishing.

Russ, M., & Jones, J. K. (2011). Knowledge Management's strategic dilemmas typology. In D. G. Schwartz & D. Te'eni (Eds). *Encyclopedia of Knowledge Management* (2nd edn) (pp. 804–821). Hershey, PA: IGI Reference.

Siegel, S. M. (2015). *Let there be water: Israel's solution for a water-starved world*. New York, NY: Macmillan.

Sivapalan, M., & Blösch, G. (2015). Time scale interactions and the coevolution of humans and water. *Water Resources Research, 51*, 6988–7022. doi:10.1002/2015WR017896.

Sivapalan, M., Savenije, H. H., & Blöschl, G. (2012). Socio-hydrology: A new science of people and water. *Hydrological Processes*, *26*(8), 1270–1276.

Stember, M. (1991). Advancing the social sciences through the interdisciplinary enterprise. *The Social Science Journal*, *28*(1), 1–14.

Thomson, C., El-Haram, M., Walton, J., Hardcastle, C., & Sutherland, J. (2007*). The role of Knowledge Management in urban sustainability assessment.* Submitted to SUE-MOT International Conference on Whole Life Urban Sustainability and its Assessment, 27–29 June 2007. Downloaded August 19, 2013 from http://download.sue-mot.org/Conference-2007/Papers/Thomson.pdf

Tosey, P., Visser, M., & Saunders, M. N. (2012). The origins and conceptualizations of 'triple-loop' learning: A critical review. *Management Learning, 43*(3), 291–307.

Tsoukas, H. (2016) Don't simplify, complexify: From disjunctive to conjunctive theorizing in organization and management studies. *Journal of Management Studies*. Accepted article. doi: 10.1111/joms.12219.

U. S. Army Corps of Engineers. (2014). *Building strong collaborative relationships for a sustainable water resources future: Understanding integrated water resources management.* Available at http://aquadoc.typepad.com/files/iwrm-report-jan2014.pdf

Vacik, H., & Lexer, M. J. (2001). Application of a spatial decision support system in managing the protection forests of Vienna for sustained yield of water resources. *Forest Ecology and Management*, *143*(1), 65–76.

Vogt, J. M., Epstein, G. B., Mincey, S. K., Fischer, B. C., & McCord, P. (2015) Putting the "E" in SES: unpacking the ecology in the Ostrom social–ecological system framework. *Ecology and Society, 20*(1), 55. 10.5751/ES-07239-200155

Wallington, T. J., Maclean, K., Darbas, T., & Robinson, C. J. (2010). *Knowledge – Action systems for integrated water management: National and international experiences, and implications for South East Queensland.* Urban Water Security Research Alliance Technical Report No. 29. Available at http://www.urbanwateralliance.org.au/publications/UWSRA-tr29.pdf

Water JPI. (2014, June 16). *Water JPI position paper on the Horizon 2020 Societal Challenge 5 2016-2017 Work Programme: Response to the stakeholder consultation.* Available at http://www.waterjpi.eu/

Winz, I., Brierley, G., & Trowsdale, S. (2009). The use of system dynamics simulation in water resources management. *Water Resources Management*, *23*(7), 1301–1323.

Ziemba, E. (Ed.) (2016). *Towards a sustainable information society: People, business and public administration perspectives.* Cambridge, UK: Cambridge Scholars Publishing.

Zygmunt, J. (2007). *Hidden waters.* London, UK: Waterwise. Available at http://waterfootprint.org/media/downloads/Zygmunt_2007_1.pdf

PART 1

Organizational and Administrative Aspects of Knowledge Management for Sustainable Water Systems

1 Perspectives from a water research institute on Knowledge Management for Sustainable Water Management

Janet G. Hering[1,2,3], Lothar Nunnenmacher[1] and Harald von Waldow[1]

[1] *Eawag, Swiss Federal Institute for Aquatic Science and Technology, CH-8600, Dübendorf, Switzerland*
[2] *Swiss Federal Institute of Technology (ETH) Zürich, IBP, CH-8092, Zürich, Switzerland*
[3] *Swiss Federal Institute of Technology Lausanne (EPFL), ENAC, CH-1015, Lausanne, Switzerland*

Introduction

Sustainable Water Management (SWM) is a domain in which the aspects of *knowledge as a public good* (van Kerkhoff, 2013) are intuitively obvious. This derives from the fact that water management has been a core responsibility of civil society throughout history (Mays, 2010; Sedlak, 2014). Here, water management is defined broadly to encompass:

- Water supply and use (e.g., for drinking water, irrigation, aesthetics, fire protection, etc.).
- Water resources protection (e.g., for fisheries, recreation and provision of other ecosystem services).
- Management of watercourses and infrastructure (e.g., for navigation, flood protection, water and wastewater conveyance).

Handbook of Knowledge Management for Sustainable Water Systems, First Edition. Edited by Meir Russ.

With this broad definition, water management must necessarily address competing uses and interests, taking into account natural, technical and societal factors (Hering & Vairavamoorthy, 2017). Whether water management can be characterized as sustainable (in a given context) will depend on whether deficits in one type of factor, such as limited natural water availability, can be offset by other factors (e.g., demand management and/or infrastructure for water storage). Conversely, sustainability is unlikely to be realized when deficits are aggravated by, for example, limited human capacity and economic resources. Meaningful approaches to SWM can only be identified and implemented in local and regional contexts (Hering *et al.*, 2015).

A huge, even overwhelming, array of knowledge is available that is highly relevant to SWM. This knowledge, which derives not only from the natural, social and engineering sciences but also from practical experiences in water management, reflects a massive investment of resources on the part of civil society. Having access to SWM-relevant knowledge could allow society to conserve its resources by using (rather than re-inventing) existing knowledge. It might also be possible to avoid unintended, adverse consequences by understanding past failures and, as well, the factors that contributed to past successes and might limit their replication under different conditions. In addition, tapping SWM-relevant knowledge could contribute to understanding what conditions would be amenable to "leapfrogging" as compared with those that are likely to require the progressive development of systems or approaches for water management (Briscoe, 2011). Ultimately, the goal of improving access to SWM-relevant knowledge would be to provide the most useful knowledge base for decision-making (Cash *et al.*, 2003; Cornell *et al.*, 2013; Hering, 2015; Martinuzzi & Sedlacko, 2016; van Kerkhoff, 2013; van Kerkhoff & Lebel, 2006; van Kerkhoff & Szlezak, 2016). This need is especially pressing in light of the commitment of the nations of the world to achieve the Sustainable Development Goals by 2030 (UN, 2015). Water supply, sanitation and the quality of receiving waters are addressed explicitly in Goal 6 "Ensure access to water and sanitation for all" and also in the Goal 11 targets to reduce the impact of water-related disasters and the adverse per capita impacts of cities.

In this chapter, issues of Knowledge Management (KM) as applied to SWM are examined from the perspective of a single organization, the Swiss Federal Institute of Aquatic Science and Technology (Eawag). Constraining the scope of this analysis to Eawag is intended to allow specific needs and demands with regard to KM to be clearly identified and characterized, not only for issues arising within Eawag but also issues that arise for Eawag as a member of the SWM community. This approach is intended to provide a basis for assessing measures to meet KM needs as well as barriers to KM that would need to be circumvented and, ideally, for proposing new approaches that could be pursued within Eawag or in cooperation with external partners.

1.1 The setting – Eawag's funding, scope and mandate

Eawag is a publically-funded research institute that receives approximately 75% of its funding through direct support from the Swiss federal government. With this level of direct support, it is particularly incumbent upon Eawag to use its resources efficiently; effective KM is an integral aspect of efficient resource utilization.

The scope of Eawag's activities is legally defined[1] and encompasses the following topics:

- Chemistry, physics, biology and microbiology of water.
- Ecology of aquatic systems.
- Drinking water and wastewater treatment technologies.
- Sustainable management of water supply and resources and of the water environment.

Eawag's thematic scope focuses mainly on water quality, which reflects both the history of the organization (Eawag, 2011) and the abundance of water resources in Switzerland. Issues related to flooding are mainly addressed by the Swiss Federal Institute for Forest, Snow and Landscape Research (WSL) though Eawag collaborates with WSL on projects dealing with aspects of flooding and flood protection that pertain to water quality and aquatic ecology.

Eawag is one of six institutions belonging to the Domain of the Swiss Federal Institutes of Technology (ETH Domain). These six institutions share a common mandate in research, education and expert consulting. Since Eawag does not grant degrees, it fulfills its mandate in tertiary education in cooperation with the Swiss Federal Institutes of Technology in Zurich (ETH Zurich) and Lausanne (EPFL), the Swiss Cantonal Universities and Universities of Applied Sciences, and also with international, degree-granting partners.

Eawag's expert consulting in context of industrialized countries is mainly focused on Switzerland with further engagement in Europe resulting from common interests and projects. Many of Eawag's applied projects in Switzerland are funded by the federal, cantonal and local authorities and are often conducted jointly with water and wastewater utilities. Eawag conducts expert consulting and research in low- and middle-income countries (LMICs) with support from the Swiss Agency for Development and Cooperation (SDC), the State Secretariat for Economic Affairs (SECO) and charitable foundations.

Because of its engagement with ETH Zurich, EPFL and the Swiss Cantonal Universities (which include joint professorial appointments), Eawag's research benefits greatly from the participation of doctoral students as well as Masters and

[1] https://www.admin.ch/opc/de/classified-compilation/20032108/index.html (in German).

Table 1.1 Eawag statistics (2016)

Total staff (number)[a]		472
Scientific staff[b] (number)		306
with adjunct professorships	12	
with tenured or tenure-track appointments	76	
Joint professors[c] (number)		17
Supervised doctoral dissertations[d] (number)		144
Supervised Bachelor's and Master's theses (number)		136
Base funding (CHF)		61,499,000
External funding (CHF)		17,627,000
Peer-reviewed (ISI) publications (number)		408
Other (non-ISI) publications[e] (number)		146

a) This number of staff corresponds to 422 FTE due to part-time appointments.
b) The scientific staff includes those doctoral students who are employed by Eawag.
c) With the exception of Eawag's Director, all joint professors are formally employed by the partner university.
d) This number includes dissertations by doctoral students who are Eawag employees as well as those who are directly supervised by an Eawag researcher and guest students spending >50% of their working time at Eawag.
e) This number corresponds to 74 non-ISI publications (which includes non-reviewed publications in trade journals) as well as 72 publications in the category of reports, books, book chapters and proceedings.

Source: Eawag Annual Report (http://www.eawag.ch/en/aboutus/portrait/annual-report/).

Bachelors students conducting projects and thesis research. The short residencies of students (especially as compared with Eawag's permanent scientific staff) poses specific challenges for KM. Summary statistics (for 2016) are presented in Table 1.1.

1.2 Understanding SWM-related demands for KM at Eawag

Eawag, as an organization with a focus on applied research, must manage a broad range of knowledge and information. Here the focus is on demands for KM that relate directly to SWM; knowledge management relating to organizational operation and performance is excluded.[2] In addition, SWM as an area of research shares many KM needs with other applied (and even fundamental) research areas. Some KM tools, structures and/or frameworks are generally applicable (i.e., to topics beyond SWM).

[2]Eawag reports operational statistics to the Board of the ETH Domain, which publishes summary statistics for all ETH Domain institutions in its Annual Reports (https://www.ethrat.ch/en/annualreport_2016).

An important aspect of SWM is that relevant knowledge includes not only technical knowledge (which derives mainly from the natural, social and engineering sciences) but also experiential and practical knowledge, which is often tacit (Tsoukas, 2011). SWM knowledge that derives from practical experience often develops during the interaction with stakeholders or practitioners in the context of applied projects or expert consulting. The tendency for such knowledge to remain as tacit, informal knowledge (which is held by individuals or, in the best cases, within networks) limits its wider application in SWM. An important challenge (but also opportunity) for Eawag is *how to articulate and access experiential and practical knowledge gained by its researchers through interactions with practice.*

Demands for SWM-related knowledge can also have different drivers. Internally-driven demands include the need for Eawag researchers to position proposed SWM projects and/or interpret the results of SWM projects in the context of previous relevant work. Analysis of gaps in existing SWM-related knowledge is generally needed as a justification for further research. Externally-driven demands arise from needs in engineering practice, water resources management and/or policy-making that may be linked with concrete projects or with formal or informal consultation. Externally-driven demands are often highly contextual, posing the challenge of *how to incorporate context adequately while still being able to identify and extract generalizable aspects.*

1.3 Current measures to meet SWM-related demands for KM at Eawag

Eawag's capacities for knowledge management have expanded over time, in concert with changing demands and opportunities. As related technologies (mainly in IT) develop, it is a constant challenge to decide on and make appropriate investments (including the decision to develop new capacities in house vs. out-sourcing). Perhaps even more challenging, though, is fostering new attitudes and norms regarding, for example, data management and data sharing to comply in a meaningful way with changing external expectations.

1.3.1 Data management

The rapidly-developing trend toward more systematic data management is a response to evolving norms in the scientific community (Fecher *et al.*, 2015; Soranno *et al.*, 2015), which in turn reflect both the expectations of the broader society for return on its investment in scientific research and the availability of tools for data management (Joseph, 2016). The fact that much scientific research is publically funded underlies growing requirements for measures ranging from the specification of data availability to public data archiving (Kowalczyk and Shankar, 2011; Vines *et al.*, 2013).

Data Management Plans (DMPs) are increasingly required by funding agencies and DMP tools[3] are being developed to support them. An accord for "Open Data in a Big Data World" has been endorsed by national scientific academies, scientific organizations, universities and research institutes (http://www.science-international.org/). Some publishers have adopted the Joint Data Archiving Policy (http://datadryad.org/pages/jdap) or comparable policies that require the publication-supporting data to be publicly available (Michener, 2015). This is also of direct interest to researchers because of the citation benefits associated with open data (Piwowar and Vision, 2013). However, the availability of reliable public repositories for data deposition varies widely among the scientific disciplines in which Eawag is engaged and the quality of deposited data is subject to criticism (Roche *et al.*, 2015). The Registry of Research Data Repositories, re3data.org (Pampel *et al.*, 2013), listed 1805 research data repositories when this chapter was written. Zenodo (https://zenodo.org) and Dryad (https://datadryad.org) are among the most used general-purpose repositories at Eawag. Options are increasing to publish data sets as products in journals such as Earth System Science Data or Scientific Data and to assign digital object identifiers (DOIs) to data sets (e.g., https://www.datacite.org/) or to publish single observations after peer review (https://www.sciencematters.io/) or "nanopublications" (Groth *et al.*, 2010).

Since 1972, Eawag has collaborated with the Swiss Federal Office of the Environment (FOEN) and the WSL in the National Long-term Surveillance of Swiss Rivers (NADUF) program. Through this program, standardized monitoring data on water quality are made available through the Eawag website (http://www.eawag.ch/en/department/wut/main-focus/chemistry-of-water-resources/naduf/) to accompany hydrologic data available through the FOEN website. Note that other water data portals (i.e., websites with data or links to monitoring data on water resources including flow, water quality and integrated water resources management) (Hering, 2017) are an important resource for Eawag researchers.

Currently, a central research data (CRD) repository is being established at Eawag to provide intermediate- and long-term storage for datasets (Figure 1.1). This project is intended to support compliance with standards of good scientific practice for data retention and to enable the long-term archiving of unique, unrepeatable measurements. An additional anticipated benefit is increased collaboration within Eawag, including re-use of data that are discoverable through appropriate search queries. Data sets can be filtered by location, sampling periods, keywords and terms from several controlled vocabularies.

It is hoped that these ancillary benefits will help to motivate individual researchers to perform the organization, quality control and annotation of their

[3]Data Management Plan (DMP) Tool, Univ. of California (https://dmp.cdlib.org/); DMP Online, Digital Curation Centre (https://dmponline.dcc.ac.uk/).

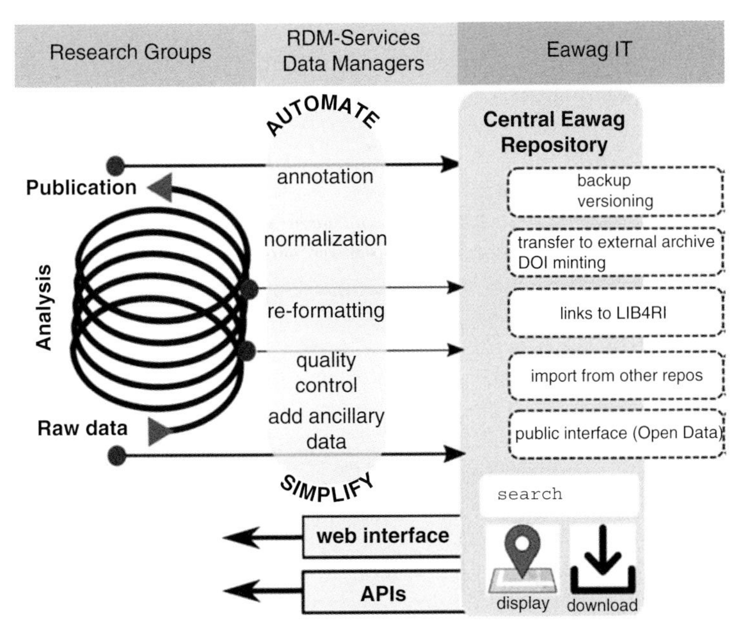

Figure 1.1 Schematic illustration of Eawag's research data management system. © Eawag. Used with permission.

data that is needed for a reliable research data repository. Additional activities related to the establishment of the repository include the provision of guidelines, training and ancillary tools and services to support researchers with their data-management and data-handling needs and thus to make usage as efficient and painless as possible.

Eawag's multidisciplinary environment (in which highly heterogeneous types of data are generated) necessitates substantial customization of its repository system. The open-source software CKAN (http://ckan.org/) was chosen largely because it is sufficiently modifiable to meet the broad spectrum of user requirements (Amorim *et al.*, 2015). Additionally, different groups require different ancillary tools to support the submission of data sets. For example, reliable uploading of large genomic datasets (10s to 100s of GB) requires a fully-featured command-line client script running on Linux (https://github.com/eawag-rdm/resup#resup), while conversion of proprietary file-formats to standardized ones is best served by MS Windows-based tools with a convenient graphical user interface (https://github.com/eawag-rdm/sav2csv#sav2csv). Eventually, a subset of Eawag's data will be made available through a publicly available server and tools will be provided to publish submitted datasets semi-automatically to external repositories such as Zenodo.

Data packages that are uploaded to Eawag's CRD repository are annotated with metadata. The design of the metadata-scheme is one of the project's central challenges. Consideration of metadata structure, extent, and implementation is essential to assure the findability, accessibility and interoperability of data in the repository with the growing number of distributed digital collections (Park & Tosaka, 2010).

Metadata can be as simple as general-purpose vocabularies such as the DCMI Metadata Terms (DCMI, 2012) or the Data Catalog Vocabulary (DCAT) (W3C, 2014), which merely target semantic interoperability. In contrast, the highest level of metadata structure incorporates formally-specified ontologies that encapsulate domain-specific knowledge in a hierarchy of and relations among the terms of a vocabulary. Data annotated in this way can be used to derive new knowledge through automated reasoning as illustrated for water quality monitoring (Jajaga *et al.*, 2015).

The degree of complexity and completeness of metadata provisioning of Eawag's CRD repository is influenced by the effort needed to annotate the data, especially given the lack of available, appropriate domain-specific ontology. This effort must be balanced against the medium-time gains that can be realized through improved interoperability with other services and the necessity to establish a sustainable solution for the life-time of the repository.

In practice, general bibliographic metadata are represented according to the DataCite Metadata Schema (DataCite Metadata Working Group, 2016) which can be largely mapped to DCAT and DCMI Metadata Terms. This facilitates metadata exchange with DataCite and the DOI registration agency and can be handled by common metadata transmission standards such as OAI-PMH (Open Archives Initiative, 2015). In addition, a limited number of domain-specific metadata fields are used, which are drawn from existing vocabularies if possible (e.g. for lists of biological taxa). In the absence of existing vocabularies, specific controlled vocabularies are developed in collaboration with Eawag researchers.

Eawag has invested in support staff to meet these technical challenges as well as to provide information on archival standards, file formats, dataset submission guidelines and vocabularies for dataset classification, to co-produce software requirements and to consult on local data management practices.

1.3.2 Management of scientific and technical knowledge

The scientific literature provides an effective means of compiling formal scientific and technical knowledge from the natural, social and engineering sciences on all topics, including SWM. The scientific literature is well indexed and Eawag researchers have broad access to search tools (e.g., Web of Science, Scopus, etc.) as well as to journals and books through a library that is operated by Eawag on behalf of the four Research Institutes of the ETH Domain (http://www.lib4ri.ch/); the LIB4RI also offers trainings (particularly for doctoral students) on searching the scientific literature. Additional access to print materials (especially books) is provided through the Network of Libraries and Information Centers in Switzerland (http://www.nebis.ch/eng/). Eawag researchers also have access to bibliographic tools (e.g., EndNote). In this regard, KM for SWM poses essentially the same challenges as arise in other technical fields.

Scientific and technical conferences (usually organized through professional societies) also provide a venue for exchange of technical knowledge, usually in advance of journal publication. The tendency of individual researchers to align their networks within specific professional societies can lead to a lack of exchange across such societies. The accessibility of conference proceedings to society non-members (or even society members who did not attend a specific conference) can be highly variable. Technical knowledge presented at conferences can, however, be expected to be published (eventually) in the scientific literature.

Various organizations, both governmental and non-governmental, also issue reports containing SWM-related technical knowledge. Examples include water-related publications released by the OECD (Organization for Economic Cooperation and Development, http://www.oecd.org/env/resources/water-publications.htm), the World Bank (http://www.worldbank.org/en/topic/water/research) or UN Water (http://www.unwater.org/publications). Eawag produces only a limited number of self-published reports but does contribute to reports issued by Swiss Federal Offices, most commonly the FOEN (http://www.bafu.admin.ch/?lang=en). Some large projects, especially those funded by the European Union (EU), post comprehensive reports on project websites; relevant examples include: Sustainable Water management Improves Tomorrow's Cities' Health (http://www.switchurbanwater.eu/), Adaptive Strategies to Mitigate the Impacts of Climate Change on European Freshwater Ecosystems (http://www.refresh.ucl.ac.uk/), Transitions to the Urban Water Services of Tomorrow (http://www.trust-i.net/index.php). The Community Research and Development Information Service (CORDIS) provides a searchable index of EU projects (http://cordis.europa.eu/projects/home_en.html). While various reports can be found through search engines or through the websites of organizations and are almost always freely available, they are not reliably indexed and the process of finding them is often haphazard. Some organizations (including Eawag) issue newsletters and/or use social media outlets to provide entry points to technical information.

Eawag archives reports to which its staff has contributed (as well as all publications in the scientific literature by Eawag authors) in a structured, searchable institutional repository (https://www.dora.lib4ri.ch/eawag/). Full access to content is available within Eawag; external users can view metadata for all publications and can access content that is not restricted by agreements with external parties (e.g., publishers).

1.3.3 Management of experiential and practical knowledge

In contrast to technical knowledge, few (if any) standardized communication channels exist for experiential and practical knowledge. Such knowledge can include both practical intuition regarding complex technical systems that is derived from the experience of operating them and vernacular knowledge held by stakeholders involved in water resources management (Simpson *et al.*, 2015).

A further aspect of experiential knowledge, involving methods and practices of stakeholder engagement, transdisciplinary and/or action research, is partly systematized through the scientific literature (Jahn *et al.*, 2012; Klein, 2008; Pohl, 2008; Polk, 2015), though this is often fragmented in terms of specific applications (e.g., translational and public health medicine, development for LMICs, etc.).

One of the principal challenges of compiling experiential and practical knowledge is that it is often highly contextual and/or specific to a particular (technical) system. Such knowledge can be partly captured in case studies, which are often included in reports, technical guidance documents or other "gray literature". A significant limitation to the further application and implementation of such knowledge, however, is the lack of a consistent and well-documented review process (quality control). In such cases, the reputation of the organization issuing the document is likely to influence the perception of its reliability. As with technical reports (described above), the lack of reliable indexing tends to result in haphazard access to this knowledge. Nonetheless, generalized searches can provide access to such material and visibility for content providers (including Eawag).

An important factor in documenting experiential knowledge is the involvement of social scientists in water management projects. Although the relevance of social sciences to water management has long been recognized (James, 1974), the recent trend of increasing cooperation among social scientists, natural scientists and engineers (which Eawag supports) has led to improved characterization of social networks and interactions, better incorporation of vernacular knowledge and more reflection on processes of integration and stakeholder engagement (Butterworth *et al.*, 2011; Hoffmann *et al.*, 2017a; Hoffmann *et al.*, 2017b; Simpson *et al.*, 2015). Without integration of the social sciences, the experience of participants in water management projects (including social learning) is likely to be held only by individuals or within networks and documented only in unsystematic records such as agendas and minutes of public meetings (though these can potentially be analyzed post hoc). Eawag is supporting a Community of Practice (CoP) for individuals employed by or affiliated with Eawag who are working at the interface of science and practice. Such a CoP, composed of individuals who share expertise and experience in a specific domain of activity (Wenger & Snyder, 2000), can facilitate the exchange of experiential knowledge within the network. It remains a challenge to make this knowledge more generally accessible.

1.4 Unresolved issues and challenges in SWM-related KM

Although Eawag's experience is certainly not unique, it provides insight into the unresolved issues and challenges related to KM, generally, and to KM for SWM in particular.

1.4.1 Information overload and fatigue

The availability (or over-availability) of information is generally perceived in the context of prior experience and/or technological innovation, making information overload a nearly timeless phenomenon. In the current context of scientific and technical KM, the overall number of records in the Core Collection of the Web of Science is estimated to be growing by about 2–3% per year (Larsen and von Ins, 2010) with estimated growth exceeding 10% per year in the field of water management (Ren *et al.*, 2013).

Associated with information overload (Eppler & Mengis, 2004) and the accompanying information fatigue are a number of coping strategies (Landhuis, 2016). For scientific and technical knowledge, these may include: over-reliance on metrics (e.g., journal impact factors or citation rates) as proxies for quality, idiosyncratic search strategies with semi-arbitrary criteria for search termination, under-use of less easily accessible information sources such as (printed) books and reliance on recommendations from personal contacts or recommender systems (Beel *et al.*, 2016).

1.4.2 Open access

The debate over open access is proceeding in a rapidly-evolving context. Publishers are coming under increasing pressure to develop business models for open access that are economically viable for authors and libraries (LERU, 2015). In late 2016, a German library consortium failed to come to an agreement with the publisher Elsevier; open access and cost were the major points of contention (Vogel, 2016). In the absence of (or in parallel to) library subscriptions, researchers are likely to rely more heavily on sites such as Google (Scholar) (https://scholar.google.com), Research Gate (https://www.researchgate.net/), Academia (https://www.academia.edu/) or Sci-Hub (http://sci-hub.io/), without regard to the questionable legal status of documents deposited on (or accessible through) such sites. Preprint servers, mainly arXiv (https://arxiv.org/) and bioRχiv (http://biorxiv.org/), are being increasingly used by Eawag researchers, not only to provide open access but also to establish (early) claims to intellectual property. Issues relating to copyright, however, remain murky and publishers have differing policies regarding whether preprints are considered as prior publication (Lawal, 2002).

Eawag does not have formal requirements for open access publication but provides financial support for "gold road" open access publication (though not for "hybrid" open access). "Hybrid" open access, particularly its vulnerability to "double-dipping" by publishers, has been a key issue in arguments with publishers. Eawag also strongly encourages "green road" open access publication of post-prints (http://www.lib4ri.ch/services/open-access.html), which makes content available to readers without institutional access to subscribed content. "Green road" open access, however, depends on authors to deliver post-prints to the repository and is, in some cases, blocked by publishers for an embargo period (which is often one to two years). These may be factors in the continuing attractiveness of "hybrid" open access to researchers in environmental science and engineering since many leading journals in this field are not "gold road" open access.

1.4.3 Quality control and collaborative editing

The proliferation of open access journals has highlighted the issue of quality control in the conventional peer review process, which was exposed in an investigation by the journal Science (Bohannon, 2013). These concerns are not limited to open access journals, as illustrated by increases in retractions (Steen *et al.*, 2013). Forums for post-publication review are provided by Faculty of 1000 (F1000, http://f1000.com/), which focuses on biology and medicine, and PubPeer (https://pubpeer.com/). F1000 has expanded its service to include an open publishing platform that offers immediate publication and transparent peer review. This model has been adopted by major funders including the Wellcome Trust (https://wellcomeopenresearch.org/) and the Bill and Melinda Gates Foundation (https://gatesopenresearch.org/). An open review process is conducted by Copernicus Publication, which provides platforms that host unsolicited public comments in parallel to reviews solicited by an editor (http://publications.copernicus.org/services/public_peer_review.html). The trend toward open and continuous peer review, in which scientific reports are considered as "living documents" (Shanahan, 2015), has the advantage of corresponding to the actual development of scientific understanding. At the same time, it poses challenges to attribution and quality control as implemented in conventional scientific publishing and peer review.

Eawag's current practices rely heavily on (external) peer review, particularly since there is no institutional requirement for internal review of papers that are submitted to peer-reviewed journals. Publication in leading peer-reviewed journals is also an important criterion for decisions on staff advancement (e.g., tenure and promotion).

More generally, there is an implicit assumption of the quality of peer-review in the growing trend toward (comprehensive) systematic reviews, which have been promoted as an antidote to individual bias (Bilotta *et al.*, 2014; Cvitanovic *et al.*, 2015; Spruijt *et al.*, 2014). This trend runs counter to the past expectation that critical reviews would provide filtering of the available material based on the expert judgment of the author(s) (Nilsen, 2015).

The issue of quality control in the peer-reviewed technical literature is even more acute for reports released by project consortia and/or institutions that are subject to varying (and often unstated) levels of review. The pressure for inclusiveness (which is often an explicit requirement of large programs funded by the EU) is not always compatible with discerning quality control. This places an even greater burden on the researcher (or other readers) who might rely on information in the gray literature.

1.4.4 Resource demands

Quality control clearly demands resources (i.e., the time and attention) of researchers. Indeed, many failings of the peer review system are attributed to the increasing and competing demands on researchers' time. Appropriate curation and archiving of knowledge (including but not limited to data) are also activities that demand both time and expertise. Even updating of websites that function as knowledge portals takes time – often there is no obvious indication when specific websites are no longer actively maintained.

Needed resources can be provided by institutions or funding agencies (Hering, 2015). When they are not, dangers arise that knowledge platforms will take on an unacknowledged promotional character (serving the self-interest of contributors) and/or that paywalls will be constructed that limit access to knowledge (Hering & Vairavamoorthy, 2017).

1.5 Future directions for SWM-related KM

Eawag attempts, through various measures, to support its researchers in effectively accessing and contributing to the knowledge base for SWM. Some of the challenges and obstacles to this, however, call not only for new tools but also for changes in attitudes and expectations within the SWM research community. As a leading research institute in this field, Eawag can hope to have some influence on these norms and attitudes.

The deluge of scientific and technical literature (including that related to SWM) has been linked with trends in metrics and incentives in academics and funding agencies (Benedictus *et al.*, 2016; Hammarfelt *et al.*, 2016). In addition, bibliometric indicators have been criticized as having an inhibitory effect on inter- and transdisciplinary research (Zilahy *et al.*, 2009), which is clearly an issue for SWM. Eawag attempts to provide recognition for publications that are directed toward non-academic audiences (e.g., appearing in trade journals) and, for academic publications, to emphasize quality over quantity. It is, however, open to debate whether quantity (as opposed or in addition to quality) affects impact; an effect of quantity was reported in a recent study in which impact was assessed on the basis of citations in the Web of Science within 3–6 years of publication (Sandstrom & van den Besselaar, 2016). Nonetheless, Eawag supports the view that a shift to focusing on content, with the understanding that metrics are (at best) surrogates, and attention to impact outside academia are needed (Benedictus *et al.*, 2016).

A change in attitudes at various levels is also needed to combat "platform proliferation syndrome", which reflects the "allure of setting up new initiatives, rather than building on what's already there" (Barnard, 2013). A key issue is the need to adopt common standards and open data principles, which would also eventually support data mining and machine learning (i.e., the semantic web). This approach is being fostered in the development (https://www.okhub.org/) and climate adaptation (https://www.weadapt.org/) communities. The issues of attribution and quality control also need to be addressed in this context. Eawag would be interested in working with other organizations to develop such models for SWM-related knowledge and welcomes input to the Open Science Framework project "Freshwater Knowledge Hub" (https://osf.io/28rhn/).

The lack of accessibility of publications and technical content (e.g., located behind paywalls of publishers or professional societies) is a major impediment to text mining (Peters *et al.*, 2014; Shemilt *et al.*, 2014). Text and data-mining (TDM) for non-commercial research purposes is explicitly allowed under UK copyright law if the researcher has lawful access (i.e., through a subscription or license) to the material (https://www.jisc.ac.uk/guides/text-and-data-mining-copyright-exception). Collections of open access papers (e.g., https://www.scienceopen.com/) offer a substantial advantage for TDM but this is, of course, limited to the papers in the collection. Other repositories (e.g., PubMed Central) allow papers to be downloaded and read individually, but connection to natural language processing tools needed for effective TDM is limited to the open access subset of the collection (Doring *et al.*, 2016).

The useful application of machine learning to the pharmaceutical patent literature (which has the advantage of being publically accessible) has been demonstrated (Schneider *et al.*, 2016). This approach could be extremely useful in SWM, but the SWM community would need to coordinate efforts to make this possible.

1.6 Concluding comments

With the SDGs (particularly Goal 6), the SWM community faces both enormous challenges and exciting opportunities. Simply collecting and compiling the monitoring data on indicators to support SDG 6 is a significant task being coordinated through several agencies of the United Nations (UN Water, 2016). Actually achieving the targets under SDG 6, particularly by 2030, will require that existing knowledge is marshaled and used to implement solutions (Hering *et al.*, 2016; SDSN, 2015). Improving knowledge management is an essential component of this task.

The knowledge management issues faced by the SWM community reflect more general issues relating to the overwhelming abundance of scientific and technical knowledge. The SWM community would do well to endorse and support the 12 Vienna Principles – accessibility, discoverability, reusability, reproducibility, transparency, understandability, collaboration, quality assurance, evaluation, validated progress, innovation and public good (http://viennaprinciples.org/) – or the "fair principles" that also address machine learning (https://www.force11.org/fairprinciples).

In this context, Eawag serves as a microcosm illustrating current practices, challenges and outlooks for SWM-related KM. Eawag's perspective is that of a research institute (i.e., mainly a knowledge provider). Knowledge production, however, also entails the discovery and integration of existing knowledge. Furthermore, it is in the interest of all knowledge providers that the knowledge produced is accessible for implementation and as the basis for further research. No single institution can overcome the barriers to the effective sharing and utilization of knowledge, but together, institutions engaged in SWM can hope to move the SWM-community in this direction.

References

Amorim, R. C., C. J. A., J. R. da Silva, & C. Ribeiro (2015). A Comparative Study of Platforms for Research Data Management: Interoperability, Metadata Capabilities and Integration Potential. In *New Contributions in Information Systems and Technologies. Advances in Intelligent Systems and Computing*, edited by A. Rocha, A. Correia, S. Costanzo and L. Reis, pp. 101–111. Springer, Cham.

Barnard, G. (2013). FEATURE: Portal proliferation syndrome responding to treatment. *Climate and Development Knowledge Network*, http://cdkn.org/2013/08/feature-portal-proliferation-syndrome-responding-to-treatment/?loclang=en_gb (January 6, 2017).

Beel, J., B. Gipp, S. Langer, & C. Breitinger (2016). Research-paper recommender systems: a literature survey, *International Journal on Digital Libraries*, *17*(4), 305–338.

Benedictus, R., F. Miedema, & M. W. J. Ferguson (2016). Fewer numbers, better science. *Nature*, *538*, 453–455.

Bilotta, G. S., A. M. Milner, & I. Boyd (2014). On the use of systematic reviews to inform environmental policies. *Environmental Science & Policy*, *42*, 67–77.

Bohannon, J. (2013). Who's afraid of peer review? *Science*, *342*(6154), 60–65.

Briscoe, J. (2011). Making Reform Happen in Water Policy: Reflections from a Practitioner, paper presented at OECD Global Forum on Environment: Making Water Reform Happen, Paris, 25–26 October, http://www.oecd.org/env/resources/48925318.pdf.

Butterworth, J., P. McIntyre, & C. da Silva Wells (Eds.) (2011). *SWITCH in the city: putting urban water management to the test*. IRC International Water and Sanitation Centre, The Hague, The Netherlands.

Cash, D. W., W. C. Clark, F. Alcock, N. M. Dickson, N. Eckley, D. H. Guston, J. Jager, & R. B. Mitchell (2003). Knowledge systems for sustainable development. *Proceedings of the National Academy of Sciences of the United States of America*, *100*(14), 8086–8091.

Cornell, S., F. Berkhout, W. Tuinstra, J. David Tabara, J. Jaeger, I. Chabay, B. de Wit, R. Langlais, D. Mills, P. Moll, I. M. Otto, A. Petersen, C. Pohl, & L. van Kerkhoff (2013). Opening up knowledge systems for better responses to global environmental change. *Environmental Science & Policy*, *28*, 60–70.

Cvitanovic, C., A. J. Hobday, L. van Kerkhoff, S. K. Wilson, K. Dobbs, & N. A. Marshall (2015). Improving knowledge exchange among scientists and decision-makers to facilitate the adaptive governance of marine resources: A review of knowledge and research needs. *Ocean & Coastal Management*, *112*, 25–35.

DataCite Metadata Working Group (2016). DataCite Metadata Schema Documentation for the Publication and Citation of Research Data. *Version 4.0*. DataCite e.V., http://doi.org/10.5438/0012 (May 3, 2017).

DCMI (2012). DCMI Metadata Terms. http://dublincore.org/documents/dcmi-terms/ (April 7, 2017).

Doring, K., B. A. Gruning, K. K. Telukunta, P. Thomas, & S. Gunther (2016). PubMed-Portable: A Framework for Supporting the Development of Text Mining Applications. *Plos One*, *11*(10).

Eawag (2011). *Eawag: past, present and future*, 95 pp. Eawag, Dübendorf, http://www.eawag.ch/fileadmin/Domain1/About/Portraet/Geschichte/Eawag_1936-2011.pdf.

Eppler, M. J., & J. Mengis (2004). The Concept of Information Overload: A Review of Literature from Organization Science, Accounting, Marketing, MIS, and Related Disciplines. *The Information Society*, *20*(5), 325–344.

Fecher, B., S. Friesike, & M. Hebing (2015). What drives academic data sharing? *Plos One*, *10*(2).

Groth, P., A. Gibson, & J. Velterop (2010). The anatomy of a nanopublication. *Information Services & Use*, *30*(1–2), 51–56.

Hammarfelt, B., G. Nelhans, P. Eklund, & F. Åström (2016). The heterogeneous landscape of bibliometric indicators: Evaluating models for allocating resources at Swedish universities. *Research Evaluation*, *25*(3), 292–305.

Hering, J. G. (2015). Do we need "more research" or better implementation through knowledge brokering? *Sustain Sci*, 1–7.

Hering, J. G. & M. Leuzinger (2017). Water Data Portals: An Annotated List, v. 4, https://osf.io/sgfz6/.

Hering, J. G., & K. Vairavamoorthy (2017). Harvesting experience to support sustainable urban water management. In *Assessing Water Megatrends*, edited by A. K. Biswas and C. Tortajada. Springer, Berlin, preprint available at: https://osf.io/zuwnr.

Hering, J. G., S. Maag, & J. L. Schnoor (2016). A call for synthesis of water research to achieve the sustainable development goals by 2030. *Environmental Science & Technology*, *50*(12), 6122–6123.

Hering, J. G., D. L. Sedlak, C. Tortajada, A. K. Biswas, C. Niwagaba, & T. Breu (2015). Local perspectives on water. *Science*, *349*(6247), 479–480.

Hoffmann, S., C. Pohl, & H. J. G. (2017a). Methods and procedures of transdisciplinary knowledge integration: empirical insights from four thematic synthesis processes, *Ecology and Society*, *22*(1), 27.

Hoffmann, S., C. Pohl, & J. G. Hering (2017b). Exploring transdisciplinary integration within a large research program: Empirical lessons from four thematic synthesis processes, *Research Policy*, *46*(3), 678–692.

Jahn, T., M. Bergmann, & F. Keil (2012). Transdisciplinarity: Between mainstreaming and marginalization, *Ecological Economics*, *79*, 1–10.

Jajaga, E., L. Ahmedi, & F. Ahmedi (2015). An expert system for water quality monitoring based on ontology. In *Metadata and Semantics Research: 9th Research Conference, MTSR 2015, Manchester, UK, September 9–11, 2015, Proceedings*, edited by E. Garoufallou, R. J. Hartley & P. Gaitanou, pp. 89–100, Springer International Publishing, Cham.

James, L. D. (Ed.) (1974). *Man and Water: The Social Sciences in Management of Water Resources*. University Press of Kentucky.

Joseph, H. (2016). The evolving U.S. policy environment for open research data. *Information Services & Use*, *36*, 45–48.

Klein, J. T. (2008). Evaluation of interdisciplinary and transdisciplinary research – A literature review. *American Journal of Preventive Medicine*, *35*(2), S116–S123.

Kowalczyk, S., & K. Shankar (2011). Data sharing in the sciences. *Annual Review of Information Science and Technology*, *45*, 247–294.

Landhuis, E. (2016). Information overload How to manage the research-paper deluge? Blogs, colleagues and social media can all help. *Nature*, *535*(7612), 457–458.

Larsen, P. O., & M. von Ins (2010). The rate of growth in scientific publication and the decline in coverage provided by Science Citation Index, *Scientometrics*, *84*(3), 575–603.

Lawal, I. (2002). Scholarly Communication: The Use and Non-Use of E-Print Archives for the Dissemination of Scientific Information, *VCU Libraries Faculty and Staff Publications, Paper 4*.

LERU (2015). "Christmas is over. Research funding should go to research, not to publishers!", *League of European Research Universities*, http://www.leru.org/index.php/public/news/christmas-is-over-research-funding-should-go-to-research-not-to-publishers/ (November 26, 2015).

Martinuzzi, A., & M. Sedlacko (Eds.) (2016). *Knowledge Brokerage for Sustainable Development*, 330 pp., Greenleaf Publishing, Saltaire.

Mays, L. W. (Ed.) (2010). *Ancient Water Technologies*, 280 pp. Springer, Dordrecht.

Michener, W. K. (2015). Ecological data sharing. *Ecological Informatics*, *29*, 33–44.

Nilsen, P. (2015). Making sense of implementation theories, models and frameworks. *Implementation Science*, *10*(53).

Open Archives Initiative (2015). The Open Archives Initiative Protocol for Metadata Harvesting. https://www.openarchives.org/OAI/openarchivesprotocol.html (April 7, 2017).

Pampel, H., P. Vierkant, F. Scholze, R. Bertelmann, M. Kindling, J. Klump, H.-J. Goebelbecker, J. Gundlach, P. Schirmbacher, & U. Dierolf (2013). Making research data repositories visible: The re3data.org Registry. *Plos One*, *8*(11).

Park, J. R., & Y. Tosaka (2010). Metadata Creation Practices in Digital Repositories and Collections: Schemata, Selection Criteria, and Interoperability, *Information Technology and Libraries*, *29*(3), 104–116.

Peters, D. P. C., K. M. Havstad, J. Cushing, C. Tweedie, O. Fuentes, & N. Villanueva-Rosales (2014). Harnessing the power of big data: infusing the scientific method with machine learning to transform ecology. *Ecosphere*, *5*(6).

Piwowar, H. A., & T. J. Vision (2013). Data reuse and the open data citation advantage, *Peerj*, *1*.

Pohl, C. (2008). From science to policy through transdisciplinary research, *Environmental Science & Policy*, *11*(1), 46–53.

Polk, M. (2015). Transdisciplinary co-production: Designing and testing a transdisciplinary research framework for societal problem solving. *Futures*, *65*, 110–122.

Ren, J.-L., P.-H. Lyu, X.-M. Wu, F.-C. Ma, Z.-Z. Wang, & G. Yang (2013). An Informetric profile of water resources management literatures. *Water Resources Management*, *27*(13), 4679–4696.

Roche, D. G., L. E. B. Kruuk, R. Lanfear, & S. A. Binning (2015). Public data archiving in ecology and evolution: how well are we doing? *Plos Biology*, *13*(11), 12.

Sandstrom, U., & P. van den Besselaar (2016). Quantity and/or quality? The importance of publishing many papers. *Plos One*, *11*(11).

Schneider, N., D. M. Lowe, R. A. Sayle, M. A. Tarselli, & G. A. Landrum (2016). Big data from pharmaceutical patents: a computational analysis of medicinal chemists' bread and butter. *Journal of Medicinal Chemistry*, *59*(9), 4385–4402.

SDSN (2015). Getting Started with the Sustainable Development Goals, 38 pp. Sustainable Development Solutions Network, http://unsdsn.org/wp-content/uploads/2015/12/151211-getting-started-guide-FINAL-PDF-.pdf.

Sedlak, D. (2014). *Water 4.0: The Past, Present, and Future of the World's Most Vital Resource*, 332 pp. Yale University Press New Haven.

Shanahan, D. R. (2015). A living document: reincarnating the research article. *Trials*, *16*, 151.

Shemilt, I., A. Simon, G. J. Hollands, T. M. Marteau, D. Ogilvie, A. O'Mara-Eves, M. P. Kelly, & J. Thomas (2014). Pinpointing needles in giant haystacks: use of text mining to reduce impractical screening workload in extremely large scoping reviews, *Research Synthesis Methods*, *5*(1), 31–49.

Simpson, H., R. de Loe, & J. Andrey (2015). Vernacular knowledge and water management – towards the integration of expert science and local knowledge in Ontario, Canada. *Water Alternatives-an Interdisciplinary Journal on Water Politics and Development*, *8*(3), 352–372.

Soranno, P. A., K. S. Cheruvelil, K. C. Elliott, & G. M. Montgomery (2015). It's good to share: why environmental scientists' ethics are out of date. *Bioscience*, *65*(1), 69–73.

Spruijt, P., A. B. Knol, E. Vasileiadou, J. Devilee, E. Lebret, & A. C. Petersen (2014). Roles of scientists as policy advisers on complex issues: A literature review. *Environmental Science & Policy*, *40*, 16–25.

Steen, R. G., A. Casadevall, & F. C. Fang (2013). Why has the number of scientific retractions increased? *Plos One*, *8*(7).

Tsoukas, H. (2011). How should we understand tacit knowledge? A phenomenological view. In *Handbook of Organizational Learning and Knowledge Management*, Second Edition, edited by M. Easterby-Smith and M. A. Lyles, pp. 454–476. John Wiley & Son, New York.

UN (2015). *Transforming our World: The 2030 Agenda for Sustainable Development*, 41 pp., https://sustainabledevelopment.un.org/content/documents/21252030%20Agenda%20for%20Sustainable%20Development%20web.pdf.

UN Water (2016). *Integrated Monitoring Guide for SDG 6: Targets and global indicators*, 25 pp., UN Water, http://www.unwater.org/fileadmin/user_upload/unwater_new/docs/SDG%206%20targets%20and%20global%20indicators_2016-07-19.pdf.

van Kerkhoff, L. (2013). Knowledge governance for sustainable development: A review. *Challenges in Sustainability*, *1*(2), 82–93.

van Kerkhoff, L., & L. Lebel (2006). Linking knowledge and action for sustainable development, in *Annual Review of Environment and Resources*, edited, pp. 445–477.

van Kerkhoff, L., & N. A. Szlezak (2016). The role of innovative global institutions in linking knowledge and action, *Proceedings of the National Academy of Sciences of the United States of America*, *113*(17), 4603–4608.

Vines, T. H., R. L. Andrew, D. G. Bock, M. T. Franklin, K. J. Gilbert, N. C. Kane, J. S. Moore, B. T. Moyers, S. Renaut, D. J. Rennison, T. Veen, & S. Yeaman (2013). Mandated data archiving greatly improves access to research data. *Faseb Journal*, *27*(4), 1304–1308.

Vogel, G. (2016). Thousands of German researchers set to lose access to Elsevier journals. *Science*, doi: 10.1126/science.aal0552.

W3C (2014). *Data Catalog Vocabulary (DCAT)*. World Wide Web Consortium, https://www.w3.org/TR/vocab-dcat/ (April 7, 2017).

Wenger, E. C., & W. M. Snyder (2000). Communities of practice: The organizational frontier. *Harvard Business Review*, *78*(1), 139.

Zilahy, G., D. Huisingh, M. Melanen, V. D. Phillips, & J. Sheffy (2009). Roles of academia in regional sustainability initiatives: outreach for a more sustainable future, *Journal of Cleaner Production*, *17*, 1053–1056.

2 Information transfer and knowledge sharing by water user associations in China

Dajun Shen, Xuedong Yu and Ali Guna

School of Environment and Natural Resources, Renmin University of China, Beijing 100872, China

Introduction

China's irrigated land produces 75% of the country's total grain, over 90% of which are economic crops. Irrigation development plays a significant role in national strategy (Wang, 2013). However, water scarcity and inefficient irrigation severely affect the sustainability of China's agricultural production and economic development, and even the world agriculture market (Lohmar, Wang, Rozelle, Huang, & Dawe, 2003). In order to improve irrigation efficiency and to overcome resource constraints, the Chinese government has been promoting reforms in agricultural irrigation management since 1990 (Riedinger, 2002). The hierarchical organizations are gradually withdrawing from rural irrigation management and their major forms of irrigation management are replaced by contract, leasehold, stock-cooperation, and the water user association (WUA) (D. Li, 2002). Presently, WUAs have been proven as effective systems that support irrigation management of the country and overcome the long-lasting problems of management agency "absence" and "dislocation" in grass-root irrigation infrastructure construction and management (Y. Li, 2009). WUAs also play an exemplary role in irrigation management (Xin, 2009) and have been the most common organizational form in China's agricultural irrigation management (Gao, Wang, & Li, 2003). According to the 2014 "Water Resources Development Statistics Bulletin" of the Ministry of Water Resources (MWR), since the first WUA was established in 1995, the total number of WUAs in China have reached 83,400,

Handbook of Knowledge Management for Sustainable Water Systems, First Edition. Edited by Meir Russ.

managing an area of about 189 million hectares, accounting for 29.2% in total cultivated land.

Information transfer and knowledge sharing play an important role in the organization and management of WUAs, and are the cornerstones of a WUA (Zhou, Weng, & Su, 2015). Information transfer connects all related stakeholders and external organizations (Colvin *et al.*, 2009). Practices prove that only through information transfer and knowledge sharing can a WUA play its required role in irrigation management by coordinating complex interests among WUA participants (Kazbekov, Abdullaev, Manthrithilake, Qureshi, & Jumaboev, 2009). Whether during set-up or operation, effective information transfer is one of the necessary conditions for a well-functioning WUA. The mobilizing and organizing to set up a WUA, the associated management of financing the infrastructure, water use, and daily management are all based upon information transfer and knowledge sharing. For all stakeholders, the participation in irrigation management is realized through access to, processing of and responding to the information and knowledge shared. Therefore, it is obvious that information transfer and knowledge sharing are the prerequisite to accomplish the functions expected from a WUA.

However, the footprints of hierarchical management are still left in China's WUAs and are affecting their development. Without effective water users' participation, the WUAs fail to establish a smooth information transfer and knowledge sharing path. Many governmental organizations were involved in the WUA pilot and promotion, and are now gradually withdrawing (Liu & Shi, 2009). Furthermore, a WUA is rarely built by water users themselves (Tong, 2005); rather they are largely formed under governmental "top–down" promotion (Wang, Liu, & Li, 2013).The bureaucratic structure seems to be common in WUA development (Liang & Hao, 2014). Such reform (listed above) could hardly encourage the grass-root users fully participating into management (Zhang, Huang, & Rozelle, 2002; Wang, Xu, Huang, & Rozelle, 2005). A hierarchical management transfers information from government to water users, but the feedback rarely works effectively. It is difficult for water users to obtain the information they require and, as a result, it is impossible to achieve effective information analysis and feedback. Many problems arose from the lack of appropriate information transfer and knowledge sharing. Combined with other adverse effects, this led the WUAs in China to function improperly and to not fit local circumstances, with only one-third under normal operation, one-third under poor management, and the rest existing only in name (Y. Li, 2009; H. Li, 2007; He, 2010).

2.1 Literature review

Domestic researchers in China rarely focus on WUA information transfer and knowledge sharing, but many studies on rural information services, could be used as references. The rural information service is an organized activity to collect, analyze, develop, store and transmit the information needed, from both rural users and the government, in order to promote agricultural and rural development (Zhao, 2006). Though it is different from WUA information transfer and knowledge sharing in terms of definition, the way in which information is transferred or knowledge is shared, is fundamentally the same. The information transfer has many costs in time, funds, benefit evaluation, and outcomes, etc. The rural information is transmitted mainly in organizational, interpersonal and public patterns. Each of the three patterns has advantages and disadvantages (Qi & Wang, 2009). Each professional association of farmers has developed its own pattern. The characteristics of information dissemination in poverty-stricken regions are mainly human-based (Ji, Gao, & Zuo, 2009). Information is transferred through character, voice, theatrical performance and training courses, etc. (Du, Qian, Zhu, & Zhang, 2010). Information need is critical in guiding the communication (Liu & Sun, 2008). To effectively deliver information to users is the precondition necessary to achieve better allocation and full utilization of the knowledge (Zhang & Li, 2011). This is also the principle and prerequisite of WUA information transfer and knowledge sharing.

There are some studies of WUA information transfer and knowledge sharing, but comprehensive studies on the whole transfer process are rare. Information asymmetry is a key factor in WUA external linkages, resulting in information barriers and less-efficient water resource allocation (Wang, 2008; Meng, 2009). The more layers in hierarchical system information transfers, the more the signals are distorted (Li, Wang, & Fan, 2006). Flattening the management structure could effectively reduce the information gap and distortion between managers and WUAs (Liang, Wang, & Qui, 2009).

Users' participation in water resource management could reallocate irrigation water rights and management powers. But under the governmental regulation, and because of their "top–down" approach, the farmers know little about water resource management information and lack the opportunity to participate in decision-making. Therefore, measures to strengthen the information transfer and knowledge sharing should be taken to mobilize farmers (Wang, 2010) and encourage their participation (Han & Zhao, 2002). Further, information asymmetry sets a huge barrier to information transfer outside of WUAs. Some water users and organizations collude and distort the actual information to cope with the administrative supervisors and, subsequently, it is difficult for the government to manage water resources (H. Li, 2007). Additionally, WUAs cooperate to focus on the common problems in order to have stronger voices in regional water resource allocation and regulation (Meinzen-Dick, Digregorio, & McCarthy, 2004).

Studies of internal information transfer focus on technical information, water use information, water tariff information, and affecting factors. The government plays an important role in the promotion of agricultural water saving technology and is the key investor in technological improvement in collective organizations. However, technologies, both learned from parents and imitated from other farmers, are important too, and the costs to access these technologies are mainly borne by the farmers (Wang *et al.*, 2013). Water use and tariff information includes information such as: direct and indirect tariff collections, tariff standards (Zhu & Zhang, 2007), water tariff broker management systems (Peng, Peng, Luo, & Sun, 2010), management of water tariff collection, and credit evaluation (Chen, 2016), accurate and quantified WUA information management systems, etc. (Liu, Xu, Zhang, & Hou, 2014; Huang & Xie, 2002). Additionally, there are many studies evaluating the current situation of, and relevant factors to, the use of and sharing of information through questionnaires and interviews (Lu, Hu, & Zheng, 2013; Huang, Weng, & Zhou, 2013; Zhou, Deng, & Weng, 2013; Wang & Wang, 2014; Zhou, Deng, & Weng, 2015). The key factors impacting information transfer and knowledge sharing are mainly farmers' education; information provision methods of managers; WUA organization, bureaucratic structure and management; and policy support.

Therefore, many aspects related to WUA information management are indeed investigated, but a systematic analysis on the whole process of WUA information transfer and knowledge sharing is lacking. This chapter will analyze information transfer and knowledge sharing in WUAs, including information transfer and knowledge sharing among the stakeholders in the set-up period and management activities in the operation period, involving water use, infrastructure, financing, daily operation, and water rights trade.

2.2 WUA set-up and operation in China

Based on the registration in the civil affairs department, a WUA's life could be divided into the set-up period and the operation period. The government guides the WUA set-up with a top–down approach (Table 2.1). First, the regional water administrative department researches the WUA reform and establishes a leading group which is responsible for set-up affairs. Then, after consultation, the preparatory group promotes advocacy and organizes training. Finally, farmers become WUA members and are familiarized with WUA charters and rules; and the WUA is registered in the civil affairs department. A complete WUA structure consists of the farmer (water user), the water use group (a group with a number of water users according to the irrigation canal system or production units in the village), the members' representative congress, an executive committee, and a supervisory board. Additionally, some large associations might set up an infrastructure management division, irrigation management division, and finance division, etc. (Figure 2.1). In operation, the external parties related to a WUA are defined as the water source management authority, the local irrigation management entity or the market-oriented water supply company; while the internal parties are work units. Information and knowledge are transmitted and shared among both internal and external parties in the set-up and operation periods.

Table 2.1 Functions of agencies in WUA set-up

Procedure	Level	Stakeholder	Functions
Proposal	Regional water administrative department	Related governmental departments	Conducting WUA reform
Plan	WUA leading group	Related governmental departments, professional agencies	Consultation, guiding, supporting
Action	WUA preparatory group	Township/village cadres, representatives of professional agencies and water users	Organizing, advocating, training
Set-up	Member	Water users and water use group	Participating, suggesting, electing

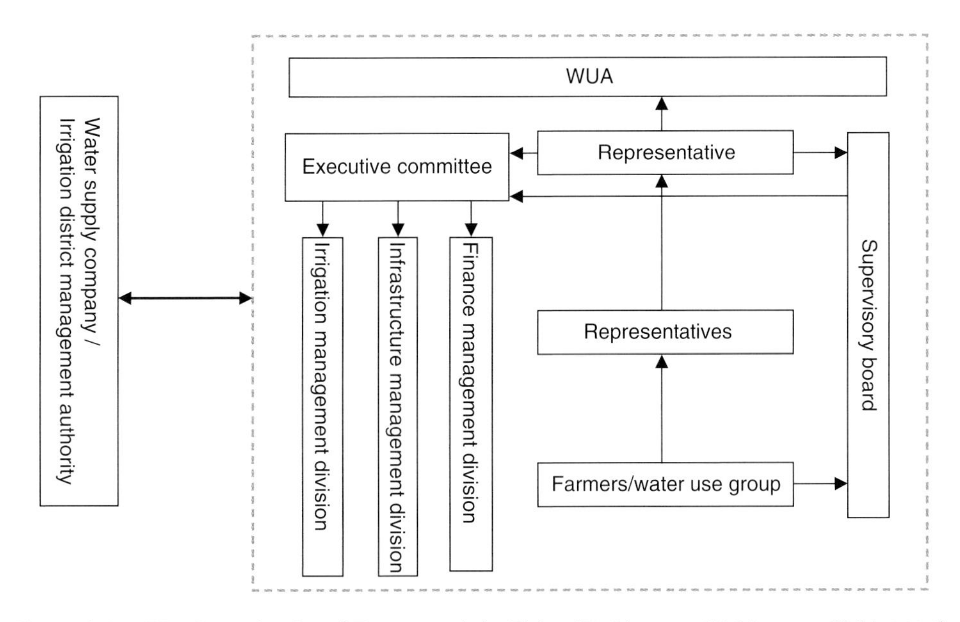

Figure 2.1 WUA Organizational Framework in China (D. Li, 2002; H. Li, 2007; Y. Li, 2009).

2.3 WUA information transfer and knowledge sharing

According to activities of WUAs during the set-up period and the operation period, information transfer and knowledge sharing can be grouped into five categories: basic, water use, finance, infrastructure, and trade information (Table 2.2).

Table 2.2 Information transfer and knowledge sharing in WUAs in China

Information type	Basic information	Water use information	Financial information	Infrastructure information	Trade information
Set-up period	• Knowledge of WUAs • Rights and obligations of members • Elect and be elected • Charters and regulations	• Water use plan • Water use management	• Financial management rules • Water tariff related obligations and rights • Water tariff standards and collection methods • Financing and expenditures	• Ownership rights • Transfer of irrigation infrastructure • Facility list • Water saving technique and management	• Water rights clarification • Water trade rules
Operation period	• Daily management information • Information access • Information analysis • Information feedback	• Cropping plan and area • Water demand plan • Water supply plan • Irrigation plan • Water release • Water delivery • Metering • Use • Balance • Trade	• Cost analysis • Water tariff standard formulation • Pay for water tariff • Water tariff collection • Public financial reports • Supervision and auditing	• Maintenance, update, and new-build of the irrigation infrastructure	• Water budget • Trade requirement • Trade time • Trade place • Trade mechanism • Control of transaction cost

The basic information transfer is the knowledge structure of the WUA, which includes advocacy and training, awareness of members, organization and staff, election, formulation and explanation of charters in the set-up period; and management, access, analysis and feedback of daily information by the members in the operation period. Water use management consists of the formulation of water use plans in the set-up period and the implementation of annual/seasonal water use plans in the operation period. The finance management formulates the financial regulations and ensures that they are understood and accepted by the members, especially water tariff and collection in the set-up period; and in the operation period, it refers to implementation of the financial activities. The infrastructure management is the transfer of the irrigation infrastructure from the local irrigation agency to the WUA in the set-up period; and the maintenance, update and new-build of the infrastructure in the operation period. The water trade management in the set-up period clarifies irrigation water rights and trade rules, and conducts water rights trading in the operation period.

Next, we will detail each one of the five aspect listed above.

2.3.1 Basic information

Basic information is used in both WUA set-up and daily operation. In the set-up period, the information of advocacy and training, and the election and formulation of rules and regulations are very important. In the operation period, these are the access to, analysis of and feedback on activities, and are used by participating members and supervisors.

In irrigation management systems adopted from other countries, water users and grass-root irrigation management organizations need to fully understand the basic information of a WUA. Therefore, during the set-up period (see Figure 2.2), the governmental departments will invest a lot of resources

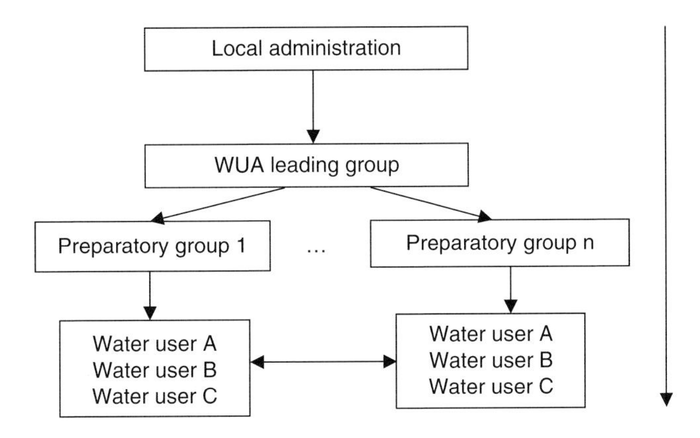

Figure 2.2 WUA information transfer and knowledge sharing in the set-up period.

into disseminating knowledge and organizing training in order to make the stakeholders understand WUA systems and encourage active participation. The information and knowledge tending to spread gradually from top to bottom in hierarchical systems. Consider an example of a county: if the county-level administration decides to set up a WUA, professional institutions and related departments form a leading group; under the promotion of the leading group, the WUA preparatory group is set up in the irrigation district; and then the group transfers and shares information and knowledge with water users and users co-share with each other as well.

Information is also conveyed during the process of electing the WUA personnel, taking a bottom-up path as follows: from water users to water use groups, to representatives of water users to the water user representative congress/water user congress, and finally to the WUA executive committee- supervisory board (Figure 2.3). A certain number of water users, according to irrigation canal systems and/or village structures, form the water use groups. Representatives will be elected from each group, which can then form the representative congress (a small-scale WUA can directly form the water user congress). The executive committee and supervisory board will be elected in the congress. The congress will pass the rules and regulations of the WUA. The process described above is a unique system in which political processes and knowledge sharing are intertwined and the knowledge seems to be shared effectively.

The formulation and revision of regulations and rules are conducted during the establishment of the WUA representative congress which may include the procedures of draft, discussion, examination, approval, temporary implementation, implementation, and revision of charters. Information is transferred and

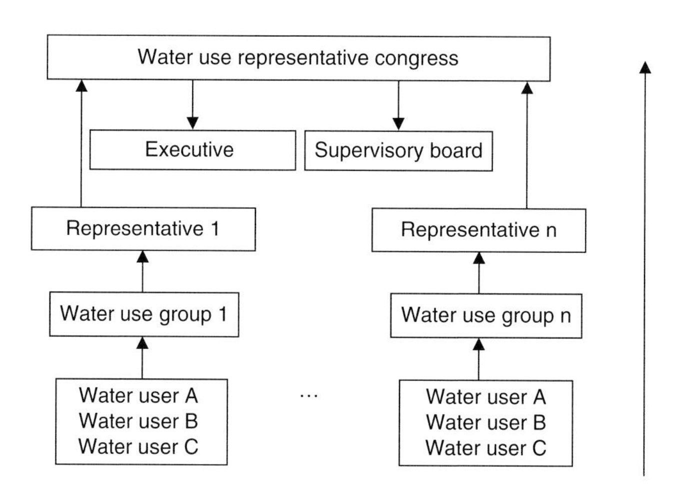

Figure 2.3 Information transfer of personnel election in WUA set-up.

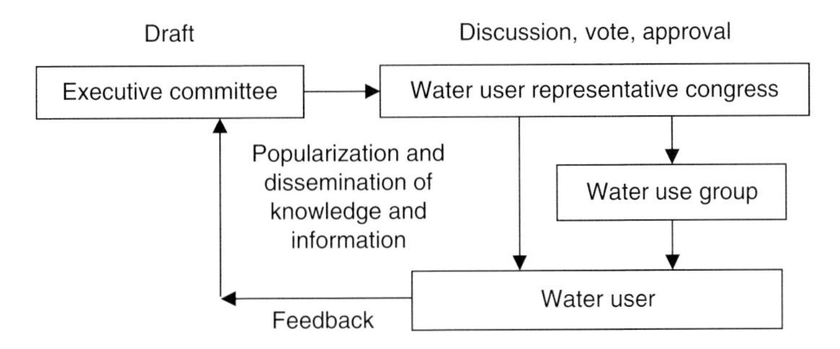

Figure 2.4 Information transfer and knowledge sharing within WUA rule formulation.

shared during open discussions and meetings and then delivered to water users. The questions from water users can be sent back as a feedback (Figure 2.4).

In daily operation and management, the executive committee and supervisory board work according to their duties as identified by their charter and information is communicated about business and activities and is shared by relevant stakeholders. Water use management, financial management and infrastructure management are three main duties of the executive committee. The detailed path of information transfer is analyzed below: here the WUA members' participation and implementation of supervision rights are used as an example to demonstrate how information is transferred in the operation of a supervisory board (Figure 2.5). WUA members acquire information from the WUA implementation plan and its daily operation and understand the irrigation services provided by the WUA. Then, the member analyzes whether the service meets their irrigation

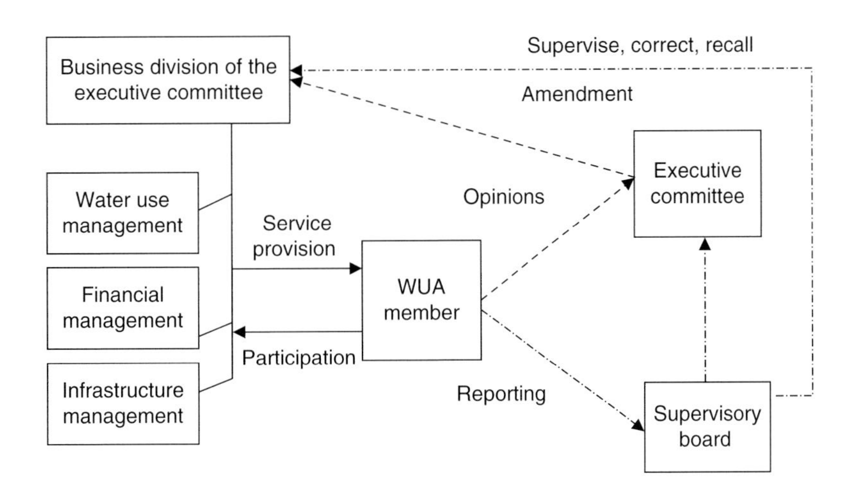

Figure 2.5 Information transfer within WUA members' participation and supervision.

needs and whether the measures that the executive committee takes comply with the rules or not. Members express opinions and decide to actively participate in the irrigation plan or propose a different opinion and report their dissents.

2.3.2 *Water use management*

Water use management consists of the irrigation plan formulation and irrigation water use. The information is used to formulate and implement irrigation plans, and processed in the following order in WUAs in China: 1. cropping pattern and cultivated area; 2. water demand plan; 3. water supply contract; 4. water supply plan; 5. water supply; 6. water delivery; 7. water metering; 8. water use; 9. balance and trade (Figure 2.6).

The executive committee formulates the water use plan according to members' demands at the start of each year. The committee measures and checks members' actual irrigation areas, records cropping patterns, and formulates a water demand plan and submits it to the water supply entity of the irrigation district. Then, the water supply entity negotiates and formulates the water supply plan in light of the supplying capacity and signs a water supply contract. Consequently, the executive committee comprehensively considers members' opinions and suggestions, and formulates and declares the detailed irrigation plan, according to the water supply contract. Thus, the information used includes the cultivated area, cropping pattern, irrigation water demand, water resource condition, water supply plan, and contract.

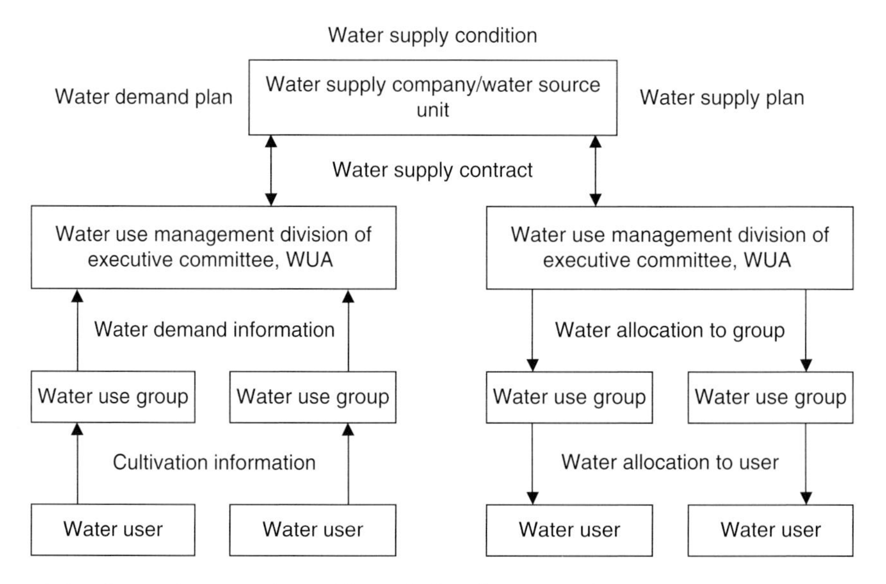

Figure 2.6 Information transfer and knowledge sharing within water use management.

During the irrigation season, the water supply entity informs the WUA to prepare, release and deliver water at the scheduled time. The water use management division is responsible for detailed implementation according to the plan in a process of: water delivery, water metering, water use, to manage integrated water supply and use. The information transferred and shared during this time includes water supply time, water metering method, water quantity, water supply order, and transit method between neighboring water users.

2.3.3 Financial management

As an independent entity, a WUA needs to set up a professional financial management division to conduct their financial work, follow regulations, and to be self-financing. The financial information is transferred and shared among the related parties in financial business. The financial management division formulates the annual financial plan, manages water tariff collection and cash flow, drafts the financial report and the budget report, provides the financial information to the members, and is audited. According to the activities of the association, the financial information could be grouped as irrigation tariff, water tariff collection, and the maintenance of the infrastructure.

Information about irrigation tariffs and water tariff collection is largely of concern to water users. The irrigation tariff consists of tariff level and structure, and the sum of other relevant expenditures. The water tariff revenue directly affects the normal and sustainable operation of the water supply agency and the WUA. The irrigation tariff and collection system is defined during WUA set-up, and normally written in the WUA charter. The executive committee drafts water tariff proposal, consults with members, and submits the proposed water tariff to the representative congress for approval. The user participation in tariff formulation is helpful for them to understand water service costs and expenditures, to consider members' willingness and ability to pay, and for tariff collection. The tariff is collected according to regulations by either levying a portion before watering and the remaining balance after watering or after irrigation according to volume consumed. The members are obligated to pay the tariff on time. Direct and indirect water tariff collection patterns are applied in large irrigation districts in China (Figure 2.7). The direct collection is what the water supply entity collects directly from water users. The indirect procedure entrusts the collection to towns, banks, and WUAs. The process of tariff collection is as follows: water users to the water use group, to the WUA, and then to the water supply entity (water supply company). The advantage of this process is to reduce the high cost in tariff collection from a large number of water users, and to avoid the additional charges from other agencies not related to irrigation activities.

The infrastructure maintenance fund can come from the support of government and other social organizations, and/or be collected from WUA members internally. The members can provide resources such as capital, labor and other inputs accordingly. At the start of each year, the WUA formulates an

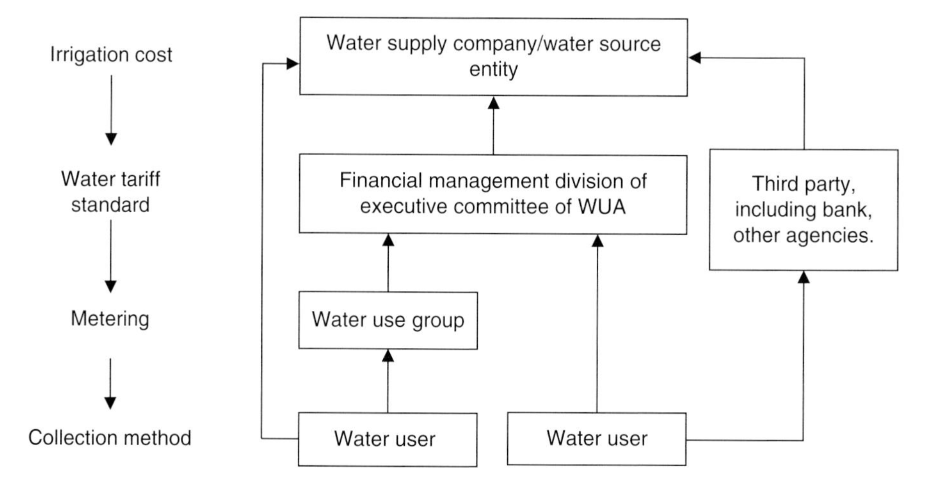

Figure 2.7 Information transfer and knowledge sharing in water tariff collection.

annual financial plan according to infrastructure maintenance and irrigation demands, and submit this to the representative congress for approval. All funding shall be used for the identified purposes. The financial management division should implement the financial plan strictly according to the rules of fund-raising and use, and accept the member supervision over implementation. The financial management division formulates financial sheets and reports and accepts supervision and auditing.

2.3.4 Infrastructure management

Infrastructure information is transferred and shared among related parties when the infrastructure management division conducts irrigation facility management. The infrastructure division is responsible for maintenance, protection and management of the irrigation infrastructures in the WUA. The division formulates the annual management and maintenance plan and organizes the implementation, which may include metering facility calibration and examination, channel cleaning and reinforcement, irrigation infrastructure renewal, revision and expansion.

According to the irrigation process, infrastructure management can be categorized as examination before irrigation, inspection during irrigation, and maintenance after irrigation. Some WUAs with better supports could renew and expand irrigation engineering (Figure 2.8). Generally, the infrastructure management responsibility is decentralized to water use groups, and then further to water users, as it is helpful for the direct beneficiary to participate in engineering maintenance. The testing of the infrastructure is conducted before irrigation and repair is conducted for those facilities not in good condition. The infrastructure management division collects information on problems, as

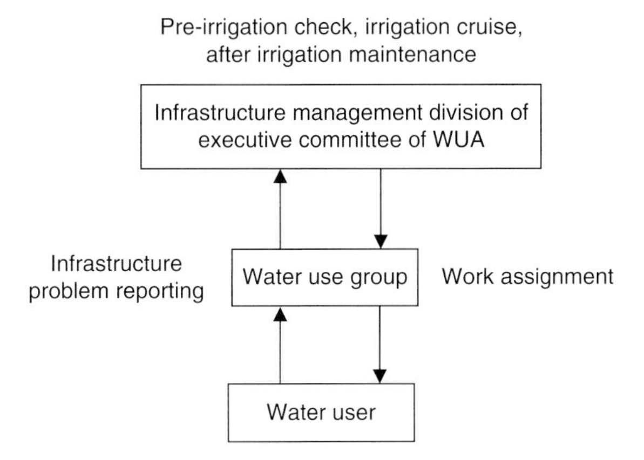

Figure 2.8 Information transfer within infrastructure management.

well as suggestions, from members, in the formulation of the quality check plan and allocates the check tasks. Then, the water use group checks the channels within the group and solves problems like leakage or blockage. Meanwhile, the members are required to regularly check and maintain the irrigation inlets in their own cultivated fields. During irrigation, a professional worker takes charge of water release and metering, and cruises the channels to avoid natural or man-made problems. After irrigation, the facilities are clearly counted and properly preserved, and fixed if needed. In terms of new-building, renewal or expansion of the infrastructure, extensive consultation with members and related parties is essential. The detailed project plans are required to be formulated, submitted to the representative congress for vote and approval, and implemented by the appointed professional staff.

2.3.5 Water trade

The farmer has the rights to use water after the rights clarification and registration in the WUA. When a water user changes the cropping pattern or reduces water use by improving technology, the saved water can be traded and becomes a new source of supply for users with more water demand. Water trade is a market behavior and both parties should provide true, complete, and valid trade information. The seller is required to submit the application to the trade platform with volume available, register and be listed. The buyer needs to submit a purchase application to the trade platform and provide demand information, register, and be listed as well. The trade platform then matches the sellers and buyers within the market (Figure 2.9). The related parties can negotiate further or attempt a re-match. If a trade agreement is reached, the platform records the trade information. Both parties then trade water. Water rights trade must be implemented on

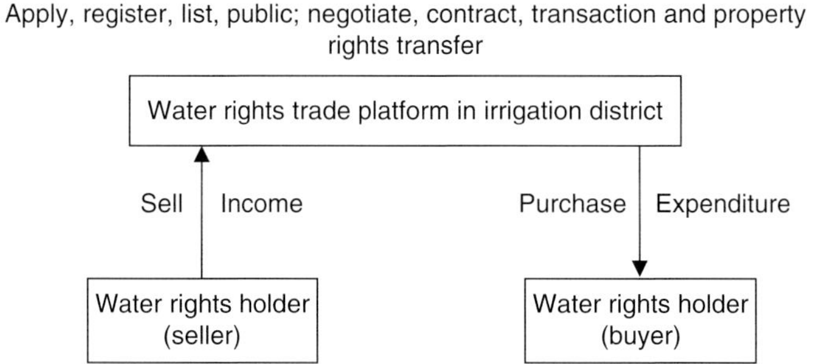

Figure 2.9 Water rights trade among water users in a WUA.

the basis of clear, deliverable and metered water rights (Shen, Yu, Zhang, Li, Shi, & Liu, 2016).

So far, five categories of WUA information transfer and knowledge sharing in China have been analyzed. Although presented separately, all of them are interconnected and function with each other in WUAs. For a successful WUA, all kinds of information must be transferred and shared properly and treated correctly. Any problem in information transfer and knowledge sharing will impact the sustainable and efficient management and operation of the WUA.

2.4 WUA in Shiyang River basin

There are more than 80,000 WUAs that have been set up in China since their introduction in the mid-1990s. Although the operation of these WUAs is impacted by many factors, water resources and irrigation requirements are the fundamental issues. If more water or watering is required by users, more activities, together with more information, will be conducted in a WUA. Therefore, given water resource circumstances in China, WUAs in semi-arid and arid regions, such as northwestern China, operate and function better. So, this chapter selects the Shiyang River basin in Gansu Province in Northwest China as an example.

The Shiyang River basin is located in the east part of the Hexi Corridor in Gansu Province with annual precipitation of 222 mm. The water resources of the basin come from upstream mountainous regions. The middle part of the river basin has extensive agricultural and urban development. The basin has suffered a severe water shortage problem and ecosystem deterioration. After a large-scale water infrastructure development, the river basin explored water rights management and actively developed a WUA (Li, 2013) in order to promote integrated river basin management and water-saving society construction.

2.4.1 *Water rights allocation*

Surface water is directly allocated from the irrigation district to the farmer and the groundwater is allocated from the irrigation district to the branch irrigation channel. The WUA is responsible for water resource regulation to, and tariff collection from, the farmers (Gao, 2007). Therefore, in order to implement a water rights system, a WUA was developed to conduct irrigation management and support water rights reform.

2.4.2 *Stakeholders of WUA*

Combined with the water rights system, the WUA in Shiyang River basin has some unique characteristics. It represents not only an independent legal entity but also participates in hierarchical management, and thus, information transfer and knowledge sharing structures exist with special features (Figure 2.10). The major responsibilities of the WUA in the river basin are to collect water tariffs and maintenance fees, to manage and maintain the irrigation area, and to manage water rights trade within the irrigation district management unit. At the same time, the WUA participates in water rights management which is the extended/residential mission of the township water resource management agency and county water affairs bureau. That is different from WUAs in other places.

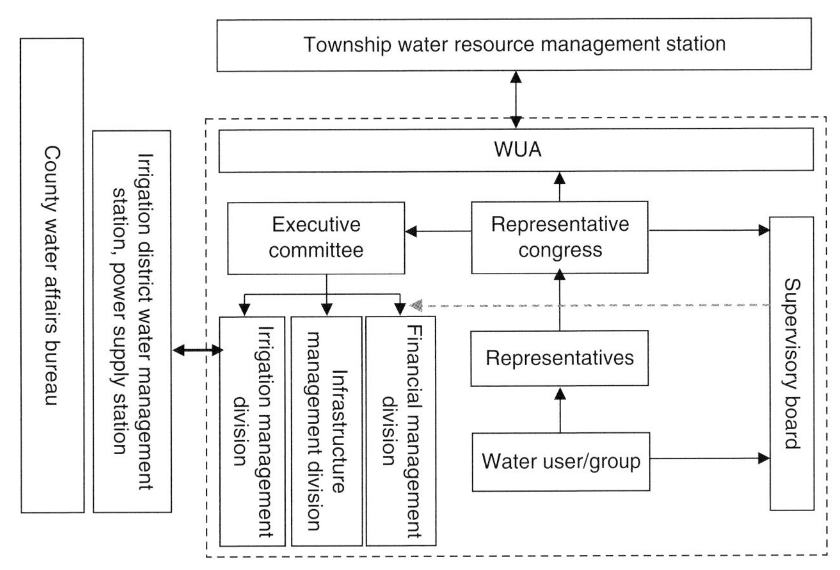

Figure 2.10 WUA organizational structure and external relationships in Shiyang River basin.

2.4.3 *Information transfer and knowledge sharing in water use management*

(1) Planned water use

The water usage in the river basin includes that for domestic, agricultural and ecological sectors, and also industrial sectors in some areas (Table 2.3, Table 2.4). In March of each year, and before each watering, every water use group requires its members to fill in two applications: an annual water use application form (which is for the whole year) and the round water use application (which is for each watering). The irrigation management division of the executive committee sums and formulates the annual water demand plan and the round plan, and then submits the water use applications to the township water resources management station. The township water resources management station, according to the annual water resource availability and water demand, formulates water supply plans and signs contracts with the WUA. Finally, the WUA notifies users of the annual water use plan and the round water use plan. The water users purchase water and electricity from the township water resources management station based on the plan.

The annual infrastructure maintenance plan is formulated at the same time as the annual water use plan. The water use group is responsible for checking, maintaining and dredging the irrigation channels and metering facilities.

(2) Water use

Before irrigation, the WUA confirms that water users have completed watering transfer according to the round water use plan; declares the irrigation timing and the order of the round for each user; and guides users to use water. Water users produce and irrigate according to the cropping plan (Table 2.5).

The water use information is measured by a metering facility. The quota and exceeding-quota volumes are developed. The quota volume is the average water demand for the normal growing of a certain type of crop. The exceeding-quota volume refers to the volume that exceeds the quota volume, probably occurring due to improper planting and management, or due to natural factors such as drought.

2.4.4 *Information transfer and knowledge sharing in water tariff management*

In order to promote efficient water use and water-saving technologies, the exceeding-quota volume is charged by the increasing block tariff, which will be charged more on each unit volume. After watering, the farmer needs to confirm the actual water consumption information with the WUA for related tariff collection.

Table 2.3 Agricultural water allocation of a WUA

No.	Water user name	Farmland area with irrigation water right	Annual water allocation for farmland (m^3)	Forest and fruit irrigation area (m^2)	Annual water allocation for forest and fruit (m^3)	Total annual water use (m^3)
Sum						

Table 2.4 WUA water allocation

WUA			Domestic water volume			Agricultural water volume			Ecological water volume			Industrial water volume
Group	User name	No. of Water Rights	Population	Livestock	Volume	Cultivated land area	Quota	Volume	Area	Quota	Volume	

Table 2.5 Actual water use form of a WUA in an irrigation district

WUA	No. of irrigation machines	Crop A				Crop B				…	Total water use volume	Signature
		Area	Quota	Exceeding-quota volume	Subtotal	Area	Quota	Exceeding-quota volume	Subtotal	…		

The county water resources management agency is responsible for formulating the annual irrigation water tariff (Table 2.6). The WUA is responsible for the interpretation and implementation of water tariff standards. The township water resources management station allocates the initial water rights and issues the water rights certificates. The basic water tariff is collected during the issuance, and other tariffs are collected based on the metering. Water tariffs are consistent between water users, the water use group and the township water resources management station.

2.4.5 Information transfer and knowledge sharing of water rights trade

In other river basins in China, market-based water rights trade has not been implemented yet; but the Shiyang River basin has introduced and implemented such a system, that entrusts more functions and activities to the WUA. The rules and regulations that deal with trade in the river basin are defined. The tradable water rights are limited to water use for industry and agriculture with economic benefits. These water rights could be traded without significant impact on third parties, such as the ecosystem and environment, public interests, and others. The basic domestic water rights are used to guarantee basic living demands and are not encouraged for trade. The ecological water rights are forbidden to be traded in order to maintain the river basin ecosystem. Additionally, water trade is restrained by the physical infrastructure to deliver water and also forbidden to be transferred for low-efficiency use.

The parties involved in the trade include the seller, buyer, WUA, township water resources management station, water rights trade center, and the irrigation district water management station. The cross-sector, cross-region, and permanent trades require more approvals because of the impacts mentioned above (Table 2.7, Figure 2.11).

The trade tariff must be strictly based on water tariffs verified by the pricing department. The agricultural water rights trade price should not exceed 3 times the water tariff, and the trade income subsequently goes to the sellers. The water rights trade price could be different according to the purpose, and the differential pricing is applied.

The trade contract includes the basic information of both parties, trade volume and purpose, trade pattern, price, trade time, and valid time of the rights. Six copies of the contracts are held as follows: one copy each with the seller, buyer, township water resources management station, and the irrigation district water management station, and two copies with the water rights trade center.

It could be inferred that such water rights trades are neither simple nor efficient. The responsibilities of the relevant parties are not clearly defined, which may cause additional information transfer obstacles and confusions.

Table 2.6 Irrigation water tariff of an irrigation district

Crop	Irrigation pattern (flood irrigation, drip irrigation)	Irrigation quota (m^3/mu)	Under quota (RMB/mu)			Exceeding-quota (RMB/mu)		Water Resource fee (RMB/m^3)	Water prices at The end channel (RMB/m^3)
			Basic fee	Volumetric tariff	Subtotal	Block tariff	Exceeding-quota tariff		

Table 2.7 Water rights trading account

No.	Seller	Buyer	Trading scope	Volume	Trade pattern	Trade time	Remarks

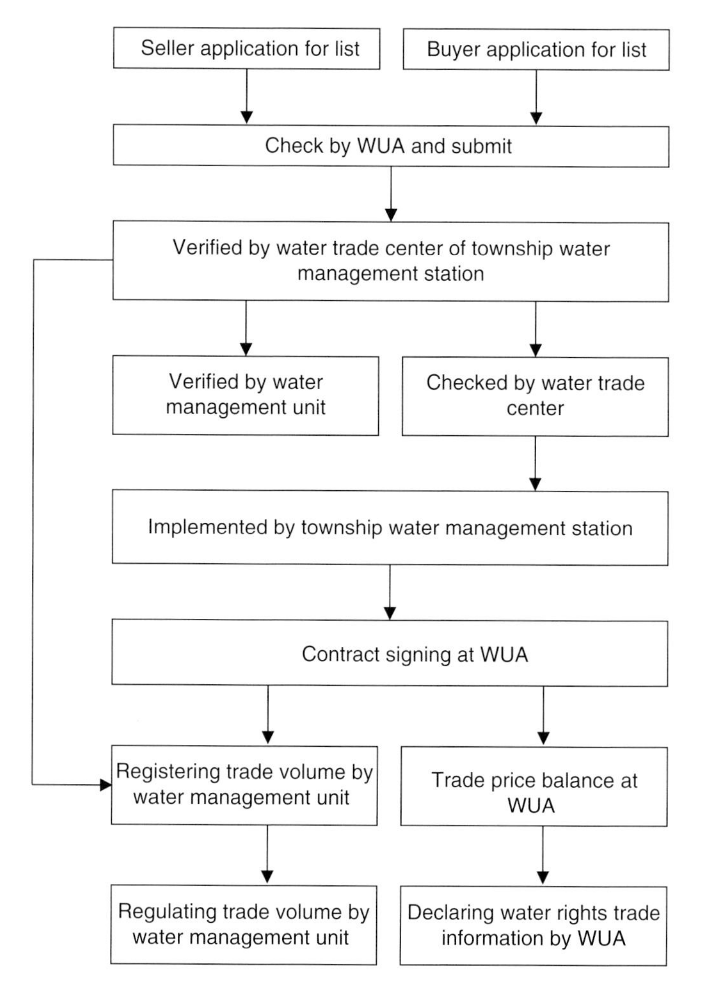

Figure 2.11 Water rights trading procedure in Shiyang River basin.

Only after several approvals can the water rights go into the trade market, which significantly increases the transaction costs. Additionally, the regulations on trade price and trade scope limit the function of markets.

So, under a centralized system and with significant water resource differences among regions in the country, the WUA in Shiyang River basin follows the national guidance in terms of a WUA internal structure, but the external relationships are different. The WUA in the basin functions as both a water user self-governing organization and a village-level water resources management agency, which is a public agency. Additionally, water rights trade, although involving too many approvals, is a unique case, compared to other WUAs in the country.

2.5 Suggestions

Based on the analysis above, we can draw the conclusion that there is a gap between arrangement and practice, with regard to WUA information transfer and knowledge sharing. It is much less effective in actual operation than what it is expected to be. In addition to geographical differentiation, the lack of necessary support in operation and the deviation from the expected goals are also significant issues. Therefore, recommendations on improving WUA information transfer and knowledge sharing are proposed from institutional development, operational mechanism, and supporting perspectives.

(1) Institutional development

The principal goals in setting up a WUA are to settle the disputes in tariff collection between the irrigation district management agency and farmers and to manage the difficulties in maintenance expenditure collection. As an independent legal entity, a WUA is formed by water users with willingness, and is in charge of grass-root irrigation management. However, the WUA in the Shiyang River basin has mixed and confused functions: it is involved within a hierarchical water resources management institution and appears more like a subordinate agency of the irrigation district management bureau than an agent of the farmer. In terms of information transfer and knowledge sharing, a top–down approach is applied and information expression and participation of water users are always insufficient and unequal. The farmer is more negatively than positively affected when participating in water use activities.

Therefore, reform of the present multi-level connecting and management method in the institution should be implemented to reduce the information flow units and costs by adopting concise specific management units and digital management units. The current process can be fully operated by internet and mobile technology, such as the internet-based water trading center in the Shiyang basin[1]. At the same time, the WUA should be clearly defined as an independent legal entity and increase water user participation in order to make the WUA function better as a grass-root irrigation management body, to convey accurate and necessary irrigation information, and to enable knowledge sharing by the stakeholders.

[1] http://www.water-trading.net/nwt/sys/login_index.action

(2) Operational mechanism

Detailed and clear working rules and mechanisms are required for the WUA set-up in order to guide and regulate operations. In practice, it turns out that the charters of WUAs, and the mechanisms for meeting arrangement, water use, water tariff collection, and infrastructure management, etc., are too general and cannot fit local circumstances. Guided by the national regulation, the majority of WUAs in the country have few significant differences between them in terms of designs and arrangements.

Another common problem is inadequate communication of information transferred and knowledge shared between management activities in a WUA. Although the activities are grouped and siloed, the information and knowledge of these activities, by their nature are related, such as between water use planning, water tariff regulation, and infrastructure. Now, most information and knowledge is transferred and treated separately in a single path. This reduces the integrated efficiency and synergistic effectiveness of the information treatment, as well as the WUA operation.

Therefore, given the institutional setting, mechanism improvement has become the key for well-functioning WUAs. Based on the activities, the participatory rights of all stakeholders in WUA operations should be redefined to clarify the responsibility, communication methods, and consulting measures of each party. Only in that case can related parties have a better understanding of their position, function, and responsibilities. The information integrated treatment mechanism should be introduced. The information transfer paths in water trade should be simplified by reducing unnecessary approval procedures and decreasing information costs for both parties.

(3) Supporting

Water allocation plans are the foundation of management and trade in WUAs. Water rights trade should be based upon certain conditions. The key is the initial allocation of water rights, which should be in accordance with the water allocation plan. The purpose of water use management is to allocate annual use based on water users' demands. However, with regard to existing WUAs, there is no opportunity for users to take part in the decision-making process to formulate the water allocation plan and annual water use plan. In fact, the plans are formulated by the local water administration and the WUA follows their decision. Herein the information transfer process about formulating water allocation plans should be open and transparent to water users. Otherwise it would lead to imbalanced distribution, perhaps even cause conflicts.

Besides the system development, the premises of water rights trade is that water shall be measured and delivered. In a large irrigation district, the costs to obtain water information through labor measurement are very high. Digital and internet technologies can reduce information collection and sharing costs considerably, and reduce overall trade costs. The developed website for the water rights trade center in the Shiyang River basin has significantly reduced the information costs involved in trade and has simplified procedures including the list, negotiation, and payment.

References

Chen, J. (2016). Insert the balance information of water arrearages into the big data platform for credit evaluation system. *Modern Economic Information*, (*01*), 66–68.

Colvin, J., Chimbuya, S., Everard, M., Goss, J., Ballim, F., Klarenberg, G., Ndlovu, S., Ncala, D., & D. Weston (2009). Building capacity for co-operative governance as a basis for integrated water resource managing in the Inkomati and Mvoti catchments, South Africa. *Water S A*, *34*(6), 681–689.

Du, G., Qian, J., Zhu, W., & Zhang, P. (2010). The actualities and the thinking in main communication medium in rural information services. *Agriculture Network Information*, (*08*), 98–100.

Gao, E. (2007). *China water right system development*. Beijing: China Water Power Press.

Gao, H., Wang, Y., & Li, Z. (2003). The connotation and development of participatory irrigation management. *China Rural Water and Hydropower*, (*08*), 27–29.

Han, H., & Zhao, L. (2002). Game-theoretic analyses of farmers' co-operative behaviors in irrigation districts. *China Rural Survey*, (*04*), 48–53.

He, X. (2010). Why WUA does not fit the local circumstances? *China Rural Discovery*, (*1*), 83–86.

Huang, B., & Xie, C. (2002). Practical irrigation management information system in ground water well-irrigation fields. *China Rural Water and Hydropower*, (*03*), 6–8.

Huang, Y., Weng, Z., & Zhou, L. (2013). Farmers' willingness to participate in the water user association and its influencing factor in major grain producing areas: A survey based on 602 farmers. *Chinese Agricultural Science Bulletin*, *29*(14), 92–97.

Ji, M., Gao, C., & Zuo, T. (2009). Patterns, characteristics and policy suggestions on farmer professional associations in poverty regions in China: A case study of Guangxi. *Journal of Anhui Agri Sciences*, *37*(6), 2747–2749.

Kazbekov, J., Abdullaev, I., Manthrithilake, H., Qureshi, A., & Jumaboev, K. (2009). Evaluating planning and delivery performance of water user associations (WUAs) in Osh province, Kyrgyzstan. *Agricultural Water Management*, *96*(8), 1259–1267.

Li, C, Wang, X., & Fan, W. (2006). Design of auction system of license of using water due to information asymmetry. *Operations Research and Management Science*, *15*(06), 139–144.

Li, D. (2002). China irrigation management and water user participation in irrigation management. *China Rural Water and Hydropower*, (*05*), 1–3.

Li, H. (2007). *Research and analysis of the reasons of rural community participate in water resource management from rights perspective: A case study of City B*. China Agricultural University.

Li, P. (2013). Continental river basin water right system reform. *China Rural Water and Hydropower*, (*11*), 57–59.

Li, Y. (2009). China WUAs development status and direction. *China Water Resources*, (*21*), 15–16.

Liang, H., Wang, H., & Qiu, L. (2009). Extraction of information rent and operation management modes for water resources of East Route of South-to-North Water Transfer Project. *Journal of Economics of Water Resources*, *27*(04), 7–10.

Liang, Y., & Hao, Y. (2014). Reviews of WUA construction research in China. *Water Resources Development Research*, *14*(5), 28–31, 51.

Liu, D., & Sun, Y. (2008). Information need for guidance of countryside information service. *Information Science*, *26*(07), 1003–1006.

Liu, F., & Shi, J. (2009). A study on the administrative hierarchy of farmer's cooperation from the view of organization relationships: Based on the construction and development of water users' association. *Issues in Agricultural Economy (Month)*, *30*(09), 30–37.

Liu, Y., Xu, H., Zhang, L., & Hou, L. (2014). Research on WUAs' quantitative management information technology. *Inner Mongolia Water Resources*, (*04*), 62–64.

Lohmar, B., Wang, J., Rozelle, S., Huang, J., & Dawe, D. (2003). China's agricultural water policy reforms: Increasing investment, resolving conflicts, and revising incentives. *Agricultural Information Bulletins*, 197–201.

Lu, C., Hu, R., & Zheng, S. (2013). Research on the present situation of water user association (WUA) in Zhejiang Province. *China Rural Water and Hydropower*, (*11*), 157–159.

Meinzen-Dick, R., Digregorio, M., & McCarthy, N. (2004). Methods for studying collective action in rural development. *Agricultural Systems*, *82*(3), 197–214.

Meng, G. (2009). Incomplete information's effect on the efficiency of water resources centralized allocation mechanism. *Journal of Wuhan Institute of Technology*, *31*(09), 27–30.

Peng, P., Peng, S., Luo, Y., & Sun, Y. (2010). Exploration of large irrigation area water supply charges agent system: A case study of Gaoyou irrigation area. *China Rural Water and Hydropower*, (*10*), 136–138.

Qi, D., & Wang, W. (2009). Research on the information transfer pattern on farmer demand. *Jiangxi Social Science*, (*5*), 218–221.

Riedinger, R. (2002). Participatory irrigation management reform in China: Self-management irrigation and drainage area. *China Rural Water and Hydropower*, (*06*), 7–9.

Shen, D., Yu, X., Zhang, M., Li, P., Shi, J., & Liu, J. (2016). Study on water right trade conditions. *Water Conservancy and Hydropower Technology*, *47*(09), 117–121.

Tong, Z. (2005). The farmer-consumer association and rural development. *The Chinese Cooperative Economic Review*, *12*(2), 25–37.

Wang, F. (2008). Urban water supply enterprises asymmetric information management. *Science and Management*, (*03*), 39–41.

Wang, H. (2010). Long-lasting mechanism of sustainable development for water users' association. *China Rural Water and Hydropower*, (*12*), 42–45.

Wang, J., Liu, K., & Li, Y. (2013). Adoption of agricultural water saving technologies: Information channels, investment sources and constraints. *Journal of Economics of Water Resources*, *31*(02), 45–49.

Wang, J., Xu, Z., Huang, J., & Rozelle, S. (2005). Water management reform, agricultural production, and poverty reduction. *China Economic Quarterly*, *5*(1), 189–202.

Wang, Y. (2013). The reform of the water user association in China: A close examination from the perspective of the policy implementation. *Management World*, (*06*), 61–71.

Wang, Y., & Wang, C. (2014). Rational design and grassroots practice of institution: Based on WUAs investigation in northern Jiangsu Province. *Journal of Nanjing Agricultural University (Social Sciences Edition)*, *14*(04), 85–93.

Xin, J. (2009). Running and management of WUAs in Duanjiaxia irrigation area. *Shaanxi Water Resources*, (*06*), 147–148.

Zhang, J., & Li, X. (2011). Guided by information needs of rural information service research. *Information Research*, (*05*), 54–56.

Zhang, L., Huang, J., & Rozelle, S. (2002). Growth or policy? Which is winning China's war on poverty. Working Paper, Center for Chinese Agricultural Policy, Chinese Academy of Sciences.

Zhao, J. (2006). Characteristics and mode option of rural information service in China. *Journal of Library and Information Sciences in Agriculture*, *18*(11), 26–29.

Zhou, L., Deng, Q., & Weng, Z. (2013). The empirical study on the influencing factors of peasant households' participation in water user association: Based on logistic-ISM. *Journal of Huazhong University of Science and Technology (Social Science Edition)*, *27*(05), 107–115.

Zhou, L., Deng, Q., & Weng, Z. (2015). Consistency of empirical research of farmers' preference and realistic choice to participate in WUA. *Journal of Agrotechnical Economics*, (*01*), 93–101.

Zhou, L., Weng, Z., & Su, H. (2015). Assessment of effectiveness of the functioning of water users' association (WUA) from the perspective of the heterogeneity of rural household income. *Journal of China Agricultural University*, *20*(04), 239–247.

Zhu, J., & Zhang, L. (2007). Agricultural irrigation water price policy and its impact on water saving. *China Rural Water and Hydropower*, (*11*), 137–140.

3 Knowledge Management Systems for urban water sustainability: Lessons for developing nations

Vallari Chandna and Ana Iusco*

University of Wisconsin–Green Bay, Green Bay, Wisconsin, USA

Introduction

Sustainability is not just an environmental problem but also a developmental one. It primarily deals with acknowledging the limits to our planetary resources, understanding that the people and the planet are connected and that resource use and distribution must be equitable. The statement made two decades ago by Larsen and Gujer (1997, p. 4) that "sustainability is only possible where there are no extreme poverty conditions" has fostered much research in the area. There is thus a significant role to be played by the development level of the nation in terms of determining how water systems in general and water sustainability issues in particular, are dealt with. So long as developing nations lack the technology and/or other resources to ensure their needs for a clean water supply, managing the same from a sustainability standpoint poses an even greater challenge. The mass exodus from rural areas to urban regions has further exacerbated the problem. While financial resources help alleviate the problem, it is not a cure-all and a system-wide approach needs to be taken to manage the basics of water supply, so as to truly advance sustainable water systems. The significance of knowledge has far outgrown the significance of its more tangible counterparts such as labor or land with knowledge management being central to any organization's success whether they be businesses, not-for-profits

*Both authors contributed equally and are listed in alphabetical order.

Handbook of Knowledge Management for Sustainable Water Systems, First Edition. Edited by Meir Russ.

or governmental bodies (Durst & Wilhelm, 2012). Knowledge Management Systems (KMSs) are information systems that can help organizations manage knowledge creation, multiple knowledge sources, its dissemination, storage, transfer and application (Alavi & Leidner, 2001). We use the example of South Africa to illustrate how a developing nation that seeks to address their water system issues, can emulate best practices from Sweden while utilizing knowledge management systems to resolve multiple water system issues which range from highly micro-level issues such as theft to the loftier goals of sustainability.

3.1 Population trends towards urbanization

With a significantly large amount of the world population living in urban areas, issues such as purification of water supply, managing water pollution, proper drainage, demand for fresh water and sustainability of water management systems are but some of the issues being faced by countries. Developing nations are at a further disadvantage as much of their urban growth and expansion is unplanned therefore making it harder for them to tackle these issues. Population growth, urbanization and globalization are the underlying reasons for the ever-growing demand for water, energy, food and land (Wilderer & Huber, 2011). According to the United Nations (UN) statistics (2014), more than of the half of the world's population lives in urban areas and the trend shows that developing countries will continue to move towards further urbanization.

Current UN predictions indicate that by 2050, the planet will be home to 9 billion people. With the constant migration to urban areas seemingly inevitable, 2050 projections indicate that depending on the region, 40–80% of the world population will be living in urban areas by that time. Numerous studies have shown that this is beyond sustainability threshold levels. Nevertheless, efforts must be made to safeguard our existing water supplies and make certain that they are satisfactory to sustain these population levels. This can be done by ensuring the widespread adoption of sustainability endeavors when it comes to water management systems.

Thus, these trends of urbanization and population growth in emerging economies indicate that there is and there will continue to be a very high demand of fresh water in those specific countries making sustainability a truly critical concern. While those are future projections, even presently we are faced with the same challenges regarding sustainability in high-demand countries. A lack of clean water is the ultimate poverty which needs to be prioritized to make sure that all people can access clean water supply in the growing urban areas, sanitation needs are met, wastewater nutrients are recycled as well as urban agriculture can avail of its irrigation needs (Niemczynowicz, 1999).

3.2 Water issues plaguing South Africa

South Africa is no stranger to rapid population growth. Overall, the population of Africa is expected to double by 2030 (United Nations Report, 2015). Rapid urbanization in the Sub-Saharan (including South African region) between 1990 and 2012 was accompanied by fewer people having access to improved drinking water supplies, nor could the water sanitation systems keep pace with this rapid urbanization (UN report, 2014). Therefore, currently less than 1% of the people in urban areas in South Africa have access to a new and improved drinking water infrastructure.

The water availability problem is quite concerning. South Africa is already using 98% of their available water supply (Thelwell, 2014). Urban South Africa depends on both groundwater and surface water for its day-to-day functioning, and as they expand, urban areas drain ever-larger landscapes (Showers, 2002). Poor water sustainability practices have contributed to the shortage of water in South Africa. Additionally, climate change has had a big impact on the regional water resources. Temperature levels around South Africa are predicted to increase between three to six degrees Celsius by the end of the century, and the country's rainfall is predicted to experience a great deal of variability as well (Ziervogel *et al.*, 2014). Other recent studies have proven that in terms of rainfall, Western and Southwestern South Africa will become drier, while the east of the country may become wetter over the next few decades (Muller, 2007).

Another issue is that 40% of the South African wastewater treatment systems are in a critical state (Thelwell, 2014). This poor state of the wastewater treatment systems is referenced in South African government reports as well, where it is stipulated that "only 7% of wastewater treatment systems complied with international standards in 2009" (Lincoln, 2011, p. 5). A transition to the worldwide accepted standards of water quality would indeed be both time consuming as well as require financial resources that the country lacks and yet, the problem is one that needs to be addressed urgently from both a utility and sustainability standpoint. South Africa is losing approximately 1.58 billion kiloliters of water per year due to faulty piping infrastructure that has outlived its lifespan ("South Africa lost 1.58 billion kiloliters", 2013). Yet not all the loss can be attributed to the poorly functioning pipes. Fifteen percent of the water loss was caused by water-theft. In South Africa, farmers steal water from these faulty systems for irrigation purposes given the dry conditions created by climate change (Barlowe & Clarke, 2002). In such extreme poverty conditions where theft is almost a necessity, it is indeed difficult for farmers to care about sustainability.

The largest Water Authority Board in South Africa is Rand Water which extracts the water from the Vaal Dam and Vaal Barrage and some additional underground sources. Before supplying water to their constituents, Rand Water purifies the water going through several steps like coagulation, flocculation, sedimentation, stabilization, filtration, disinfection and chlorination (Lincoln, 2011). Rand Water complies with the drinking quality standards adopted by

South African National Drinking Water Standard SANS 241 (Rand Water website). However, the Water Board has concerns about the water in both their main sources of extraction: Vaal Dam and Vaal Barrage. Vaal Dam is contaminated with various dissolved solids, chlorine, sulfate and increasing salinity. This contamination is a problem they are struggling with despite following required local standards. Their other water extraction source, Vaal Barrage is highly contaminated with fecal bacteria and toxic chemicals from industrial production such as sulfates, sodium, iron, manganese and heavy metals. Given that Rand Water is the biggest water authority in South Africa, this is indeed quite a concern since despite their local compliance and constant monitoring, the water quality continues to deteriorate. Thus, while all the steps taken by Rand Water are in line with local legislation that seeks to ensure clean water, there is still a great deal of room for improvement. Additionally while water purity is a concern that the legislation seeks to address, environmental sustainability has not historically been as crucial an issue. Treated and untreated industrial waste and some sewage has been known to contaminate the ecosystem in South Africa as they often end up in streams and oceans (Showers, 2002). The biggest water pollution problem is perhaps the industrial wastewater that contaminates both surface and groundwater with heavy metals like lead, nickel and molybdenum and the runoff from landfills (Showers, 2002). This is a threat to the fresh water supply that is already under attack by climate change, lack of supporting infrastructure and limited resources.

Because of poor sanitation, the water resources in South Africa contain five times the acceptable level of coliforms and bacteria and these continue to reproduce and spread within the pipes. Research on Limpopo, one of the poorest provinces of South Africa suggests that the water in the region was rarely disinfected and chlorinated, even through the municipal water infrastructures (Mellor, Smith, Samie, & Dilingham, 2013). This lack of attention can lead to illness, fatalities and damage the ecosystem irreparably. Therefore, overall, the quality of water in South Africa is regarded as subpar.

3.3 Evaluating South Africa

Even though the domestic amount of water consumption by country is stabilizing, the projected urban population growth will continue to place a great deal of pressure on natural water resources. This is especially true for the urban areas in developing countries. Wastewater treatment plants not only have to fulfil the current needs of the people for a clean, safe and stable supply but also secure the needs of future generations by ensuring the sustainable use and reuse of water. Lundin and Morrison (2002) propose a classification system evaluating the sustainability of urban water systems. Their lettered classification system is from a low D to a high A. Thus, regions could be lacking in their current water supplies, lacking focus on sanitation and universal water supply and have issues with maintenance even. At the middle levels would be those that meet quality standards and sustainability needs to a minimal or moderate level while at the highest level would be those much more focused on technology and sustainability. Using all tools at their disposal such as best practices, continuous improvement, infrastructures, new technologies, etc. countries at a lower level of urban water sustainability can and should improve their water systems to reach exemplary levels while those at higher levels should continue in their efforts to reach the very top and maintain those levels.

Another model that looks at the evolution of urban water systems as proposed by Brown and colleagues (2008) recommends that cities can move from lower stages to higher stages of water systems management. The lowest stages such as Water Supply City and Sewered City are those wherein water supply systems and sewage systems are established and most cities moved through these stages by the 20th century. However, in many developing countries, there are substantial sub-sections that are still failing in this respect. There are parts of South Africa where drinking water for all is still a pipe dream. The middle stages of Drained City and Waterways City are those where drainage systems are more evolved and pollution management is given credence. For many regions, the lack of focus on sustainability can be attributed to a lack of social awareness, of regulations or of funds. Urban cities in South Africa are only now becoming more aware of the significance of sustainability but their regulations are still minimal and there are other cities eager to become sustainable but lacking the necessary resources. Finally, the highest levels are the Water Cycle City and the Water Sensitive City. A Water Cycle City is one wherein issues of water supply security, public health protection and flood control are addressed while a Water Sensitive City is the highest level and can be perhaps regarded as the pinnacle with their use of cutting-edge technology and research. There is a great focus on sustainability, continuous learning and ensuring that there is a social evolution in this stage.

Figure 3.1 has been modified and adapted by combining both these models to indicate four levels from least to most-geared towards urban water sustainability. Each level greater, includes all that the previous level encompasses but builds up on that and does more for sustainability.

As can be seen, "Exceptional" depicts the highest level attainable, wherein clean technologies are used and the recycling processes exceed worldwide acceptable standards. Meanwhile at the other end of the spectrum we have "Insufficient" representing the lowest level of development, where some or all of the population is unable to receive the minimum necessary amount of clean water and sustainability does not receive its due recognition.

When evaluating South African water systems much of what characterizes the bottom of the spectrum, is true. The existing water systems and maintenance thereof are thus quite inadequate from an efficiency, safety and sustainability standpoint. "Sustainable development of surface and groundwater must involve a clear cost-benefit assessment and an optimal level of interaction between social

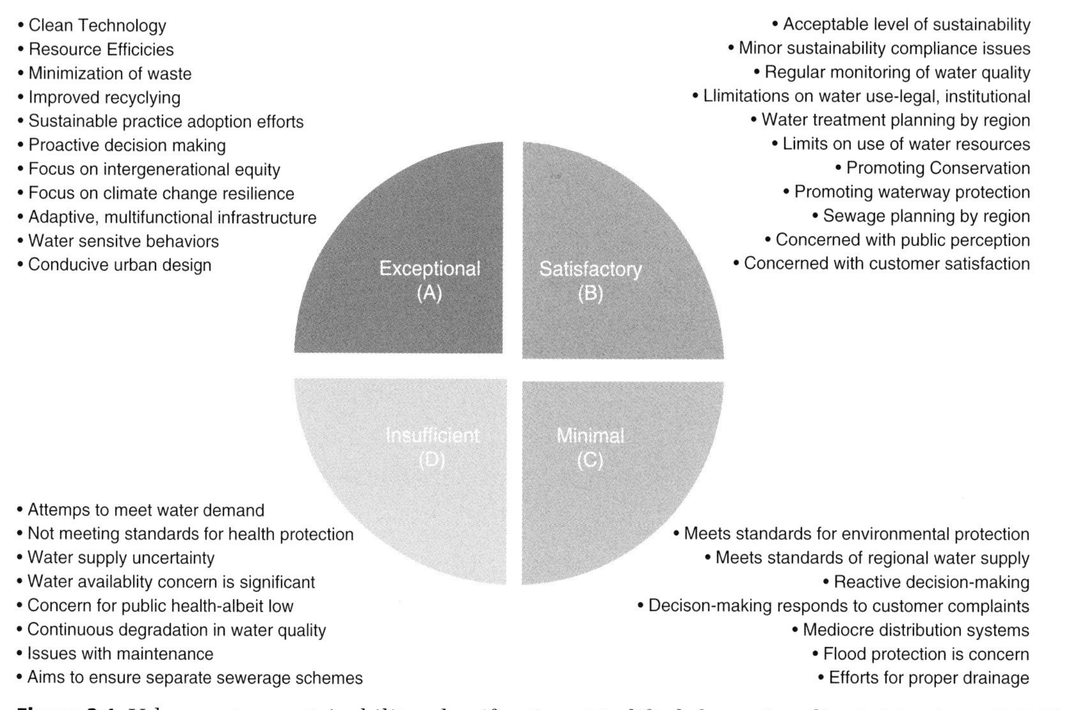

Figure 3.1 Urban water sustainability classification. Modified from: Lundin & Morrison (2002) and Brown *et al.* (2008).

and natural resource concerns" (Barbier, 1987, p. 102). For developing nations there is a dilemma when it comes to the cost-benefit aspect in that they may not be able to conserve and sustain safe water supplies while addressing both these social and natural concerns. Hence, it becomes imperative to utilize every tool at their disposal to ensure that they are supplying current needs in a sustainable fashion while doing so as efficiently as possible from an economic standpoint.

The entire South African system is thus in need of an overhaul as it lies in the "Insufficient" quadrant while showing promise to move into the "Minimal" quadrant with some efforts. The significance of implementing a knowledge management system that streamlines the entire process and that includes both individuals and organizations along with technology, cannot be overstated. However, before that, the outcomes sought need to be determined. As with industry, water systems too can benefit from institutional theory lessons (DiMaggio & Powell, 2000). Mimetic forces "encourage" regions to turn to their counterparts in different parts of the world and by looking at other countries, they can use their water systems as benchmarks which could help speed up the process of improving the local, regional system. This benchmarking could help determine what outcomes are desirable as well gain insight into the potential of the KMS.

3.4 Sweden – the aspirational model

Currently Sweden is regarded as one of leading countries in terms of sustainable water practices. Their geographical location near coastal islands and many lakes has motivated them to take water management and sustainability, quite seriously. Countries like South Africa can significantly benefit from existing Swedish sustainability knowledge. For example, the transformation of sludge into construction material is a unique practice that could be adapted and utilized in South Africa. Sweden with its broad hydrological resources, makes substantial attempts to conserve water. They have recognized that desalinization is a great method to maintain an ample fresh water supply. Every summer, cities like Gotland face the problem of very low levels of groundwater, which drove Sweden to search for efficient alternatives like desalination.

While South Africa is vulnerable to water supply issues due to its predominantly arid and semi-arid regions, it has a significant coastline and could benefit from Sweden's experiences and technologies to create similar desalinization plants. The continuous decrease in drinking water quality in South Africa would contribute to catastrophic issues if unchecked affecting the current generation and their failure to ramp up water sustainability would have consequences for future generations as well.

Until 1940s, Sweden had faced numerous problems related to the release of nutrients into water that were non-native to the water supply that caused hypoxia, fish death and waterborne epidemics (Ljunggren, Ericksson, & Unger, 2014). Other major issues that Sweden has had to contend with include flooding and challenges in recycling the agricultural nutrients (Lundin & Morrison, 2002; Nyberg, Evers, Dahlstrom, & Petterson, 2014). A Swedish EPA report from 2006 suggests that agriculture is the source of 44% of nitrogen discharges and 45% of phosphorus discharges (Ljunggren *et al.*, 2014). Despite their sufficiently large water resources, contamination was an issue due to toxic waste from municipal wastewater treatment plants. However, over the last few decades they have consistently worked towards improving their water management systems and sustainability endeavors. The collective consciousness for preserving the Swedish resources has evolved and the urban water systems in Sweden have moved to a more sustainable status, albeit slowly (Lundin *et al.*, 1999).

The biggest contribution perhaps to the water sustainability practices in Sweden, came from the existence and functioning of SEPA (Swedish Environmental Protection Agency), which was formed in 1967 and has continuously worked on improving the water quality ever since. The extensive measures taken by SEPA include the improvement of the urban wastewater treatment, which has continued to evolve over time. The wastewater treatment system has been modernized multiple times since 1945 from simple sludge removal to secondary biological treatment in 1950, tertiary biochemical treatment in 1980 and finally it is currently at the phase of special nitrogen removal that started in 1990s. These methods took many years as well as resources to evolve as a result of SEPA imposing limits to concentrations of different substances and chemicals, managing wastewater, creating and enforcing rules of inspection and timely sampling of water which were all managed in an organized manner (Ljunggren *et al.*, 2014). Managing the water systems in Sweden for current use and from a sustainability perspective, necessitates the use of knowledge management systems that help actors in a multitude of ways such as to navigate the legal-political frameworks or assist in the management of ad hoc projects (Hellström, Jeppsson, & Kärrman, 2000).

Sweden not only endeavors to protect their current water supply for consumers, ensure sustainable water use for the future and protect the ecosystems, but also ensures that by utilizing tools, technology, and including all actors involved, that it has an integrated system in place to manage knowledge in the urban water schema. Significant legislation that favors sustainability is in place that underlies their efforts such as the Urban Waste Water Directive, which imposes water quality standards that have to be met by various industries. Sweden adopted 16 environmental quality objectives to attain by 2020, five directly related to water, some others indirectly related. Among these measures are the requirements for a non-toxic environment, zero eutrophication, natural acidification only, flourishing lakes and stream, a balanced marine environment, thriving wetlands, a good built-environment and good quality groundwater. Each of these objectives is accompanied by specific measurable goals and deadlines. All water systems have their own guidelines even small-scale local installations for wastewater properties that are not connected to the municipal wastewater treatment plants, are provided specific recommendations (Ljunggren *et al.*, 2014). The conscientiousness of Sweden as a whole towards water issues is reflected in their adoption of a complex net comprised of laws, national and regional objectives and sustainability recommendations.

Sweden engages in a lengthy process to ensure that all water is treated correctly using mechanical, biological and chemical treatments following the high standards set (Ljunggren *et al.*, 2014). Steps such as nitrogen removal are also taken to ensure that the water is safe for consumption as well as for when it is released back into the ecosystem. Other steps are also included as part of the process. In Goteborg, for example, like other regions, the city of Goteborg has very specific challenges and objectives set, including those regarding the recycling of nutrients, decrease in hazardous substances and increased knowledge of material flows. The city's wastewater treatment plant produces about 50,000 tons of sludge, which is re-used rather than dumped. The resultant sludge from this process is used as construction soil on golf courses or as noise barriers alongside roads since it consists of peat and bark (Novotny & Brown, 2014).

Sweden has thus, made great strides when it comes to (1) cleaning their existing water supply; (2) establishing legislation and guidelines for the present and future; (3) establishing methods to manage knowledge about the systems in place; (4) positively impacting the health of its citizens; (5) protecting their aquatic resources such as lakes and streams; and (6) ensuring the sustainable use of water all over the country including in urban areas. Through these endeavors, by referencing Figure 3.1, it can be seen that Sweden firmly plants its feet in the "Satisfactory" level though it is working towards the "Exceptional" quadrant through constant improvements.

3.5 Urban water sustainability

Sustainability is crucial to achieving a good quality of life while conserving the environment for future generations. Urban water management is a system that encompasses four major areas – water supply, solid wastes, urban drainage and flood management, and sanitation (Bahri, 2012). All these elements interact with each other and lead to environmental issues if proper management is ignored. These components are highly sensitive to variables like institutional legislation, people in management and urban planning. Similar to the business analogy, urban water sustainability should address issues of the product (urban water in this instance), processes, actors involved and implication for stakeholders (Robinson, Anumba, Carrillo, & Al-Ghassani, 2006). In the sustainability context of water management this would mean involving the stakeholders like consumers, society, water treatment institutions, individuals at all levels of government, and executive committees that will enforce sustainable management of these elements in order to achieve the final desired product: safe water today and a sustainable future tomorrow.

One of the key ways of achieving those end goals is to coordinate the water services to embrace innovative water treatments that will create uses of unnecessary and toxic waste and transform it into a useful final product. Sweden presents an excellent model to follow when it comes to innovative water treatment systems. As mentioned earlier, the removal of sludge and reutilization in the construction industry is one such instance of innovative water use. Legislation is another important way to move closer to the end goals, which enforces what norms must be followed, what quality standards must be met, as well as what technology should be adopted for sustainable development. Elected officials and administrative services create and enforce the laws that permit the use, reuse and recovery of damaged water at the local level. Additionally, there is a socio-cultural aspect that needs to be assessed. The constituents of various municipalities need to be informed by various sources regarding the existing problems. Their opinions need to be accounted for and local customs and traditions must be acknowledged as well. The cultural aspect will always affect the institutional legislation because the embedded "water values" differ dramatically from one country to another. Social denial of existing sustainability problems creates a burden on much-needed reform and drive the water quality into stagnation and degradation.

It must be remembered that implementing knowledge management systems for sustainability are difficult to achieve and costly. Moreover, the periodic screening of water quality in different locations requires human capital that is knowledgeable in the sustainability field. A nation like South Africa, with a lot of uncertainty about the day-to-day supply for physical needs, has much to contend with. Their local wastewater systems are not sustainable nor clean – the contamination levels and lack of proper cleaning technologies, make them conducive to the spread of bacteria. Nations that suffer from a lack of funds and resources still struggle to meet the basic water demands by failing to keep pace with the urbanization and demographic changes. At the same time, their sustainability endeavors and practices are inconsistent. Under these circumstances, it is critical that experts educate people about how to conserve their water resources better. This would be possible by sending international water sampling researchers from more developed to developing countries as well as encouraging local experts to visit developed countries and see their urban water management systems in use. Thereafter, it is important to ensure that all these conservancy efforts and the significance about the quality of their water and its preservation are communicated to the local population.

Funding urban water sustainability education programs in developing countries could also impact the public consciousness, local efforts and the sustainability narrative in general. The developed countries need to be cognizant of the issues faced by developing nations and provide timely assistance on humanitarian grounds as well, because sustainability affects us all. In order to prevent the deterioration of the quality of our water, developed countries should support international funds that target urban water sustainability.

3.6 Knowledge Management Systems (KMSs)

At the outset, it bears remembering that data, information and knowledge are distinct from one another with information being more in the nature of transactional data being organized in a meaningful way while knowledge is the next higher level of understanding that also involves the people and social interactions (Bose, 2003; Sveiby, 2001). Knowledge is regarded as one of the most unique and important resources possessed by an organization. Knowledge is crucial for better decision-making that encourages greater participation by all actors and a lack of knowledge could lead to unsustainable resource use (Lemos, Bell, Engle, Formiga-Johnson, & Nelson, 2010). Knowledge Management Systems ensure that organizations use their tools, technologies and the people of the organization as well as their connections to other actors to benefit the organization and optimize all of the knowledge-related functions.

Knowledge Management Systems (KMSs) have three major purposes which are primarily the codification and sharing of a variety of best practices, creating knowledge directories and databases and the creation of knowledge networks that enable the organization to achieve their desired end-goals in the most efficient manner possible (Alavi & Leidner, 2001). It is crucial that the information systems are not just looked at as a quick technological fix but rather as complementary to the knowledge activities of individuals and organizations as a whole, organizational culture, structure and processes (Xu & Quaddus, 2005). Organizations can facilitate knowledge accumulation and knowledge sharing through the KMS. KMSs are thus crucial to documenting tacit knowledge and keeping track of explicit knowledge and internal knowledge experts. KMSs provide a computerized decision system that helps address specific challenges and fulfil specific needs of the actors by providing access to knowledge throughout the system.

Frequently KMSs are only thought of as tools for business but they can play a crucial role for any organization and the public sector can greatly benefit from utilizing a KMS to streamline their processes for efficiencies and increased effectiveness. Whether or not a KMS can be successfully adopted by an organization, depends on multiple factors. Technology Innovation needs such as (1) benefits from the KMS; (2) complexity of the KMS; (3) how compatible is it with the local needs; as well as the organization's situation; (4) whether the organization has sufficient resources; (5) technology competencies of the actors involved; (6) support systems in place, all impact the KMS system chosen, adopted and implemented (Wang & Wang, 2016).

3.7 Knowledge Management for urban water sustainability in South Africa

Due to climate change, the increased number of pollutants, chemical run-off, industrial accidents, human waste and lack of investment in better water treatment systems, urban water sustainability is a crucial issue that should be at the forefront of every country's environmental efforts.

Knowledge Management Systems (KMSs) are crucial from a long-term perspective- an uncertain future requires a level of flexibility that would help municipalities across South Africa at different stages of modernization and with actors having different levels of expertise, to respond to the emerging changes and threats to water safety and sustainability. There is a great need for more focused efforts in adopting and implementing a KMS that not only manages existing supplies of water, prevents waste and ensures that water standards are being met, but does so while promoting water sustainability. There is a growing awareness in South Africa about the importance of knowledge and KMSs and that local contextual elements such as diversity of actors, socio-political conditions and cultural elements, all need to be heeded (Kruger & Johnson, 2013).

To optimize the process of developing and adopting a KMS, it is important to ensure that all elements of a KMS are focused upon. The creation of knowledge itself is a crucial element. An in-depth analysis of social, technical and management fronts will facilitate understanding the context of that particular urban area. Knowledge creation can only be developed by building the necessary knowledge networks. Bringing the most knowledgeable local experts together with experts in the field of sustainable performant water systems and KMS experts. Thus, it is important to bring together all the concerned actors, local and regional water authorities, environment-related authorities, consultants, water treatment/sanitation experts and even the consumers from business and agriculture. South Africa has shown a tendency to be willing to build and use online, social platforms for knowledge sharing in the form of communities of practice that are already quite widespread across the public sector (Mkhize, 2015). These can be integrated into a KMS or similar internal system made available as part of the KMS, while also including other stakeholders.

It is thus possible to co-create value in the KMS, i.e. the providers and the consumers/beneficiaries collaboratively create unique value for everyone during the process (Prahalad & Ramaswamy, 2004). Co-creation involves the providers of a service and the recipients as well as any other stakeholders, to all come together and work on creating value in a way that all parties are able to benefit and move forward collectively (Prahalad & Krishnan, 2008). Bringing these actors together will add multiple perspectives and bring new insight and interactions as well, ultimately leading to the creation of knowledge management networks. A KMS includes the technological components as well as the social components which are the various actors involved in the process. Thus, another outcome from these meetings is the cyclical creation of explicit knowledge from tacit and of tacit knowledge from explicit. Throughout this urban water sustainability knowledge creation process, the knowledge will flow from some groups of experts to others, which will create intranets and extranets of knowledge.

Figure 3.2 depicts the process of adopting a KMS. Based on what is known about KMSs in general as well as looking at the South African urban water sustainability system in particular, we develop and depict a streamlined structure

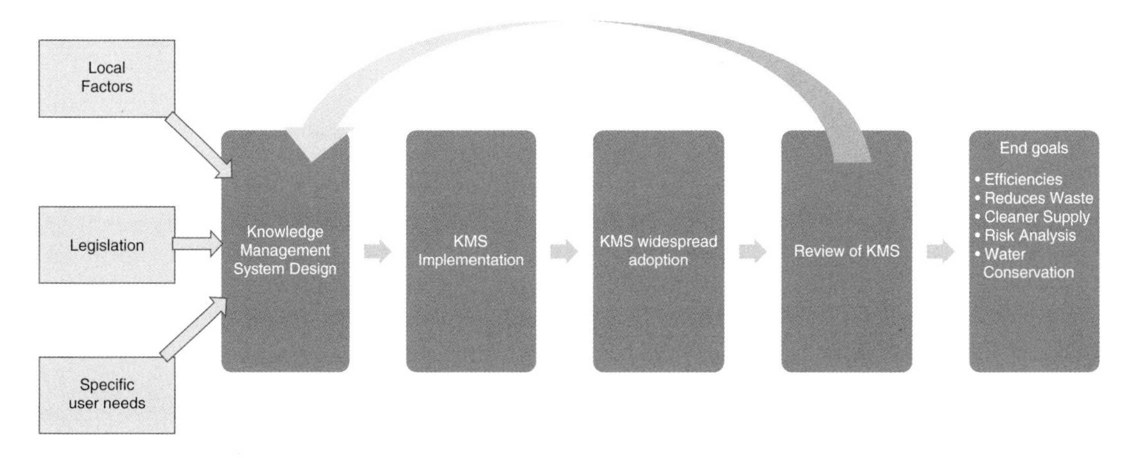

Figure 3.2 Adoption and implementation of KMS for urban water sustainability in South Africa.

of what this process entails. The end-goals must be determined by the country in question (in this instance, South Africa). This can be done by determining the issues currently faced by the country as well as by looking at what other nations (in this instance, Sweden) have been able to achieve. The effect of the initial research phase is that particular issues that a region is facing, can be identified. Context-specific environmental factors must be kept in mind as it is within the context of local rules and regulations that the KMS must operate. For South Africa, there are other local issues when it comes to knowledge management such as a large number of languages being used and different communication styles (Kruger & Johnson, 2013). The urban water planning bodies have the important role of selecting the appropriate alternative resources of water, like various types of freshwater and wastewater (black, yellow, brown, grey etc.) for the right purposes, while following the legislation in place. The urban water sustainability framework is highly dependent on specific goals, legislation and urban planning. Government agencies must possess a high level of understanding of new and safe water treatment technologies available. Meanwhile, periodic assessment of the goals and achievements are necessary to learn from mistakes and improve upon them as can be seen at the review stage which can then lead back to redesigning the KMS resulting in a feedback loop and continuous improvement through ongoing learning.

Another notable aspect, is that actors will be able to rely on the created knowledge once the KMS is used to build databases which will store the accumulated local urban water sustainability knowledge. After successfully creating knowledge and ensuring storage, the next step would be to efficiently transfer that knowledge to the relevant organizations of experts. It is very important that the knowledge databases are complete and relevant and that modes of knowledge transmission are appropriate. Thus, the main role of this comprehensive stored knowledge, is to provide access to a retrievable and useful database for all the interested institutions, organizations and individuals concerned with knowledge transfer and dissemination. Transparency and knowledge sharing enable the knowledge management process to be more impactful. The final step is to make sure that this knowledge is successfully applied. Frequently the knowledge that is made available could be overwhelming without real life application experience. It is important to understand how to successfully extract the particular knowledge needed for specific issues to then target specific outcomes. There is no better application of knowledge than the real experience in applying that knowledge. In the long run, the learning and knowledge application process will allow even more knowledge flow and creation, flowing like a spiral upwards, enhancing the knowledge horizons of sustainable development.

Figure 3.3 shows the four main stages of knowledge creation, knowledge structuring, knowledge dissemination and knowledge application (Bose, 2003) as they are applicable to the urban water sustainability situation of South Africa. By conducting a detailed review into the urban water sustainability issues plaguing South Africa, we were able to classify the much-needed solutions into these four steps of the knowledge management process. The cyclical process

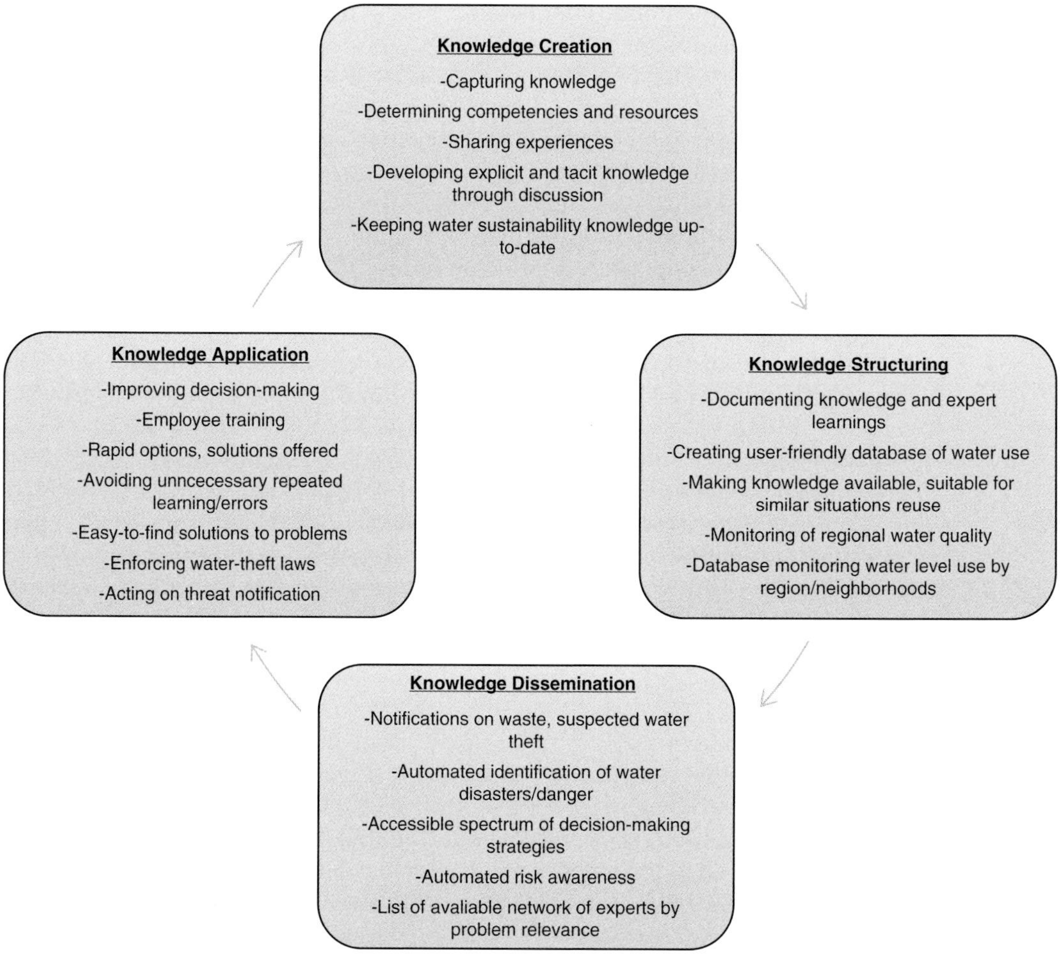

Knowledge Creation

-Capturing knowledge
-Determining competencies and resources
-Sharing experiences
-Developing explicit and tacit knowledge through discussion
-Keeping water sustainability knowledge up-to-date

Knowledge Structuring

-Documenting knowledge and expert learnings
-Creating user-friendly database of water use
-Making knowledge available, suitable for similar situations reuse
-Monitoring of regional water quality
-Database monitoring water level use by region/neighborhoods

Knowledge Dissemination

-Notifications on waste, suspected water theft
-Automated identification of water disasters/danger
-Accessible spectrum of decision-making strategies
-Automated risk awareness
-List of avaliable network of experts by problem relevance

Knowledge Application

-Improving decision-making
-Employee training
-Rapid options, solutions offered
-Avoiding unncecessary repeated learning/errors
-Easy-to-find solutions to problems
-Enforcing water-theft laws
-Acting on threat notification

Figure 3.3 Knowledge management stages applicable to urban water sustainability in South Africa. Modified from: Bose (2003).

starts at knowledge creation and moves through the remaining stages. However, as is always true for a sound KMS, the process is a continuous and on-going one. Thus, at the knowledge application stage, a feedback loop is present that evaluates and allows valuable lessons to be taken back to the knowledge creation point. Feedback loops are crucial for system-wide learning. While at the outset these may be single-loop, i.e. for incremental learning, in time they should move towards being double-loop wherein the learning is more in line with the systems-level of thinking and is generative, leading to fundamental changes in assumptions (Rubenstein-Montano, Liebowitz, Buchwalter, McCaw, Newman, Rebeck, & The Knowledge Management Methodology Team, 2001).

As stated earlier, the significance of financial resources cannot be understated. Obtaining funding to support the KMS is crucial. However, the financially vulnerable urban areas in developing nations do indeed have difficulty accessing funding. They have very limited access to monetary resources and will struggle to find funding when it comes to ensuring that the KMS adoption, implementation and continuance remains well-funded. One option could be to seek funding from international, non-governmental agencies, non-for-profit or water funding programs like the "Sustainable Water Fund Grant" or the "World Bank Water Partnership program" which are funded by the World Bank for supporting global water sustainability.

Numerous researchers have examined water sustainability in multiple contexts while others have looked at urban supplies of water (Bahri, 2012). However, our attempt here is to take their efforts further and look at how a KMS can be utilized to address current and future issues faced by developing nations with a particular emphasis on urban water sustainability. Urban water sustainability is a dynamic issue comprised of many unpredictable elements. For a more efficient adoption of sustainable water management practices, it is necessary that various countries create their own personalized water knowledge management systems. From a practical perspective, it is not always feasible to implement a KMS as-is, but countries that have faced similar issues in the past and now have similar goals, offer a template that can then be adapted for the local context. Before implementing a KMS, it is important to be cognizant of how the system will enable the different actors to achieve the goals sought.

Table 3.1 provides a detailed summary of the major issues related to urban water sustainability faced by South Africa along with the more specific ways that a KMS may deal with said issues. The issues were identified after an extensive review of literature pertaining to water issues of the region.

Thus, there is much that a KMS can achieve in terms of urban water sustainability in South Africa. By taking a systems perspective (Von Bertalanffy, 1972) of the entire urban water suitability issue, knowledge can be managed with the inclusion of all stakeholders in an optimal manner.

3.8 Conclusion

When sustainability first rose as a concern, it brought with it an effort to manage water supplies. Coupled with technological improvements, the use of water purification systems began to arise. We have come a long way from the first purification technologies in medieval times wherein removal of particulate matter was achieved by means of sedimentation; yet there is still a way to go. At the same time our populations have risen, people have moved from rural regions to urban cities and with them urban water systems have begun to have their own unique issues and problems. With the increased awareness of the importance

Table 3.1 Synopsis of issues and KMS's potential

Main urban water sustainability goals	Specifics within the goal
Reduce water theft	• Meter all water connections • Detect areas and amount of theft • Send local authorities for investigation and punishment
Reduce water degradation and maintain water quality	• Frequently sample regional water quality • Determine the contaminated water sources • Identify the origin of toxic substances • Create underground dams to manage wastewater spillage.
Supply municipalities with the necessary amount of water	• Create municipal dual distribution systems: potable and treated water • Use the desalination technologies to cover the increasing water demand • Encourage people to use reclaimed water • Demand management
Water conservation	• Determine impact of improved fixtures on water conservation • Facilitate education regarding water use, repairs • Manage desalination systems • Manage extraction, treatment and the use of low-quality ground water

of conservation and sustainability in urban water systems, developing countries have continued on a quest to determine the best way of managing knowledge. Through this chapter we sought to lay out the significance of an appropriate knowledge management system that can enable a developing country (like South Africa) end its struggle with urban water sustainability issues and move towards the standards set by more developed nations (such as Sweden).

References

Alavi, M., & Leidner, D. E. (2001). Review: Knowledge management and knowledge management systems: Conceptual foundations and research issues. *MIS Quarterly*, 107–136.

Bahri, A. (2012). Integrated urban water management. *GWP Tec background papers. Global Water Partnership, Stockholm, 16.*

Barbier, E. B. (1987). The concept of sustainable economic development. *Environmental Conservation*, *14*(02), 101–110.

Barlow, M., & Clarke, T. (2002). *Blue gold: The fight to stop the corporate theft of the world's water*. The New Press, New York.
Bose, R. (2003). Knowledge management-enabled health care management systems: capabilities, infrastructure, and decision-support. *Expert Systems with Applications*, *24*(1), 59–71.
Brown, R., Keath, N., & Wong, T. (2008). Transitioning to water sensitive cities: historical, current and future transition states. In *11th International Conference on Urban Drainage* (Vol. *10*). Edinburgh, UK.
DiMaggio, P. J., & Powell, W. W. (2000). The iron cage revisited institutional isomorphism and collective rationality in organizational fields. In *Economics Meets Sociology in Strategic Management* (pp. 143–166). Emerald Group Publishing Limited, Bingley, UK.
Durst, S., & Wilhelm, S. (2012). Knowledge management and succession planning in SMEs. *Journal of Knowledge Management*, *16*(4), 637–649.
Hellström, D., Jeppsson, U., & Kärrman, E. (2000). A framework for systems analysis of sustainable urban water management. *Environmental Impact Assessment Review*, *20*(3), 311–321.
Kruger, C. N., & Johnson, R. D. (2013). Knowledge management according to organisational size: A South African perspective. *SA Journal of Information Management*, *15*(1), 1–11.
Larsen, T. A., & Gujer, W. (1997). The concept of sustainable urban water management. *Water Science and Technology*, *35*(9), 3–10.
Lemos, M. C., Bell, A. R., Engle, N. L., Formiga-Johnsson, R. M., & Nelson, D. R. (2010). Technical knowledge and water resources management: a comparative study of river basin councils, Brazil. *Water Resources Research*, *46*(6).
Lincoln, J. (2011, December). Water Treatment technologies in SA. *Swiss Business Hub South Africa*, 1–9. Retrieved November 16, 2016, from http://www.s-ge.com/sites/default/files/private_files/Water Treatment technologies South Africa_SBHSA_December-2011.pdf
Ljunggren, R., Eriksson, M., & Unger, M. (2014). *Wastewater Treatment in Sweden*. Arkitektkopia, Stockholm.
Lundin, M., Molander, S., & Morrison, G. M. (1999). A set of indicators for the assessment of temporal variations in the sustainability of sanitary systems. *Water Science and Technology*, *39*(5), 235–242.
Lundin, M., & Morrison, G. M. (2002). A life cycle assessment based procedure for development of environmental sustainability indicators for urban water systems. *Urban Water*, *4*(2), 145–152.
Mellor, J. E., Smith, J. A., Samie, A., & Dillingham, R. A. (2013). Coliform sources and mechanisms for regrowth in household drinking water in Limpopo, South Africa. *Journal of Environmental Engineering*, *139*(9), 1152–1161.
Mkhize, P. L. (2015). A knowledge sharing framework in the South African public sector: original research. *South African Journal of Information Management*, *17*(1), 1–10.
Muller, M. (2007). Adapting to climate change water management for urban resilience. *Environment and Urbanization*, *19*(1), 99–113.
Niemczynowicz, J. (1999). Urban hydrology and water management–present and future challenges. *Urban Water*, *1*(1), 1–14.
Novotny, V., & Brown, P. (2014). Cities of the Future-Towards integrated sustainable water and landscape management. *Water Intelligence Online*, *13*, 9781780405308.
Nyberg, L., Evers, M., Dahlstrom, M., & Pettersson, A. (2014). Sustainability aspects of water regulation and flood risk reduction in Lake Vänern. *Aquatic Ecosystem Health & Management*, *17*(4), 331–340.

Prahalad, C. K., & Krishnan, M. S. (2008). *The New Age of Innovation*. McGraw-Hill Education, New York.

Prahalad, C. K., & Ramaswamy, V. (2004). Co-creation experiences: The next practice in value creation. *Journal of Interactive Marketing*, *18*(3), 5–14.

Rand Water and Infrastructure Management. (n.d.). Retrieved November 16, 2016, from http://www.randwater.co.za/WaterAndInfastructureManagement/Pages/WaterQuality.aspx

Robinson, H. S., Anumba, C. J., Carrillo, P. M., & Al-Ghassani, A. M. (2006). STEPS: a knowledge management maturity roadmap for corporate sustainability. *Business Process Management Journal*, *12*(6), 793–808.

Rubenstein-Montano, B., Liebowitz, J., Buchwalter, J., McCaw, D., Newman, B., Rebeck, K., & Team, T. K. M. M. (2001). A systems thinking framework for knowledge management. *Decision support systems*, *31*(1), 5–16.

Showers, K. B. (2002). Water scarcity and urban Africa: An overview of urban–rural water linkages. *World Development*, *30*(4), 621–648.

South Africa lost 1.58 billion kiloliters - report. (2013, June 23). Retrieved November 11, 2016, from http://www.iol.co.za/news/south-africa/sa-lost-158-billion-kilolitres---report-1536163#.UpqXQMRDtzU

Sveiby, K. E. (2001). A knowledge-based theory of the firm to guide in strategy formulation. *Journal of Intellectual Capital*, *2*(4), 344–358.

Thelwell, E. (2014). *South Africa's looming water disaster*. Retrieved November 16, 2016, from http://www.news24.com/SouthAfrica/News/South-Africas-looming-water-disaster-20141103

United Nations Report (2014). World Urbanization Prospects: The 2014 Revision, Highlights. Department of Economic and Social Affairs. *Population Division, United Nations*.

United Nations Report (2015). *The United Nations world water development report 2015: Water for a sustainable world* (Vol. *1*). UNESCO Publishing.

Von Bertalanffy, L. (1972). The history and status of general systems theory. *Academy of Management Journal*, *15*(4), 407–426.

Wang, Y. M., & Wang, Y. C. (2016). Determinants of firms' knowledge management system implementation: An empirical study. *Computers in Human Behavior*, *64*, 829–842.

Wilderer, P. A., & Huber, H. (2011). Integration of water reuse in the planning of livable cities. *Intelligent Buildings International*, *3*(2), 96–106.

Xu, J., & Quaddus, M. (2005). Exploring the perceptions of knowledge management systems. *Journal of Management Development*, *24*(4), 320–334.

Ziervogel, G., New, M., Archer van Garderen, E., Midgley, G., Taylor, A., Hamann, R., Stuart-Hill, S., Myers, J., & Warburton, M. (2014). Climate change impacts and adaptation in South Africa. *Wiley Interdisciplinary Reviews: Climate Change*, *5*(5), 605–620.

4 A Knowledge Management model for corporate water responsibility

Fabien Martinez

EM Normandie, Métis Lab, Dublin Campus, 19–21 Aston Quay, Dublin 2, Ireland

Introduction

This chapter explores how Knowledge Management can play a part in enhancing the capacity of business organizations to contribute to the sustainable management of water resources. Industry leaders in different sectors (e.g. General Motors, Ford, Toyota, Intel, Nestlé, Unilever and Coca-Cola) recognize fresh water as the Earth's most valuable and fastest depleting resource, and its availability a more critical problem than energy conservation (Caplan, Dutta, & Lawson, 2013). Their production sites are often located in regions currently enduring, or forecasted to endure, water stress, water scarcity or water flooding which represents a direct threat on their expanding operations, production levels, profit margins, and even 'license to operate' (Barton, 2010; Chalmers, Godfrey, & Lynch, 2012). Despite the risks, businesses are generally argued to have limited knowledge about how to embed water management within core business activities (Egan, 2015). Only a few academic studies have focused attention on the critical role and impact of business activities on water (Kurland & Zell, 2010; Lambooy, 2011; Martinez, 2015). Until then, scholarly interest had typically focused on the management of water resources by water utilities (e.g., B. Harvey & Schaefer, 2001; Ogden & Watson, 1999). The term "corporate water responsibility" was recently coined and a framework was articulated that identifies the management competences and organizational capabilities that are necessary for the sustainable management of water by companies – e.g., willingness-to-act, trust-building, partnering, innovating, long-termism, going beyond compliance, integrating stakeholders' views (Martinez, 2015). These competencies and capabilities are however rarely found, or to some extent

Handbook of Knowledge Management for Sustainable Water Systems, First Edition. Edited by Meir Russ.

disregarded, in traditional business-led, technology-driven innovation, in which water management is assumed to represent a subsidiary concern. What is more, stakeholders' participation and bottom-up approaches at the micro-level are often unlikely to contribute effectively to sustainable water management without the involvement of the business community. The powerful macro- and meso-level processes and challenges that characterize water management (Biswas, 2008) mean that companies are often seen as the only institutions in the modern world that are large, powerful and pervasive enough to foster sustainability progress (Hawken, Lovins, & Lovins, 2002).

The lack of specific knowledge about how to manage water as critical for sustainable development is highlighted in Martinez (2015) as a major obstacle to corporate water responsibility. The historical disregard of water as physically contained and isolated from people has tended to facilitate the widespread adoption of unsustainable techniques of management. Individuals and organizations have largely overseen the idea that water can be changed as a result of human usage and placed all the responsibility for maintaining our relations with water to experts (Linton, 2010). The introduction of the integrated water resources management framework by the Global Water Partnership (GWP, 2000) was a step toward a different conception of water management. It fundamentally relied on the assumption that water and people constitute great potential for changing each other in ecologically healthy and socially just ways by considering the existence of water systems composed of a variety of water users (e.g., domestic, industrial, agricultural, recreational, and navigational). The integrated water resources management framework promoted collaboration between water users as a means to fostering the development, allocation and monitoring of water resource use in the context of sustainable development (Cuickshank & Grover, 2012). Despite its laudable intentions, epitomized by a public call by the World Bank and UNDP for the creation of a global water partnership at the Stockholm Symposium in August 1995, the framework has generated a number of criticisms, mainly in regard with its tendency to disregard knowledge that does not meet positivist definitions of validity and generalization. The consequence is a general failure to account for the social causation processes involved in (and social constructionist explanations on) water management (Cook & Spray, 2012; Wissenburg, 2013). From a practitioner and academic perspective, Knowledge Management is an interesting way of exploring the capacity of business to address water-related issues because it is equally concerned with the challenge of handling positivist and socially constructed data.

Quintas, Lefrere, and Jones (1997) defined Knowledge Management as the process of critically managing knowledge to meet existing needs, to identify and exploit acquired knowledge assets and to develop new opportunities. The process is framed by McAdam and McCreedy (1999); and Boisot (1987) as a matter of accounting for both codified and un-codified knowledge. What this model entails that is useful to our understanding of corporate water responsibility is that it fundamentally depends on a firm's capacity to exploit both the knowledge that is accessible to all, and readily prepared for transmission purposes (codified), and the knowledge that is embedded in the experience of individuals, and cannot be readily prepared for transmission purposes (un-codified). In this chapter, it is hypothesized that water abstraction, usage and disposal patterns are critically underpinned by a complex nexus of codified and un-codified knowledge that are hardly combinable. Progress in relation to corporate water responsibility may well depend on a firm's capacity to manage these different categories of knowledge, notably by reaching out to the individuals and communities who are affected by water quality and quantity variations, and may hold specific information about how to solve the related problems. An argument in this sense is that companies may manage water in sustainable ways if they operated as systemic entities – i.e., within a network (or community) of interdependent parts – and if water was no longer perceived merely as a quantifiable commodity, easily accessible and exploited to achieve economic objectives. The idea is to develop a model that explains how codified and un-codified knowledge can be simultaneously managed to help the firm reduce information asymmetries (Montiel, Husted, & Christmann, 2012) about the water that is withdrawn by various actors from rivers, lakes and aquifers. An extended, and distinctively critical realist, model of this kind may help corporate managers to substantially improve their understanding of the complex socio-hydrological realities and challenges of our time, on the basis of which they will be equipped to develop a more "ethical" and equitable way of abstracting, using and disposing of water resources.

In practice however, the general reliance of firms on Knowledge Management systems that foster competitive advantage implies a narrow conception of Knowledge Management as a process that mobilizes the stakeholders who are able to suit the concept to the economic interests of the firm rather than the wider social interests of neighbouring individuals and communities. It follows that firms submitting to a narrow conception of Knowledge Management, in which water is likely to be merely regarded as a free good not worthy of strategic consideration (Jones, 2010), are not likely to perceive the full scope of opportunities in water management. The study of corporate water responsibility might thus greatly benefit from a reconsideration of Knowledge Management processes. This chapter contributes in a preliminary way to discussions of the core components of a Knowledge Management framework for corporate water responsibility. The model is likely to be useful to corporate managers who consider water as essential to sustaining life rather than merely an economic good yielding immediate reputational and financial benefits, and are willing to facilitate, and perhaps accelerate, the transition toward a more sustainable water management system.

4.1 Corporate water responsibility as a socially oriented process

Martinez explains that "the ultimate goal of corporate water responsibility is that companies contribute to ecological integrity via the efficient and equitable abstraction, usage and disposal of water resources" (2015, p. 141). There is no shortage of reports documenting firms' competencies at enhancing water efficiency on the basis of water footprint[1] analysis as one way of reducing water use intensity and mitigate business water-related risks (e.g., DEFRA, 2011; SABMiller & WWF, 2009; The Coca Cola Company and The Nature Conservancy, 2010). Because water scarcity is typically caused by natural (arid climate, intermittent drought years) and man-induced (e.g., desiccation of the mindscape driven by land degradation, population-driven water stress) phenomena (Falkenmark, Lundqvist, & Widstrand, 1989), the idea of putting efficiency and productivity at the heart of any water management systems has seemed credible and legitimate as a positivist and quantitatively "sound" solution (Tacconi, 1998). That also meant that firms have typically undertaken at most slight adjustments of existing metrics of "efficiency" to address water-related issues. They have tended to operate within the comfort zone of long-held business competencies, dominated by technical "end-of-pipe" solutions to monitor water quality, as well as technological innovation (e.g., water-saver flushing toilets, low-flow drip-irrigation systems) and market mechanisms (e.g., pricing water) to foster demand reduction. These trends were somewhat supported by the emergence of a quantitative view of water as part of the process that enabled science to pursue and entrap "nature as a calculative coherence of forces" (Heidegger, 1977, p. 291). Quantification yielded estimates of the stock of water, captured the limits of supply, presupposed the prospect of scarcity, and promoted the exercise of allocative power over what became, in this context, a finite resource (Linton, 2010). This provided water managers with a substantial amount of codified knowledge upon which conceptions of water as a commodity, and of water management as an efficiency-based process, have prospered. But this also created an overemphasis on utilitarian and managerial attitudes toward water resources that, in an era of globalized corporate activities, have fostered a view of water as a deterritorialized, dematerialized and metaphorically abstract resource. What this implies is a general denial of the reality of local, specific human–environmental relationships and relational dialectics that create, sustain or undermine water consumption patterns (D. Harvey, 1996; Linton, 2010). Pahl-Wostl, Mostert, and Tàbara (2008) explain that in many places, the natural dynamics of the river environment have been destroyed as a consequence of an overly instrumental "prediction and control" approach to water management.

Many companies have realized that "end-of-pipe" solutions and efficiency metrics are not sufficient on their own to warrant sustainable water management in the face of both the geo-spatial complexity of water management (Money, 2014) and the potential disastrous hydrological consequences of climate change (Linton, 2010). The recent water governance challenges faced by Coca Cola (e.g., Hoffman & Howie, 2010; Karnani, 2012) and Nestlé (e.g., Mehta, Veldwisch,

[1]The water footprint concept was introduced in 2002 as an analogue of the ecological footprint concept originating from the 1990s (Hoekstra, 2009). Water footprint and ecological footprint, according to Hoekstra (2009), should be regarded as complementary in the sustainability debate.

& Franco, 2012; Sojamo & Larson, 2012) demonstrate that corporate water responsibility requires a much sharper understanding of local contingencies and knowledge structures influencing access to clean water, food security, basic sanitation, and ecological integrity. That implies the need to consider the overall societal context – in particular the critical role of the socially constructed forms of management of water resources. Here, I notably include the normative imperatives of encouraging collective action that involves local communities in order to overcome unsustainable water management at various levels – e.g. individual, family and organizational levels (Rist, Chidambaranathan, Escobar, Wiesmann, & Zimmermann, 2007). Water management may therefore benefit from the formulation of a more socially oriented process of knowledge production that links the firm with an extended stakeholder community, and seeks to establish dialogue among all actors affected by the issue of water quality and quantity variations. This approach, known as post-normal science (Funtowicz & Ravetz, 1995), promotes a view of water systems as dynamic and complex, and of water management as based on the assumptions of unpredictability, incomplete control and a plurality of legitimate perspectives. By contrast, water management practices have traditionally relied on experts using technical means based on designing systems that can be predicted and controlled (Pahl-Wostl, Tàbara *et al.*, 2008), with little analytical substance to account for the wider environmental, social and cultural contexts from which water-related issues emerge.

One important aspect that makes water management highly susceptible to socially constructed environmental dynamics is that most of the "used" water is not used up but reused or recycled, thus forming complex hydrological systems connecting a variety of human entities in specific geographical areas (Seckler, Molden, & Sakthivadivel, 2003). In other words, the water that is used by the firm may be reused by other actors for domestic, industrial, agricultural, recreational and environmental purposes, as well as hydropower generation (Biswas, 2008). Despite the existence of a complex socio-hydrological cycle, organizational science has largely concerned itself with human mediated transactions between organization boundaries (Peloza, 2009), and therefore ignored the human–environmental relationships that underlie, and crucially fuel, these transactions, perhaps because these relationships require management competencies and innovative capacities that are not (or rarely) found in conventional ("technocentric" and instrumental) ways of organizing and thinking of the role of business in society. The spatial complexity of the hydrological dynamics in which the firm is implicated demands the creation of multi-stakeholders dialogues to produce locally specific forecast (Cash *et al.*, 2003) and create opportunities for social learning (Pahl-Wostl, Tàbara, *et al.*, 2008). Stakeholders at different scales are thus to be connected in flexible networks allowing them to develop the capacity and trust they need to collaborate in a wide range of formal and informal relationships – ranging from formal legal structures and contracts to informal, voluntary agreements. Effective water management systems, it is argued, ought to apply a variety of mechanisms that facilitate the development of formal and informal relationships across organizational and spatial boundaries, through enhanced communication, translation and mediation intended to increase a firm's knowledge of the business risks induced by water quality and quantity variations (Cash *et al.*, 2003). The existence of a substantial share of un-codified knowledge in this area represents a serious obstacle to the advancement of corporate water responsibility. There appears to be a necessity to attend to the knowledge development mechanisms that challenge the hegemony of conventional business models, and their underpinning instrumental logics, by revealing the importance of the mixture of social processes and hydrological dynamics that characterize how water is managed, distributed, valued and used in a specific social context (Linton, 2010). In the corporate context, water has typically been treated as an economic resource, allowing firms to use it as a means to whatever ends they have the economic and technological capacity to effect. A leap of faith is arguably required in the area of water management that institutes a view of water as essential to sustaining life and livelihood, likely to be held by many communities and social actors who are affected by the impact of global and local business activities on water quality and quantity variations.

More than likely, improvements in the area of corporate water responsibility will require that corporate managers raise to the same level of gravity and concern. Martinez, O'Sullivan, Smith, and Esposito (2017) portray business managers as capable of making a step toward a more engaging standpoint on the management of social problems. They highlight the features of a (human-centric) social innovation perspective that, I allege, is well suited to explain how firms can establish processes to acquire new (and perhaps "un-codified") knowledge about water-related issues. The first feature is that business managers engaged in social innovation are self-directed and self-organized around the moral purpose of fostering social progress. They may hence assent to a view of water as essential to sustaining life rather than a mere commodity that is managed to enable many of us to survive without having to think much about it (Linton, 2010). The second feature is that social innovation is a process driven by human relations and creative capacity breaking routines and path dependencies. That is arguably linked to the idea of straying from old conceptions of water management as an efficiency-based process (Seckler *et al.*, 2003) toward a broader focus on the socio-hydrological nexus that characterizes it. The third important feature of social innovation is that it fundamentally relies on the socially constructed dynamics between business and social actors who carry ideas, focus their energies, mobilize competences and create new complementarities to tackle social problems. These dynamics may notably contribute to reinstate the particularity of water to, and the relation of interdependency between water and, the individuals and groups involved. They may act to bring together and confront (diverging) stakeholder perspectives, identify water-related issues and make feasible collective choices (Pahl-Wostl, Tàbara *et al.*, 2008). The result may be the emergence of an adjusted distribution of water resources in ways that flexibly and equitably responds to local needs, spatial specificities and affordability conditions. The fourth feature of social innovation highlighted by

Martinez *et al.* (2017) is that it best manifests itself as an informal social process that comes into existence at the margin of conventional ways of thinking and organizing business activities. This aspect throws light on the particular social circumstances that may lead to the advancement of water management as a socially constructed process. The possibility of an informal social process means that business agents can be engaged in networks or 'communities of practice' that are influenced by the governance structure in which these communities are embedded (Pahl-Wostl, Mostert *et al.*, 2008). They may therefore participate, formally or informally, in governance structures that are specifically designed to fit the societal context of water management. For example, a firm can be called to collaborate with authorities and other stakeholder groups from water-related sectors (e.g., spatial planning, flood protection, water supply management) with whom it may not be accustomed to interact. This component of social innovation may trigger a move away from the power relationships that characterize established corporate governance structures and are possibly too exclusive, rigid and inflexible in their general configurations to constitute a viable catalyst for corporate water responsibility.

4.2 Insights from Knowledge Management theory

The concept of corporate water responsibility developed by Martinez (2015) is extended above as a socially oriented process triggered at the margin of conventional business models. This process essentially tends toward the simultaneous management of "codified" and "un-codified" knowledge related to how water is abstracted, used and disposed of by both the firm itself and various stakeholders in a specific spatial context. The objective of this section is to explore how existing Knowledge Management models can be made to cohere with this process and constitute one of the fundamental underlying pillars of corporate water responsibility.

In the corporate context, Knowledge Management is typically concerned with identifying, developing and leveraging knowledge that contributes to secure and sustain competitive advantage (Alavi & Leidner, 2001; Easterby-Smith & Prieto, 2008). The role of corporate managers in this process is mainly confined to secure ongoing synchronicity, ordering and certainty in day-to-day business activities (Eisenhardt & Zbaracki, 1992). This is associated with a general emphasis on the management of knowledge that is easily handled and a neglect of more tacit (or perhaps "esoteric") forms of knowledge that are harder to codify, yet potentially more significant for competitive advantage. The postulate that corporate water responsibility offers opportunities for competitive advantage is discussed in this chapter to critically depend on a firm's capacity to combine new or "un-codified" and codified knowledge, plausibly boosted by the emergence of a different view of the corporation as an open system with greater reliance on transparency, co-creation and dialogues with stakeholders (Sai Manohar & Pandit, 2014). The Schumpetarian view of competition holds that companies ought to achieve the dual objective of "creatively destructing" existing resources and developing "novel combinations" of new functional competences to boost their competitiveness. Corporate water responsibility is arguably one area of competitiveness (Pahl-Wostl, Tàbara *et al.*, 2008) in which both creative destruction and the development of novel combinations are necessary. But, unlike the Schumpetarian view, these actions are not exclusively focused on the financial aspects of performance. They intend to enhance the social coherence and long-term economic viability of corporate activities – in line with the core tenets of the concept of social innovation in business (Martinez *et al.*, 2017).

What remains to be examined in this context is what Quintas *et al.* (1997) refer to as the strategies for developing, acquiring and applying knowledge, as well as monitoring and evaluating knowledge assets and processes for an effective management of water resources. There is no shortage of Knowledge Management models in the literature that explain how knowledge is constructed within the organization (e.g.; Clarke & Staunton, 1989; Demarest, 1997; Hedlund, 1994; Jordan & Jones, 1997; Kruizinga, Heijst, & Spek, 1997; McAdam & McCreedy, 1999). These models are particularly useful to the study of Knowledge Management for corporate water responsibility because they more or less explicitly converge on the idea of a social learning process that engages the firm with a variety of knowledge carriers. The process is commonly concerned with the following categories of knowledge: articulated, tacit, codified and un-codified. In generic terms, knowledge is carried by a variety of actors or entities at the level of the individual, group, organization and inter-organization or stakeholders. These categories are arguably broad enough to include any entities which are somehow involved in a firm's contingent water system. In this chapter, I essentially draw from the model of Boisot (1987) to identify the different categories of knowledge that are carried by these entities. The resulting framework is refined by incorporating insights from Nonaka and Takeuchi's (1995) Knowledge Management model to propose a set of levers that can be actioned by the firm to enrich its knowledge assets, in consideration of the diversity of knowledge carriers that are likely to constitute the wider environmental, social and cultural context(s) of water management.

The category of knowledge that is articulated and/or codified is framed by Boisot (1987) as public or proprietary knowledge, according to whether it is diffused or un-diffused. Public knowledge is readily prepared for transmission. It is diffused in various ways, e.g. reports, books, journals. In the area of water management, such knowledge may relate to the public availability and diffusion of quantified estimates of the stock of water, the limits of water supply, as well as the prospect of scarcity (e.g., OECD, 2012; UN WWDR-4, 2012; UNDP, 2006; United Nations (UN) – Water and Food and Agriculture organization of the United Nations (FAO), 2007; WRG, 2013). Knowledge Management for corporate water responsibility, it is argued, requires a firm to consult public knowledge and interact with the entities who produced it. Consultation provides for an appropriate lever for action because it principally constitutes a (collective) procedure of diagnosis, prognosis and treatment of accessible (and thus diffused) knowledge. Proprietary knowledge is also readily prepared for transmission; but it is un-diffused and restricted to a selectively small population. It may consist of a firm's individual talents, innovative skills and creativity. I also include the knowledge assets that are developed "in-house" (Lichtenthaler & Ernst, 2006), often patently protected and deemed critical for economic performance and competitiveness, such as specific operational and production processes, technologies, efficiency metrics, etc. It may also relate to the knowledge carried by water experts at a particular moment in time and in a specific spatial context. Knowledge management for corporate water responsibility, it is argued, requires that a firm assumes the ongoing application and exploitation of their proprietary knowledge whilst also striving to solicit and acquire the proprietary knowledge held by other important stakeholders. The contemplated pattern of Knowledge Management that corresponds to this dual objective is that of an endogenous and exogenous mobilization of (un-diffused) knowledge assets. The occurrence of public and proprietary knowledge in the field of water management reflects what critical realists (e.g., Bhaskar, 1978; Wry, 2009) refer to as the domain of the Real. In other words, these categories of knowledge represent high-level and relatively enduring sets of beliefs that provide the broad principles that shape actors' understanding of legitimate behavior. Currently, one might expect to find a general tendency to privilege positivist knowledge, essentially grounded in quantitative data about water stocks, limits of supply and prospects of scarcity. The domain of the "Real" is arguably where corporate managers satisfy their needs for prediction and control, typically translating into the deductive development of water management tools and techniques (such as water footprinting methodologies and end-of-pipe quality controls) through the combined mobilization of proprietary knowledge and consultation of public knowledge.

The second category of knowledge proposed by Boisot (1987) is unarticulated and/or un-codified. It can take two distinct forms: personal and common sense. Personal knowledge is represented in individuals' perceptions, insights and experiences. This type of knowledge is typically produced by the individuals' transactions with a social and/or institutional context. One implication for water management is that individuals may perceive and consume water differently according to whether they situate themselves in a domestic, industrial, agricultural, or recreational context; or whether they face upstream or downstream issues. A firm's consideration of the knowledge carried by these individuals may enable them to identify the parameters of a water system that need to be monitored in order to function in a sustainable manner (Biswas, 2008). The challenge is to make personal knowledge available in a language that is known and accessible to the firm. Hence, a process of integration is suggested that brings together corporate actors and carriers of personal knowledge about water management into a non-segregated community of practice. Common sense knowledge is concerned with a process of socialization, harboring customs and intuition. A connection can be made between common sense and the cultural dimension of water management since they both determine the varying ways in which water is perceived, and consumed, by individuals, groups and organizations across regions and cultures (Linton, 2010). Knowledge Management for corporate water responsibility, it is argued, ought to offer opportunities for socialization between corporate actors and important water stakeholders in ways that increase proximity between the firm and knowledge carriers and raise a firm's awareness of local customs and interpretations of water management, and therefore of the socio-hydrological realities that characterize a specific spatial context. The existence of personal and common sense knowledge in the field of water management reflects what critical realists call the domain of the Empirical (e.g., Bhaskar, 1978; Wry, 2009). In other words, these categories of knowledge consists of actions as they are understood by the actors themselves (Wry, 2009). Here, one might expect to find a general tendency to privilege socially constructed knowledge, grounded in the lives and experiences of individuals. It is essentially the cultural context, and the cognitive activity within this context, that shapes interpretations and behavior. In the domain of the 'Empirical', corporate managers are confronted with the unpredicted and uncontrolled aspects of water management. They are mobilized toward the creation of novel combinations of ideas and resources, breaking routines and path dependencies (Martinez *et al.*, 2017). The patterns of integration and socialization that are suggested above as a means to manage personal and common sense knowledge require an inductive effort of knowledge development which, according to Brown and Wyatt, "allows for serendipity, unpredictability, and the capricious whims of fate – the creative realm from which breakthrough ideas emerge" (2010, p. 33).

Figure 4.1 puts the essential aspects, and underlying mechanisms, contemplated in this chapter together into a preliminary Knowledge Management framework for corporate water responsibility, based on the tenets of critical realism. The argument is that the interpretations of business agents in the "Actual" – i.e. corporate water responsibility activities – stem from the combined filtering through the agent's (or agents') conceptualization of facts in the "Real" and actors' (or water stakeholders') understanding of lived experiences in the "Empirical". Corporate water responsibility is thus framed as a function of a firm's capacity to simultaneously capture proprietary, public, cultural and personal knowledge through four distinct patterns of Knowledge Management: mobilization, consultation, integration and socialization.

4.3 Contribution, limitations and implications

The Knowledge Management framework presented in Figure 4.1 specifically addresses the elements identified in the analysis of corporate water responsibility as a socially-oriented process. The suggested configuration accounts for contemporary (post-normal) water management trends and challenges. A fundamental contribution is made to the water management literature by explaining how

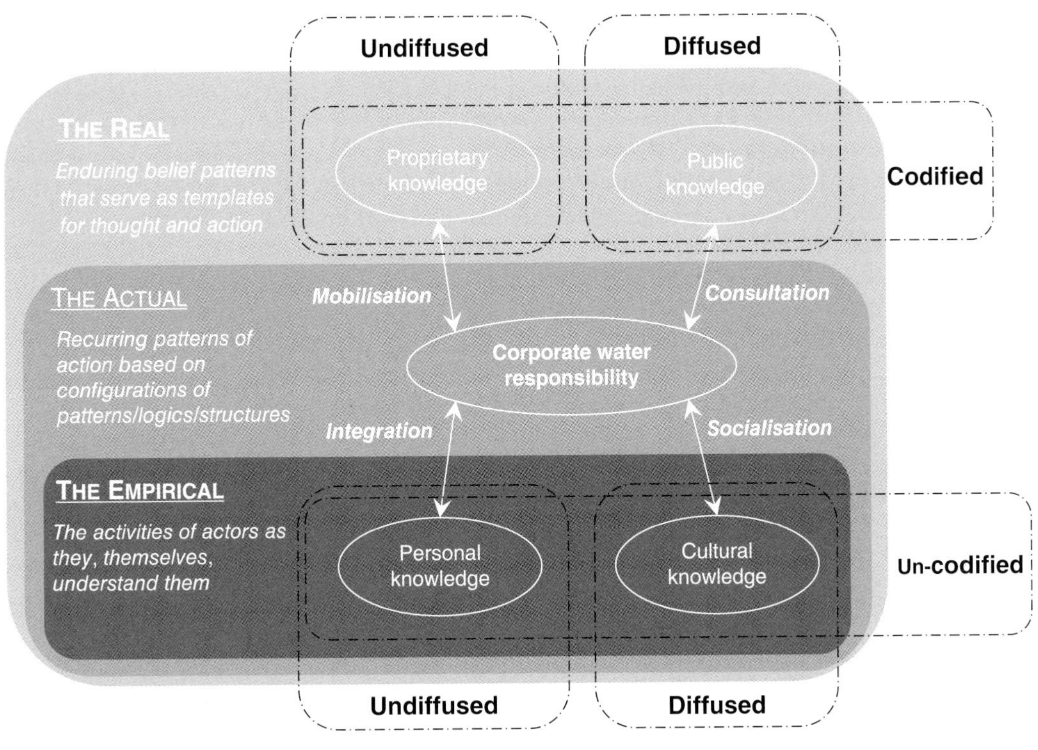

Figure 4.1 A (critical realist) Knowledge Management framework for corporate water responsibility.

business organizations may establish dialogues with an extended community of water-related stakeholders, respond to their water needs and demands, and tend toward a more holistic and inclusive management of knowledge in this area. The critical realist foundations of the framework contrasts against the general overreliance of water management on positivist knowledge claims. It attends to the tendency of the proponents of integrated water resources management to reduce a complex problem into a set of quantified interpretations, and therefore largely neglect data availability, reliability and accessibility issues (Biswas, 2008). The result is an increased level of confidence that one could have on predicting a firm's ability to manage water-related knowledge and contribute to a collective momentum of sustainable water management practice, enriched by a distinctively sharper understanding of the means by which corporate decision-makers can capture the real impact of water on human lives, the environment and other development indicators. However, several limitations are worthy of consideration in future developments of the model.

Firstly, it seems appropriate to raise some doubts about (and highlight some challenges in relation to) the extent to which the Knowledge Management model developed in this chapter results in knowledge utilization and/or improved corporate water responsibility. One challenge for the firm is to elaborate a system able to cope with the potential information-overload, recursive knowledge flows and conflicts of interests (or trade-offs) that a holistic model of Knowledge Management is likely to engender. That holds particularly true in the area of water management characterized by numerous interrelated yet conflicting needs and interests (Biswas, 2008; Martinez, 2015). The issues of decision making and conflicts management have received considerable scholarly interests in strategic management and corporate sustainability research (e.g., Hahn, Figge, Pinkse, & Preuss, 2010; Hahn, Pinkse, Preuss, & Figge, 2014, 2015; Smith, Binns, & Tushman, 2010; Smith, Gonin, & Besharov, 2013; Van der Byl & Slawinski, 2015). Most inferences are drawn from sense-making and paradox theories to identify the diversity of pathways managers take in the exercise of sustainability-oriented decision-making, and with an effort to align with the strategic concerns of the firm. The combination of informal and formal processes of Knowledge Management for corporate water responsibility contemplated in this chapter does not tend to be represented in these research streams. They are too often exclusively focused on the business case for sustainability, underpinned by a systematic search for controllable (win-)win–win solutions. The business case logic stems from a view of the relationship between business, society and the natural environment as one of "intersecting" areas of common interest, rather than the "embedded" relationship that sustainable water management is based on, and which Marcus, Kurucz, and Colbert (2010) argue frames the relationship more realistically as business existing as a construct within society, that itself is embedded within, and dependent upon, nature. The notion of a "win" is also problematic in that it suggests a final resolution and an end point in some particular endeavor, but in the case of the human–environmental relationships that characterize water management it is in reality a more complex, open-ended and constant process of knowledge development, strategic adjustment and negotiation. A key flaw in the win–win argument, therefore, is that it relegates sustainability into a particular set of strategic challenges and opportunities for companies that may prove a source of differentiation and competitive advantage, rather than recognizing it as a more fundamental challenge to the dominant economic paradigm and as an alternative approach to management thought and practice. One implication is that management tends to rely on simplified and biased information on water-related issues mainly based on statistical methods and whatever data are available (Mannor, Simester, Sun, & Tsitsiklis, 2007). In this chapter, I advocate for a more substantive consideration of elements from the concept of social innovation (Martinez *et al.*, 2017) in planning for corporate water responsibility, therefore inducing behavioral responses that have been ignored by economic analysis of optimal ecosystem management (Chen, Jayaprakash, & Irwin, 2012).

Another area that can be explored in future developments of the model is the capacity of business agents to evaluate information quality. Arguably, knowledge quality and importance varies from source to source and according to the proximity that characterizes the relationship between the knowledge carriers and the firm. If the Knowledge Management model illustrated in Figure 4.1 is to generate improved corporate water responsibility, it needs to rely on valid knowledge claims, implying the analytical capacity to filter the information stemming from a wide range of individuals and stakeholders who have, e.g., domestic, industrial, agricultural, navigational, environmental and hydropower generation needs within the water system out of which the firm operates. The complexity of this task suggests that business agents are able to stray from two important conventional facets of strategic decision-making: (1) the focus of the firm on readily transmittable (codified) knowledge coupled with a plausible lack of know-how on managing un-codified/tacit knowledge; and (2) a relatively narrow conception of stakeholder salience coupled with a potential reluctance to value the interests, and comprehend the issues faced by, actors with whom the firm is not accustomed to interact. Hopefully, further research into the dynamics of social innovation, and the "unusual" forms of business actions that are contemplated in this concept (Martinez *et al.*, 2017), will help us to develop a novel conception of the business agent as capable of building trusting relationships with water stakeholders and analytically filtering the knowledge claims that are used to inform the construct of a sound corporate water responsibility programme.

The third major challenge highlighted in this section relates to one of the essential tenet of critical realism: distinct forms of knowledge (codified and un-codified) cannot be conflated into one single pattern of Knowledge Management. Rather, the existence of varied categories of knowledge is likely to require a complex aggregation of operational, structural, technical and strategic actions that mobilize different actors inside and outside the firm to implement new trajectories without jeopardizing the performance of existing ones. This is known in the business model literature as a conflict between sustaining performance with existing models and nurturing new, potentially sustainability-oriented, models (e.g., Amit & Zott, 2010; Chesbrough, 2010; Chesbrough & Rosenbloom, 2002; Schaltegger, Lüdeke-Freund, & Hansen, 2012, 2016). The approach proposed in this chapter achieves the dual objective of preserving existing planning approaches, essentially based on physical solutions in the domain of "the Real", and contemplating new ones, grounded in complex human–environmental relationship systems in the domain of the "Empirical", without committing the epistemic fallacy of reducing "what is" to "what is known" (Bhaskar, 1978, 1998; Jackson, 2008). An important implication is that a firm ought to leverage competencies across four distinct processes of Knowledge Management (cf. Figure 4.1) that are likely to demand the mobilization of different sets of skills. It is plausible that the individuals who manage proprietary and public knowledge are not prepared to manage personal and cultural knowledge, and vice versa. Most of the contributions to the Knowledge Management literature to date have focused on IT-based tools and systems and other technical Knowledge Management infrastructures that are presented as a pipeline to knowledge codification and organization (Alavi & Leidner, 2001). The processes of integration and socialization suggested in this chapter are likely to require a distinctively more human-centric approach to Knowledge Management. One mechanism that can be explored in future research is that the detection of new knowledge can be facilitated by the establishment of an agenda for corporate water responsibility that includes an initial step toward the actors who are an evident source of information in specific water systems and industries. A firm may for example endeavor to integrate local water experts who are specialized and likely to have established solid ties with salient water stakeholders. More specifically, a food company may strive to interact with local food experts who are increasingly concerned about the realities of water availability, given the fundamental connections between food and water (Gleick, 2000). This first step may trigger a snowball "stakeholder sampling" effect, leading business agents to build an increasingly integrated network of water users within which they may strive to capture new knowledge in the ways depicted in Figure 4.1.

4.4 Conclusion

In this chapter, the conceptual underpinnings of corporate water responsibility were refined by integrating elements from Knowledge Management theory. Figure 4.1 was constructed, a suggested approach to Knowledge Management that explains how business can be more substantively involved in the management of sustainable water systems. The proposed model takes a balanced approach between the positivist and socially constructed knowledge base that is advanced to influence water management in the extant literature. It is suggested that this model has great potential as a guide for future research and literature evaluation in the area of Knowledge Management for sustainable water systems. In the following, I summarize the key contents of this chapter that contributed to the construction of the framework.

Corporate water responsibility was presented as a socially oriented process that requires unconventional modes of commitment by the business community. I argued that the formal and informal means by which business agents can participate in water management are partly articulated in the concept of social innovation. Indeed, the socially constructed dynamics that are assigned to social innovation (e.g., Martinez *et al.*, 2017) are seen as apropos for 'post-normal' water management – an important tenet of which is that the complex, and to some extent unpredictable, nature of water systems requires that the wider environmental, social and cultural contexts from which water-related issues emerge be evaluated. Knowledge management theory is generally concerned with similar challenges around the capacity of individuals to combine factual accounts and social causation processes.

Insights from Knowledge Management theory were hence explored to identify some elements that are likely to drive improved corporate water responsibility. I particularly interweaved elements from Boisot's (1987) model and critical realism (Bhaskar, 1978, 1998) to articulate a preliminary framework for the management of the codified and un-codified knowledge that are likely to play a role in the socio-hydrological systems out of which the firm operates. The main argument is that Knowledge Management for corporate water responsibility generates a combined filtering through a business agents' evaluation of (codified) proprietary and public knowledge and water stakeholders' interpretations of lived experiences, framed in Figure 4.1 as (un-codified) personal and cultural knowledge. Four distinct patterns are identified that underpin this complex process and represent how business agents are likely to be involved: mobilization, consultation, integration and socialization.

I finally drew out some important limitations and implications, all suggested to constitute elements of a research agenda in this domain. Three important challenges are discussed: (1) ensure ongoing synchronicity between formal/informal, controlled/uncontrolled, and non-economic/economic aspects of water management; (2) evaluate information quality and validity; and (3) develop tools and techniques that meet the critical realist task of simultaneously handling positivist and socially constructed knowledge. It is hoped that the efforts made in this chapter to clarify the core tenets of Knowledge Management for corporate water responsibility will be a useful resource to assemble the social actors (and perhaps essentially business agents) who seek to accelerate change toward more sustainable water systems, and avert the socially divisive and exclusivist agenda that traditional conceptions of water in business and the wider economy have tended to encourage.

References

Alavi, M., & Leidner, D. E. (2001). Review: Knowledge management and Knowledge Management systems: Conceptual foundations and research issues. *MIS Quarterly, 25*(1), 107–136.

Amit, R., & Zott, C. (2010). *Business model innovation: Creating value in times of change.* IESE Business School, University of Navarra.

Barton, B. (2010). *Murky waters? Corporate reporting on water risk: a benchmarking of 100 companies.* Retrieved from Boston, MA:

Bhaskar, R. (1978). *A realist theory of science.* Hassocks, England: Harvester Press.

Bhaskar, R. (1998). General Introduction. In M. S. Archer, R. Bhaskar, A. Collier, T. Lawson, & A. Norrie (Eds.), *Critical Realism: Essential Readings.* London: Routledge.

Biswas, A. K. (2008). Integrated water resources management: Is it working? *International Journal of Water Resources Development, 24*(1), 5–22.

Boisot, M. (1987). *Information and Organizations: The manager as anthropologist.* London: Fontana/Collins.

Brown, T., & Wyatt, J. (2010). Design thinking for social innovation. *Stanford Social Innovation Review, 8*(1), 31–35.

Caplan, D., Dutta, S. K., & Lawson, R. A. (2013). Corporate Social Responsibility Initiatives Across the Value Chain. *Journal of Corporate Accounting & Finance, 24*(3), 15–24.

Cash, D. W., Clark, W. C., Alcock, F., Dickson, N. M., Eckley, N., Guston, D. H., & Mitchell, R. B. (2003). Knowledge systems for sustainable development. *Proceedings of the National Academy of Sciences, 100*(14), 8086–8091.

Chalmers, K., Godfrey, J. M., & Lynch, B. (2012). Regulatory theory insights into the past, present and future of general purpose water accounting standard setting. *Accounting, Auditing and Accountability Journal, 25*(6), 1001–1024.
Chen, Y., Jayaprakash, C., & Irwin, E. (2012). Threshold management in a coupled economic-ecological system. *Journal of Environmental Economics and Management, 64*(3), 442–455.
Chesbrough, H. W. (2010). Business model innovation: Opportunities and barriers. *Long Range Planning, 43*(2–3), 354–363.
Chesbrough, H. W., & Rosenbloom, R. S. (2002). The role of the business model in capturing value from innovation: evidence from Xerox Corporation's technology spin-off companies. *Industrial and Corporate Change, 11*, 529–555.
Clarke, P., & Staunton, N. (1989). *Innovation in technology and organization*. London: Routledge.
Cook, B. R., & Spray, C. J. (2012). Ecosystem services and integrated water resource management: Different paths to the same end? *Journal of Environmental Management, 109*, 93–100.
Cuickshank, A., & Grover, V. I. (2012). A brief introduction to integrated water resources management. In V. I. Grover & G. Krantzberg (Eds.), *Great Lakes: Lessons in Participatory Governance*. Boca Raton, FL: CRC Press, Taylor & Francis Group.
DEFRA. (2011). *Europe 2020 Strategy: roadmap to a resource efficient Europe*. Retrieved from http://www.defra.gov.uk/publications/files/resource-efficient-europe.pdf.
Demarest, M. (1997). Understanding Knowledge Management. *Long Range Planning, 30*(3), 321–384.
Easterby-Smith, M., & Prieto, I. M. (2008). Dynamic capabilities and Knowledge Management: An integrative role for learning? *British Journal of Management, 19*(3), 235–249.
Egan, M. (2015). Driving water management change where economic incentive is limited. *Journal of Business Ethics, 132*(1), 73–90.
Eisenhardt, K. M., & Zbaracki, M. J. (1992). Strategic decision making. *Strategic Management Journal, 13*(S2), 17–37.
Falkenmark, M., Lundqvist, J., & Widstrand, C. (1989). Macro-scale water scarcity requires micro-scale approaches. *Natural Resources Forum, 13*(4), 258–267.
Funtowicz, S. O., & Ravetz, J. R. (1995). Science for the Post Normal Age. In L. Westra & 29 J. Lemons (Eds.), *Perspectives on Ecological Integrity* (pp. 146–161). Dordrecht: Springer Netherlands.
Gleick, P. H. (2000). A look at twenty-first century water resources development. *Water International, 25*(1), 127–138.
GWP. (2000). *Integrated Water Resource Management* Technical Advisory Committee Background Paper Number 4. Stockholm: Global Water Partnership.
Hahn, T., Figge, F., Pinkse, J., & Preuss, L. (2010). Trade-offs in corporate sustainability: you can't have your cake and eat it. *Business Strategy and the Environment, 19*(4), 217–229.
Hahn, T., Pinkse, J., Preuss, L., & Figge, F. (2014). Cognitive frames in corporate sustainability: Managerial sensemaking with paradoxical and business case frames. *Academy of Management Review, 39*(4), 463–487.
Hahn, T., Pinkse, J., Preuss, L., & Figge, F. (2015). Tensions in corporate sustainability: Towards an integrative framework. *Journal of Business Ethics, 127*(2), 297–316.
Harvey, B., & Schaefer, A. (2001). Managing relationships with environmental stakeholders: A study of U.K. water and electricity utilities. *Journal of Business Ethics, 30*(3), 243–260.

Harvey, D. (1996). *Justice, nature and the geography of difference*. Malden: Blackwell.
Hawken, P., Lovins, A. B., & Lovins, L. H. (2002). *Natural Capitalism. The Next Industrial Revolution*. London: Earthscan Publications Ltd.
Hedlund, G. (1994). A model of Knowledge Management and the N-form corporation. *Strategic Management Journal, 15*(S2), 73–90.
Heidegger, M. (1977). The question concerning technology. In D. Farell Krell (Ed.), *Martin Heidegger: Basic writings* (pp. 287–317). New York: Harper & Row.
Hoekstra, A. Y. (2009). Human appropriation of natural capital: A comparison of ecological footprint and water footprint analysis. *Ecological Economics, 68*(7), 1963–1974.
Hoffman, A., & Howie, S. (2010). *Coke in the cross hairs: Water, India, and the University of Michigan*. Ann Arbor, MI: GlobaLens, William Davidson Institute at the University of Michigan.
Jackson, P. T. (2008). *Back-to-back review: Ontological investigations, Colin Wight, agents, structures, and international relations*. Cambridge: Cambridge University Press, 2006, 340 pp. ISBN 0 521 67416 6. *Cooperation and Conflict, 43*(3), 341–347.
Jones, M. J. (2010). Accounting for the environment: Towards a theoretical perspective for environmental accounting and reporting. *Accounting Forum, 34*(2), 123–138.
Jordan, J., & Jones, P. (1997). Assessing your company's Knowledge Management style. *Long Range Planning, 30*(3), 322–398.
Karnani, A. (2012). *Corporate social responsibility does not avert the tragedy of the commons - Case study: Coca-Cola India*. Arm Arbor, MI: Ross School of Business.
Kruizinga, E., Heijst, G., & Spek, R. (1997). Knowledge infrastructures and intranets. *Journal of Knowledge Management, 1*(1), 27–32.
Kurland, N. B., & Zell, D. (2010). Water and business: A taxonomy and review of the research. *Organization & Environment, 23*(3), 316–353.
Lambooy, T. (2011). Corporate social responsibility: sustainable water use. *Journal of Cleaner Production, 19*(8), 852–866.
Lichtenthaler, U., & Ernst, H. (2006). Attitudes to externally organising Knowledge Management tasks: a review, reconsideration and extension of the NIH syndrome. *R and D Management, 36*(4), 367–386.
Linton, J. (2010). *What is water? The history of a modern abstraction*. Vancouver: UBC Press.
Mannor, S., Simester, D., Sun, P., & Tsitsiklis, J. N. (2007). Bias and variance approximation in value function estimates. *Management Science, 53*(2), 308–322.
Marcus, J., Kurucz, E. C., & Colbert, B. A. (2010). Conceptions of the business-society-nature interface: implications for management scholarship. *Business and Society, 49*(3), 402–438.
Martinez, F. (2015). A three-dimensional conceptual framework of corporate water responsibility. *Organization & Environment, 28*(2), 137–159.
Martinez, F., O'Sullivan, P., Smith, M., & Esposito, M. (2017). Perspectives on the role of business in social innovation. *Journal of Management Development* (in press).
McAdam, R., & McCreedy, S. (1999). A critical review of Knowledge Management models. *The Learning Organization, 6*(3), 91–101.
Mehta, L., Veldwisch, G. J., & Franco, J. (2012). Introduction to the Special Issue: Water grabbing? Focus on the (re) appropriation of finite water resources. *Water Alternatives, 5*(2), 193–207.
Money, A. (2014). Corporate water risk: A critique of prevailing best practice. *Journal of Management and Sustainability, 4*(1), 42–58.

Montiel, I., Husted, B. W., & Christmann, P. (2012). Using private management standard certification to reduce information asymmetries in corrupt environments. *Strategic Management Journal, 33*(9), 1103–1113.

Nonaka, I., & Takeuchi, K. (1995). *The knowledge creating company: How Japanese companies create the dynamics of innovation.* Oxford: Oxford University Press.

OECD. (2012). *OECD Environmental Outlook to 2050: The Consequences of Inaction*: OECD Publishing.

Ogden, S., & Watson, R. (1999). Corporate performance and stakeholder management: Balancing shareholder and customer interests in the U.K. privatized water industry. *The Academy of Management Journal, 42*(5), 526–538.

Pahl-Wostl, C., Mostert, E., & Tàbara, D. (2008). The growing importance of social learning in water resources management and sustainability science. *Ecology and Society, 13*(1).

Pahl-Wostl, C., Tàbara, D., Bouwen, R., Craps, M., Dewulf, A., Mostert, E., & Taillieu, T. (2008). The importance of social learning and culture for sustainable water management. *Ecological Economics, 64*(3), 484–495.

Peloza, J. (2009). The challenge of measuring financial impacts from investments in corporate social performance. *Journal of Management, 35*(6), 1518–1541.

Quintas, P., Lefrere, P., & Jones, G. (1997). Knowledge management: A strategic agenda. *Long Range Planning, 30*(3), 385–391.

Rist, S., Chidambaranathan, M., Escobar, C., Wiesmann, U., & Zimmermann, A. (2007). Moving from sustainable management to sustainable governance of natural resources: The role of social learning processes in rural India, Bolivia and Mali. *Journal of Rural Studies, 23*(1), 23–37.

SABMiller, & WWF. (2009). *Water footprinting: identifying and addressing water risks in the value chain.* Retrieved from http://awsassets.panda.org/downloads/sab0425_waterfootprinting_text_artwork.pdf, 29.07.17.

Sai Manohar, S., & Pandit, S. (2014). Core values and beliefs: A study of leading innovative organizations. *Journal of Business Ethics, 125*(4), 667–680.

Schaltegger, S., Lüdeke-Freund, F., & Hansen, E. G. (2012). Business cases for sustainability: The role of business model innovation for corporate sustainability. *International Journal of Innovation and Sustainable Development, 6*(2), 95–119.

Schaltegger, S., Lüdeke-Freund, F., & Hansen, E. G. (2016). Business models for sustainability: A co-evolutionary analysis of sustainable entrepreneurship, innovation, and transformation. *Organization & Environment, Online First.* doi:10.1177/1086026616633272.

Seckler, D., Molden, D., & Sakthivadivel, R. (2003). The concept of efficiency in water-resources management and policy. In J. W. Kijne, R. Barker, & D. J. Molden (Eds.), *Water productivity in agriculture: Limits and opportunities for improvement* (pp. 37–52). Oxon, UK & Cambridge, MA: CAB International.

Smith, W. K., Binns, A., & Tushman, M. L. (2010). Complex Business Models: Managing Strategic Paradoxes Simultaneously. *Long Range Planning, 43*(2–3), 448–461. doi:10.1016/j.lrp.2009.12.003

Smith, W. K., Gonin, M., & Besharov, M. L. (2013). Managing social-business tensions: A review and research agenda for social enterprise. *Business Ethics Quarterly, 23*(3), 407–442.

Sojamo, S., & Larson, E. A. (2012). Investigating food and agribusiness corporations as global water security, management and governance agents: The case of Nestlé, Bunge and Cargill. *Water Alternatives, 5*(3), 619–635.

Tacconi, L. (1998). Scientific methodology for ecological economics. *Ecological Economics, 27*(1), 91–105.

The Coca Cola Company and The Nature Conservancy. (2010). *Product water footprint assessments: Practical application in corporate water stewardship*. Retrieved from http://www.businesswire.com/news/home/20100908006074/en/Coca-Cola-Company-Nature-Conservancy-Release-Water-Footprint, 29.07.17.

UN WWDR-4. (2012). *The 4th United Nations World Water Development Report: Managing water under uncertainty and risk*. Retrieved from http://www.unesco.org/new/en/natural-sciences/environment/water/wwap/wwdr/wwdr4-2012/, Paris, 29.07.17.

UNDP. (2006). *Human Development Report*. Retrieved from http://hdr.undp.org/sites/default/files/reports/267/hdr06-complete.pdf, New York, NY, 29.07.17.

United Nations (UN) – Water and Food and Agriculture Organization of the United Nations (FAO). (2007). *Coping with water scarcity: challenge of the twenty-first century*. Rome: FAO.

Van der Byl, C. A., & Slawinski, N. (2015). Embracing tensions in corporate sustainability: A review of research from win–wins and trade-offs to paradoxes and beyond. *Organization & Environment, 28*(1), 54–79.

Wissenburg, M. (2013). What is water? The history of a modern abstraction. *Environmental Politics, 22*(2), 356–358.

WRG (2030 Water Resources Group) (2013). *Expanding our Horizon: Water Security Partnerships for People, Growth, and the Environment – 2030 WRG Annual Report*. Retrieved from https://www.2030wrg.org/work/, Washington, 29.07.17.

Wry, T. (2009). Does business and society scholarship matter to society? Pursuing a normative agenda with critical realism and neoinstitutional theory. *Journal of Business Ethics, 89*(2), 151–171.

5 How 21st century Knowledge Management can greatly improve talent management for sustainable water project-teams

Stephen Atkins, Lesley Gill, Kay Lion, Marie Schaddelee and Tonny Tonny

Otago Polytechnic of New Zealand, Dunedin, New Zealand

Introduction

Davenport and Prusak (1998) offer a very appropriate definition of this book's theoretical context stating "Knowledge Management draws from existing resources that your organization may already have in place [such as] good information systems management, organizational *change management, and human resources* management practices" (p. 163, emphasis added). In the present instance, where Davenport and Prusak (1998) refer to "existing resources that your organization may already have ... " (p. 163), they point to, in our very crucial but idiosyncratic case, an organization that is better perceived as a predominantly informal global consortium of aid-agencies with mutual interests in the *war on unsafe water* or sustainable water-aid generally. This chapter places a special emphasis on *human resource management* practices, and especially those related to optimizing the recruitment, assignment and management of human talents using contemporary and emerging information systems technologies (e.g., large government databases on workforce/work information and other related forms of big data, these then coupled with computer adaptive

Handbook of Knowledge Management for Sustainable Water Systems, First Edition. Edited by Meir Russ.

testing and its very high information-processing-needful cousin: item response theory-based mental metrics).

Human resource management (HRM) is defined as a fundamental activity of the management of work and people towards goals. Boxall, Purcell and Wright (2009) distinguish three subdomains of knowledge in HRM: Micro-HRM, Strategic HRM (SHRM) and International HRM. While Micro-HRM is concerned with the particular functions of HRM in an organization, both Strategic HRM and International HRM consider a broader and global context for HRM. The critical position of HRM to the strategic management process of an organization is supported in research. Boxall and Purcell's (2008) understanding of strategic management as a process of developing critical goals and resources firmly relies on the involvement of HRM to make this happen. The responsibility for HRM to support business strategy rests partly on the shoulders of human resource practitioners and is partly an organization-wide responsibility where all levels of management are involved, as well as involving leadership commitment (Connell & Teo, 2010).

It is widely accepted that an organization can create a competitive advantage through its human resources (HR) and the intellectual capital created by these human resources. Boudreau and Ramstad (1997) view the role of HRM as closely associated with developing intellectual capital and managing knowledge. Human resource practices need to value and support this scarce resource in the future, a perspective that is aligned with the knowledge-based view (KBV) of an organization, and assumes knowledge is the most important resource and so needs to be protected (Jashapara, 2011). Intellectual or human capital can be easily lost when valuable members of the organization leave or skills become outdated (Ployhart, Van Iddekinge, & Mackenzie, 2011).That knowledge grows from sharing, (Sveiby, 1997) is established, but the importance of collective knowledge and learning is of particular interest when developing sustainable organizational advantage, as it involves collective knowledge that is unique to an organization or team and inimitable, as opposed to individual knowledge (Ployhart *et al.*, 2011). Knowledge-sharing is dependent upon good relationships and equity (Daum, 2003; Ricceri, 2008). Both developing intellectual capital and knowledge-sharing can be fostered and maintained by good human resource practices.

What is considered to be good human resource practices is the subject of some debate in HRM circles. The best-practice approach assumes a one-size fits all universal approach (Macky, 2008). The best-fit approach in HR tends to better describe what organizations actually do as opposed to the best-practice approach (Macky, 2008). There is merit in identifying best practice, particularly when examining some of the sub-structural principles and influences on HR practices (Boxall & Purcell, 2008). High-performance HR Work Systems (HPWS) not only identify which practices lead to higher productivity but also to work satisfaction (Stone, 2014). Work satisfaction has a positive influence on discretionary effort which in turn also effects performance. While HPWS advocate that skills development, participation and job security are factors that increase performance, critics point to worker well-being as imperative to satisfaction and performance (Stone, 2014). Recruiting and selecting the right talent for a specific team for successful worker performance and well-being is a persistent HR challenge.

Understanding the connection between a team's human capital and its knowledge integration can give HR a basis for identifying competencies that may need to be deliberately recruited to a team (Newell, Adams, Crary, Glidden, LaFarge, & Nurick, 2005). Knowledge integration competency is regarded by Newell *et al.* (2005) as a potential that is created with a diversity of abilities and skills, including the dynamic interactions that result. In other words, the interactions that result in collectively being able to create solutions and achieve team objectives. Dalkir (2011) supports this idea that potential can be realized through diversity in team competencies and comments further that an imbalance could result when a team member leaves a particular role within a team. This latter notion (... of "role-within-a-team") takes on far more complexity when simultaneously considering parallel personality-based team-roles; somewhat alongside technical-specialist or specialty-competence based roles essential to successful project teams. Arguments by Belbin (2012) and Atkins (2013) expand these concerns for competency-based imbalances deriving from team-member losses. In other words, personality-based perspective-losses (or diminished richness in personality-diversity) might result from team-losses or staff-turnover. Schneider (2007 – generally) and Atkins (2014 – specific to personality) demonstrated that staff perceiving themselves as misfits are likelier to exit a line of work. These arguments can be subsumed under a broader notion of vocational competencies. Within this knowledge-management context, and especially applied to sustainable water-projects and their challenging talent requirements, we need to briefly re-visit the basic tenants of talent-acquisition, and thus the notion of what, in present times, we refer to as *competency modeling*.

5.1 Talent-requirements or competency modeling as applied to water projects

As mentioned in this chapter's introductory section, optimizing the effective use of available human talents arguably requires *knowledge-intense* work analysis which facilitates competency models and these in turn, given the extreme multi-dimensionality afforded modern knowledge-managers, yield emergent opportunities for optimally-populated project teams, and in this chapter this refers to sustainable-water project teams.

In parallel with optimizing the short-to-mid-term the staffing of project teams, the projects that a specific firm's valued talents are *assigned to* need to be consistent with career pathways or career aspirations valued by these same talented employees. Arthur, Hall and Lawrence (1989, p. 8) define a career as "the evolving sequence of a person's work experiences over time"

This definition only addresses *career* in part, but excludes other factors that shape vocational choice such as agentic-needs, environment, opportunities, and limitations (Gill, 2004). Volatilities in vocational pathways, or so-called career-ladders, have been widely-seen as increasing exponentially. Others have presented counter-evidence, suggesting time (in years) in one's principal vocation is relatively stable, or even increasing (Anwar, Barends, & Briner, 2016). Even so, self-perceptions of one's goodness-of-fit to one's own job-roles, including project-team assignments, predict one's own sustained enthusiasm for these projects.

5.1.1 Aspects of modern HR management relevant to staffing project teams

The modern HR talent-manager's role within an organization is to assimilate the data about the organization's needs *with* their data about available talent. Available talent refers to the available workforce-candidates' and workforce members' fit of skills, knowledge, qualifications and experience within a context of dynamic and competing worker-compatible alternatives (Atkins, 2013). A talent manager must also evaluate external data from competitors, partners, economies and environments for the purpose of establishing benchmarks. This data is crucial in identifying recruitment strategies, and in particular helps in the recognition of constraints and incompatibilities, as well as worker prerequisites and preferences. Rigorous work analysis (the best form of competency modeling) is crucial if effective vocational planning and talent-management is the objective.

The multi-dimensional dynamics of managing talent and work roles advances the need for better online algorithms for assessing work-roles to available-talent "fit". The sheer enormity of work-analysis and talent-requirements' data now available in government-hosted databases strongly suggest greater knowledge-manager attention (e.g., via U.S. Department of Labor alone, this includes 200+ talent dimensions crossed with a 1000+ of the most populous professions, and aggregates 45 incumbents for each dimension inside each occupation; hence nearly a million work-analysis ratings). The number of human talents (i.e., skilled workers) associated with this big data now numbers in the billions, and represents an under-managed (and frequently under-employed) talent source; unprecedented in human history. Thus, it is prudent that professional knowledge-managers consider their own firm's optimal engagement with these huge databases.

Thus, such unprecedented availability of *talents-accessible* to *talent-requirements* information strongly suggests re-consideration of multi-dimensional "person-to-job" fit algorithms. As complex as such hugely multi-dimensional algorithms can become (e.g., Drewes, Tarantino, Atkins, & Paige, 2000), 21st century Knowledge-Management systems can easily process these algorithms as such. Such web-affordances necessitate re-consideration of the abovementioned 200+ dimensional indices of *person-to-job fit*.

Person-to-job fit refers to the compatibility of the span of the job with the qualifications, interests, and vocational goals of the worker (Atkins, 2013). Models of "fit theory" have been conceptualized in vocational literature such as Work Adjustment Theory (Dawis & Lofquist, 1984; Rounds, Dawis, & Lofquist, 1987); Myers Briggs Type Indicators (MBTI; Briggs-Myers, 1980), Job Characteristics Model (JCM; Hackman & Oldham, 1980); and the classic hexagonal vocational personality model (Holland, 1997) argued with some success as being a *universal* theory (Day & Rounds, 1998).

We would be remiss not to mention there are at least two notable criticisms of "fit" theories. These include both statistical/mathematical criticisms and philosophical criticisms, and these are discussed next.

Firstly, *statistical/mathematical criticisms*, which are largely concerned with the similarity of weighted Minkowski distance indices to higher-order interaction effects, in terms of their underlying or burdening statistical assumptions. Weighted Minkowski distance algorithms (Drewes *et al.*, 2000) allow talent-managers and career-counsellors to calculate the "misfit" gaps between a job-candidate's or a career-exploring student's vocational trait profiles, and a given vocation's talent profile. The latter profiles the many features in a given vocation that require a particular ability or reinforce a particular set of worker-values, e.g., need-for-autonomy, need-for-status, need-for-creativity-outlet, need-to-provide-communal-service-beyond-self, etc.). Understandably, vocations vary widely in these profiles (e.g., consider "Water-toxicity Chemist" vs. "Real Estate Sales" vs. "Airline Pilot" vs. "Middle School Teacher" vs. "Water Project Engineer"). Not surprisingly, at least moderate performance or tenure impacts are seen from greater degrees of incongruence or misfit between what workers need to be able to do well (or convey well or resonate-with) and what they actually can do well (or convey well or resonate-with). Likewise, there are varying degrees of incongruence or dissonance between what workers value and the values a given job reinforces (Dawis & Lofquist, 2000; Atkins, 2013; Porfeli & Mortimer, 2010; Atkins, 2014; Tziner, Meir, & Segal, 2002).

Some whose expertise spans both psychometrics and statistical-mathematics (Hesketh, 2000; Tinsley, 2000) have passionately argued that if multiple vocational talent dimensions are to be considered simultaneously to *predict how* a job candidate's incongruence (or misfit) subsequently impacts on successful tenure, then such jointly-constructed misfit indices (or profile gaps) are akin to highest order interaction terms (even absent multiplicative, product crossing dimensions). In secondary-education, many of us learned that linear prediction equations including multiplicative-product terms must first include each element in such a product term as a simple main effect. Next, subordinate interaction terms are entered ad infinitum, in the case of many profile gaps being considered. Once statistical considerations of this nature are completed and included in one's model, then one can test for incremental predictive power associated with a Minkowski distance algorithm (the unleashed or generalized form of a simple squared Euclidean distance – where all relevant profile-gaps contribute simultaneously). Where *as few as* half-a-dozen predictors were to be entered as simple main effects, then if all subordinate interactions were subsequently entered, this approach to linear modeling would likely dispel statistically reliable contributions generated from the highest order term (e.g., a weighted Minkowski distance between candidate profiles and job-specific competency profiles or some-such summary metric of profile-misfit). This sort of outcome would be very likely to occur despite this latter *profile-misfit* being the main matter of interest (… and possibly, in practical terms, this *profile-misfit* being the most important predictor of successful tenure in a vocation, e.g., see Atkins, 2014).

Secondly, there are *philosophical criticisms*. These are largely concerned with "fit" theories being frequently accused of excessive reductionism and sterile determinism, and afflicted with "man-as-machine" Taylorism and simplistic Colonial-era stereotype-driven categorization. For example, Killeen (1996, p. 27) elucidates vocational career theorizing in terms of "systematically differential social environments of individuals such as coercively-maintained, gender-related social institutions which determine occupational type" in making sense of structural theory. In other words, Killeen (1996) asserts that social classification powerfully influences an individual's vocational career. It follows that such coercive processes would corrupt career theory-based algorithms while also distorting human inputs that these algorithms necessarily act upon.

In this context, Tinsley (2000) also warned of the impact, good and bad, of 21st century Knowledge-Management technologies. Most of these potentially competing frameworks for viewing a job-candidate's fit to a particular job can now be simultaneously considered via online databases, online testing and fast-acting, online algorithms (how "fast-acting" these are depends on the internal consistency of inputs, such that an algorithm's iterative cycles converge on stable solutions). In contrast, in the knowledge-management arena, much work remains to establish trust-worthiness. In the context of this chapter, the degree to which this trustworthiness is crucial depends on the life-and-death criticality of a given water project's purpose or mission.

Establishing "fit" is crucial in predicting worker commitment, performance success and tenure, while providing alternative frameworks for integrating work analysis and vocational planning. Work Adjustment Theory sums up the gaps between what workers value and what their jobs reinforce. In addition, Dawis and Lofquist (2000) highlight the job-content to job-skills fit. Thus, the degree to which a worker is satisfied with and capable to do the work along with the values that support or reinforce the job, predetermine fit or misfit, and influence vocational intentions. "Satisfactoriness" refers to the fit between how satisfactory the employee is viewed from the employer's perspective; and observable through the lens of a supervisor with regard to potential (and real or developing abilities) to succeed at the work (Dawis and Lofquist, 2000).

Increasingly, vocational management has engaged Myers Briggs Type Indicators (MBTI) which have experienced increasing acceptance in business arenas (vis-à-vis academic) relating to personality and leadership type in vocational decision-making. Holland's vocational personality theory (1997) considers three matched dimensions of vocational personality (realist–social, enterprising-investigative, and artistic-conventional) which are oppositionally paired; describing the corners of a hexagon. This model provides another view of vocational congruency (Holland, 1997; Day & Rounds, 1998). Holland's model provides a basis for generating a profile of individuals' vocational personality across the six dimensions and for profiling jobs along the same six dimensions. Proceeding along the lines of a fit-misfit paradigm requires analogous data. While this model might seem straightforward, it highlights the relative distinctiveness of individual performance, interests and stability and apparently demystifies an otherwise complex theory. MBTI's popularity is found specifically in the context of assessing the compatibility of individual personality profiles, to profiles of attributes required for success in particular occupations. R.J. Harvey (Harvey & Hammer, 1999; Atkins, 2013; Bess & Harvey, 2002; Harvey, Murry, & Stamoulis, 1995; Atkins, Carr, Fletcher, & McKay, 2006) is largely credited for advancing the state-of-the-art both in MBTI analyses and in vocational talent-requirements analysis (traditionally termed job analysis). For example, Brown and Harvey (1996) illuminated the MBTI-to-work analysis linkages via job-component validity (JCV). Until the Harvey team's pioneering work, "virtually all JCV research had been conducted using the Position Analysis Questionnaire (PAQ)" (Brown & Harvey, 1996, p. 5).

Developed by McCormick, Jeanneret, and Mecham (1972), the PAQ evaluates a job holder's abilities associated with successful tenure in different occupations. It is based on the assumption of a *gravitational hypothesis* whereby workers will "gravitate" towards jobs they believe they can succeed at as long as it provides adequate challenge, esteem and other tacit rewards.

Where personality-at-work is concerned, the most interesting developments in the current decade conceivably derive from research undertaken on Pat Raymark's *Personality-focused Performance Requirements Form* (PPRF, see Highhouse, Zickar, Brooks, Reeve, Sarkar-Barney, & Guion, 2016; Aguinis, Mazurkiewicz, & Heggestad, 2009) and on job-candidate abilities to manipulate work-related personality and *integrity* tests (Raymark & Tafero, 2009; Van Iddekinge, Raymark, & Odle-Dusseau, 2012). PPRF is valuable for three related reasons: (1) its direct mapping of worker/talent requirements in the personality domain; (2) the "open-source" nature of Raymark's pioneering work giving the humanitarian/sustainable "water-aid" community's need for an affordable uptake of these ideas; and (3) water projects will nearly always require project teams. Staffing of teams, especially if for substantial projects, requires consideration of candidate team-member personalities (e.g., as evidenced in the Belbinesque research cited multiple times in this chapter). It is for these just-mentioned three reasons that PPRF research is worthy of greater consideration in our current *HR for water-projects* context.

Expanding on these three reasons, firstly Raymark's work facilitates direct assessments of misfit in the multi-dimensional workplace-personality domain while also connecting this to the dominant Big 5 or Five-Factor Model (FFM) of human personality (Judge & Ilies, 2002 for meta-analysis; Vernon, Villani, Vickers, & Harris, 2008). Thus, the new open-source, collaboratory approach facilitated by Raymark's work and that of Highhouse *et al.* (2015) means that the above-mentioned and somewhat controversial JCV inferential linkages (Brown & Harvey, 1996) will no longer be the only publicly-available mechanism here. In other words, it will no longer be the only way to rigorously test the importance of personality in project (or other forms of) work.

Secondly, the open-source nature of PPRF research is crucial to this chapter's mission, given that in humanitarian-aid/development arenas (where much water-project work occurs), there is historically-reliable resistance to expensively-invest in HR consultancy support. Thus, a premise emerges for big-data-based research and development (R & D) efforts to optimize recruitment and assignment to water-project teams. This R & D work should be pro bono where summary-data publication rights are awarded (i.e., the benefiting organization-anonymous if requested).

Thirdly, the psychology of project teamwork is complex, and since water projects will almost always require high-levels of teamwork, affordable pathways to optimize staffing methods that form optimal water-project teams are likely valuable. While debate continues on the best ways to populate project teams, the simplest illustration of this idea has derived from Belbin's (2001/2012) work. He was apparently the first to use decades of observing (via his consultancy work with multitudes of project teams), as a primary source of evidence-distilling, complementary personality styles for high-performance teams. The highest impact criticisms of Belbin's work (e.g., Furnham, Steele, & Pendleton, 1993a; 1993b) have focused on the psychometric properties of its associated questionnaire, and thus its value as an illustration of groupthink-countering team-member diversity remains.

Some psychometric flaws in Belbin questionnaires have been partially addressed (Aritzeta *et al.*, 2007), such that online delivery for assessments of team-role predilection comparison, worker-by-worker, should soon be attainable. Generally, these analyses should predict the sorts of team-roles that project team-members will gravitate towards when finding themselves in teamwork environs (…where *success of the team* appeals to them - and for water-aid projects such appeal is reasonably safe to presume for most water-aid workers).

From studying Figure 5.1, it's rational to expect that most water-project teams will benefit from the sort of talent-mix that Belbin's work suggests. Water-project teams might need several different specialists (e.g., hydrologist, field-epidemiology/technician, plumber, builder, etc.). This obviously might leave less than eight team vacancies for covering the other Belbin-style (or Belbinesque) team-roles. Thus team-members will likely be multi-rolled. The identification of these candidates should fall within the capacity of near-future, talent-management systems. Pragmatically, ensuring adequate coverage of these nine team-member styles in each water-project team requires sufficiently large pools of candidates. This century's "war on unsafe water" may depend

	Belbin Role Strengths	Allowable Weaknesses
	**Plant:** Creative, imaginative, unorthodox. Solves difficult problems	Ignores incidentals. Too pre-occupied to communicate effectively.
	**Resource Investigator:** Extrovert, enthusiastic, communicative. Explores opportunities. Develops contacts.	Over-optimistic. Loses interest once initial enthusiasm has passed
	**Coordinator:** Mature, confident, a good chairperson. Clarifies goals promotes decision-making, delegates well.	Can be seen as manipulative. Offloads personal work
	**Shaper:** Dynamic, challenging, thrives on pressure. The drive and courage to overcome obstacles.	Prone to provocation. Offends people's feelings
	**Monitor Evaluator:** Sober, strategic and discerning. Sees all options. Judges accurately.	Lacks drive and ability to inspire others
	**Teamworker:** Co-operative, mild, perceptive and diplomatic. Listens, builds, averts friction and conflict.	Indecisive in crunch situations
	**Implementer:** Disciplined, reliable, conservative and efficient. Turns ideas into practical actions.	Somewhat inflexible. Slow to respond to new possibilities.
	**Completer Finisher:** Painstaking conscientious, anxious. Searches out errors and omissions. Delivers on time.	Inclined to worry unduly. Reluctant to delegate
	**Specialist:** Single-minded, self starting, dedicated. Provides knowledge and skills in rare supply	Contributes on only a narrow front. Dwells on technicalities

Figure 5.1 Belbin's (2001) team-roles with associated strengths and allowable weaknesses.

upon availability of thousands of new/additional water-aid workers with relevant talents. In this chapter's recommendations for future project-team field research, we will revisit the Highhouse team's (2016) open-source collaboration idea, and suggest ways that the Atkins (2012) "Generic Response Scales for Worker-Oriented Group or Team Mission Analysis" (see the Appendix below) might be applied. A goal for addressing this situation would be to have water-aid project team-talent needs explicitly quantified at team-level. The criticality of some Belbin team-roles is likely to vary across different types of water-project work (e.g., insufficient "monitor/evaluation" emphasis is repeatedly seen in media reports of aid-work, but lack of adequate shaping by a demonstrable leader has also been repeatedly seen). For present purposes the R&D needed just to simultaneously model water-project team-talent requirements, both in terms of soft-skills (e.g., Belbinesque team-work styles/team-member values) and hard-skills (e.g., specialist knowledge and abilities), will require advances in what HR-affiliated knowledge-managers currently handle.

5.1.2 Currently available HR-related online technologies in the public domain

In moving the arena of water-project talent requirements into the domain of modern knowledge-management technologies, consideration must be given to the different vocational development products available to measure vocational fit. For example, Drewes *et al.* (2000) developed a computer algorithm that allowed some attribute profile gaps (i.e. misfits) to influence the profile comparison outcomes, while it highlighted other gaps that generated minimal influence (i.e. moderate under-qualification for a particular skill, versus moderate over-qualification). Drewes *et al.* (2000), algorithm sits in the public domain, on the U.S. Department of Labor's O*Net Online (a free online service).

This increasingly comprehensive online shell or infrastructure called O*Net is based on the work of Peterson, Mumford, Borman, Jeanneret *et al.* (2001). O*Net's myriad of hyperlinked web sites and web pages is hosting the envisioned future of data-supported exploration and vocational management. It was once very time consuming, tedious and expensive for most individuals to acquire comprehensive multi-dimensional job-fit/misfit data. The government-funded O*Net system provides zero-to-low-cost products for vocational exploration, as well as work-analysis questionnaires. The U.S. government has been using these work-analysis questionnaires for over a decade, to collect and update quantified talent-requirements information for America's top 1,000 most populous occupational fields.

However, it should be noted, there is ongoing debate as to its empirical rigor. O*Net's potency lies in its large set of logically-related hyperlinks. For instance, one feature enables users to incorporate O*Net occupations and information within a customized "vocational ladder." A vocational ladder is a device that helps people visualize and learn about job options that become available through vocational progression over time. These vocational ladders usually incorporate visual representation of vertically job progression in a specified vocation. In shifting a knowledge manager's interest in water-project talent-requirements towards more "sustainable water-projects" it will be prudent to ensure water-related talent-management systems convey career progression options. Beyond such ladders, vocational lattices (a derivative) display both vertical and horizontal movement between jobs, thereby reflecting more closely the vocational paths of today's work environment (US Dept. of Labor, 2011). Vocational ladders and lattices are useful for tying work analysis insights into to vocational planning and development. They also highlight potential alignments across jobs (vocations or career-fields) – along with fluid movement upwards and/or across job types. This career progression should aid efforts to attract high school recruits (and other career explorers) into water-related vocations, where these are seen to offer both vertical and horizontal pathways. Consequently, they have potential to attract new recruits who can see vocational opportunities within a specified vocational-field laid out from inception of their water-related career exploration.

In sum, given the broader data collection now occurring in the social media era and given the increasingly-powerful data analysis tools that are becoming available for talent-management, important advances will emerge. As some of these can be anticipated, it is prudent now to address some specific applications.

5.1.3 Practices specific to sustainable water-aid

Doctors Without Borders (DWB) and Engineers Without Borders (EWB) are examples of one way to attract the necessary talents required for addressing future waterborne-diseases and safe-water needs. DWB and EWB are highly-acclaimed developments largely in the 21st century, but it remains a notably small pair of movements, given the anticipated numbers of skilled talents needed to cope with a looming humanitarian water crisis. Given the gravity of the impending water crisis we urge there be a new and fundamental engagement of the *talent-management profession* with the *knowledge-management profession*. Clearly DWB, EWB and efforts akin to these specific to water (e.g., Rotary International's WASRAG, see www.wasrag.org), have made advances in the use of social media, yet these advances fall short of what will be required for a sustainable response to the anticipated crises in human water supply. It is foreseeable that big-data in the HR arena will be notably helpful in identifying, recruiting and selecting talents needed for addressing future water crises. It is also helpful that Rotary International's WASRAG is investigating ways to greatly increase the salience, uptake and reliability of online information relevant to waterborne diseases and drinking water crises. WASRAG-supported efforts at initiating the use of talent-related big data (for *competency modeling of water-project work*) are described below.

5.2 Empirical glimpse at needed competencies for sustainable water projects via HR big data

While the information above has set the contextual stage for readers of our initial empirical work, this chapter now needs to briefly review a very common technique for data reduction (sometimes called dimension-reduction so as to reflect its use in reducing or aggregating large numbers of measured dimensions down to a more parsimonious, and thus practical, set of dimensions). Then below, this chapter describes how dimension reduction algorithms make possible the demarcation and distinction of groups or panels of like-minded raters. These algorithms play a crucial role in our water-projects' talent-research program.

5.2.1 *Fundamentals of statistical dimension-reduction*

A premise of this chapter is that HR effectiveness, near-term to long-term, means dealing with the growing world of big data. *HR-relevant* data deemed "big" involves multiple dimensions and/or categories, either co-related, interacting or confounded, so as to allow our simple comprehension of it. In many cases, its complex dimensionality can be simplified via applications of matrix algebra performed in high-speed stability-seeking iterations, performed by computerized algorithms. Some users label these systems via the terms "data-reduction" and "dimension-reduction" somewhat interchangeably. In the past decade, dimension-reduction software has become increasingly affordable and user friendly (e.g., SPSS/AMOS). Thus, for its present purpose, it may suffice to compare a philosophically orthogonal pair of increasingly popular forms of dimension-reduction, referred to as Q-methodology and R-methodology.

5.2.2 *Q-methodology contrasted with traditional R-methodology/questionnaire factor analysis*

Q-methodologies facilitate the use of commonly-available *dimension-reduction* algorithms (e.g., SPSS factor analysis) in studies of human subjectivity. R-methodology (e.g., principal factor analysis) is most commonly used to identify sub-tests or sub-scales or groupings of questionnaire items (e.g., Likert scale survey questions) measuring targeted constructs, attributes, or attitudes. Essentially, it allows researchers to empirically group "like-evoking" questions. In contrast, Q-methodology uses factor analysis algorithms to group "like-minded" raters or participants. While this requires reasonably large numbers of statements, phrases, images, or other relevant stimuli, the number of raters employed can be reasonably small. As described by Davis and Michelle (2011) this approach is:

> ... consistent with the post-structuralist view that meaning is inherently social and contextual, and that audience members must inevitably draw on discourses of the wider social world in constructing and articulating an account from their own unique location... It is this insistence on subjectivity as self-referential within a field of discourse produced by other selves that makes Stephensonian Q-methodology attractive to post-positivist, critical realist and post-structuralist scholars alike (p. 566).

The recent empirical research summarized here (e.g., to illustrate crucial "water-project" talent arguments made in this chapter) utilized Q-methodology. Nevertheless, it should be noted that stimulus-sets used in the present Q-analysis were derived directly from a big data source where the data structures had been generated from multiple applications of R-methodology (e.g., factor analysis, discriminant and convergent validity analysis).

5.2.3 Important big data sources for future water-project required talents

Historically, critical work tasks involving worker talents delineated by vocation (McCormick, 1979) were amassed and articulated via functional job analysis (Fine, 1974; U. S. Department of Labor, 1965). Ultimately, the greater variety of technical skills has caused the number of articulated skill-sets to swell enormously. In the end, over 12,000, sometimes superficially distinct, occupational titles were requiring costly updates to their listed tasks and talent requirements (Peterson, Mumford, Borman, Jeanneret, Fleishman, Levine *et al.*, 2001). Thus, a clustering of highly-similar occupations was pursued, reflecting the increasing fluidity of workers across thinner vocational boundaries. This required multi-million US dollar/decade-plus collaboration across multiple universities and the U.S. government and various American state governments. This collaboration produced the above-described Occupational Information Network or O*Net. As mentioned, this process more recently settled on a list of roughly one thousand of the "World of Work's" (www.onetcenter.org accessed Nov 2016) most populous skilled-occupations. Notably, some do prefer to think of O*Net occupations as, instead, occupational families, expressing concern that O*Net titles are too abstract and thus, associated task requirements and talent requirement ratings are distorted via the necessitated aggregations of somewhat different jobs into such occupational families (Harvey, 2009).

However, as data-collection work ceased over a decade ago via the 12,000+ occupational titles in the classic DOT, those seeking big data sources in the HR realm for task/talent requirements generally will now rely on O*Net data sources. O*Net periodically updates their 1000+ occupations for current levels on 277 task/talent/average-worker-attribute descriptors, yielding big data which could be approached with data-reduction methods. Thus, O*Net's online system is a primary online source of occupational information for building ranked or rated lists of talent requirements or competency models. The O*Net's competency models deliver a list of knowledge, skills and abilities (KSA) requirements essential for frequently required tasks, important interactions, responsibilities, common work scenarios, congruent work styles and essential mental capabilities for each of the occupations (National Center for O*Net Development, n.d.). This outcome was achieved by averaging ratings from three panels of 15 job-incumbents in each occupation.

The O*Net database yields a list of water project-related jobs, for example, water resource specialist and water/sanitation engineer, or aid work specialist. However, a specific water and sanitation aid worker is not currently available in the O*Net database. Rotary's large water-aid component required a competency model, which is specifically tailored for them. The model for our Rotary research presumes that there will be important differences between the competency model of a general aid-work volunteer compared to specific "global war on unsafe water" related volunteers (e.g., Rotary International's senior-most director of its water-aid program, and his scientific advisors, requested that our research team focus first on water-aid, as it is amongst the most technically and scientifically challenging aid sectors funded by Rotary).

5.2.4 Water-project data source for water-related talents specific to the "war on unsafe water"

The empirical research on human talents needed for *sustainable water projects* summarized in this chapter derives from expertise provided by a large charitable organization, Rotary International, which we sometimes refer to herein as our collaborator. To set the context for the data collections and analyses explained in this chapter, our Rotary host organization is briefly described. Rotary's purpose is to gather professionals from diverse backgrounds together to exchange ideas, raise funds and take actions in global humanitarian work, ensuring they work within a highly ethical framework. A primary aim of Rotary is to generate increased levels of health, social justice and peace so that conflict is reduced. Currently, this organization has roughly 33,000 branches capturing well over one million members across every continent (Rotary, 2007), and through its charitable foundation, started a century-long project focusing on water and sanitation/hygiene (WASH) work, sometimes referred to as its global 'war on unsafe water'. A primary goal is to ensure sustainable water and sanitation projects globally. Both highly skilled professionals and moderately skilled volunteers are essential to undertaking many of the projects. However, it is sometimes problematic getting the right person in the right role. Various volunteer roles are available in the collaborating organization's projects; they span general duties (administrator or manager) to more specific roles (hydrologist, public health specialist, or water resources engineer). On-site volunteers in the field and those engaging via social media, can sometimes be more problematic than helpful (Guttentag, 2009). For example, there can be a disconnect between community needs and what volunteers bring in terms of talent, attitude and experience (Atkins, 2012). The inexperienced aid-worker who is used to living in affluent contexts may suffer notable culture shock when carrying out duties in an austere rural village in a different culture. While culture shock might yield valuable personal growth in the volunteer, in the short-term, it can result in less than helpful aid-work contributions (Atkins, 2012).

To address this problem, Otago Polytechnic in New Zealand and Rotary (with laboratory support from the University of Otago) are developing matchmaking software under the working title, *SmartAid*. This technology operates within a set of algorithms by corresponding aid-volunteer-applicants' skills and experiences with new volunteer project requirements. The needed skill-sets and their relative availability will be ascertained from large public-domain databases (U. S. Department of Labor's O*Net-OnLine system). The integrated application of modern HRM and Knowledge Management (KM) technologies for volunteer recruitment and placement is expected to enrich water-project high-skills volunteerism. In turn, it will benefit our collaborating organization's projects and the global community, if better-suited aid project-teams result.

Towards this end, Bachelors in Information Technology (BIT) students at Otago Polytechnic, with Rotary IT advice, developed and tested a prototype of an online application that facilitated the upload and delivery of training videos and asynchronous cultural training materials specific to water-aid websites. Contemporary knowledge-management technologies allow the facilitation of systematic indexing and peer-review of online materials for preparing water-aid interventions by anticipating the need as well as the urgency. This collaborative student-project work (Tonny, Atkins, & Howison, 2016), with Rotary support, sought connections with O*Net Online systems and expert ratings of potentially key human talents, in this instance, human talents potentially crucial to sustainable water projects.

*5.2.5 First empirical study of O*Net competencies specific to sustainable water-aid projects*

Our Rotary-funded research team had current data from relevance-ratings of 235 volunteerism work-talents, drawn from the following sources: Personality-based Performance Requirements Form or PPRF, see Aguinis, 2009; Highhouse, 2016; and U. S. Dept. of Labor's O*Net Online, 2001. These 235 items were rated for relevance to volunteer aid-work by 21 "aid-volunteerism experienced" university students. These raters rated the importance of how significantly an ability or knowledge or attribute would benefit persons conducting general aid work, and what features (required tasks, interactions and work styles) usually occur in common volunteering scenarios. Three researchers selected 87 items (of 235) relevant to water/hygiene and sanitation (WASH) aid work. These three researchers then added, to these 87 items, 30 WASH aid-related tasks from the O*Net Online taxonomy of task lists. Thus, 117 items were administered to six of Rotary International's *most field-experienced* rural-health/hygiene/drinking-water volunteers forming the expert e-panel. This e-panel rated the "hygiene/water-aid" relevance of these 117 PPRF/O*Net work-dimensions.

Initial empirical results applying a quasi-Q-methodology to water-project competencies

It was straight-forward to factor-analyze and thus dimension-reduce rater responses in the rating task via the use of common quantitative response-option metrics. The application of a confirmatory factor analysis (CFA) paradigm allowed a direct competition between a "dust-bowl empirical/capitalization-on-chance" CFA model versus a theoretical model based on extreme water-project expertise. This relatively novel use of CFA was first presented at the 15th European Conference on Research Methods in Management Studies (ECRM/London, 2016).

Figures 5.2 and 5.3 compare Rotary's six field experts clustering, with notable tightness along a distinct dimension. In this case, it is a dimension

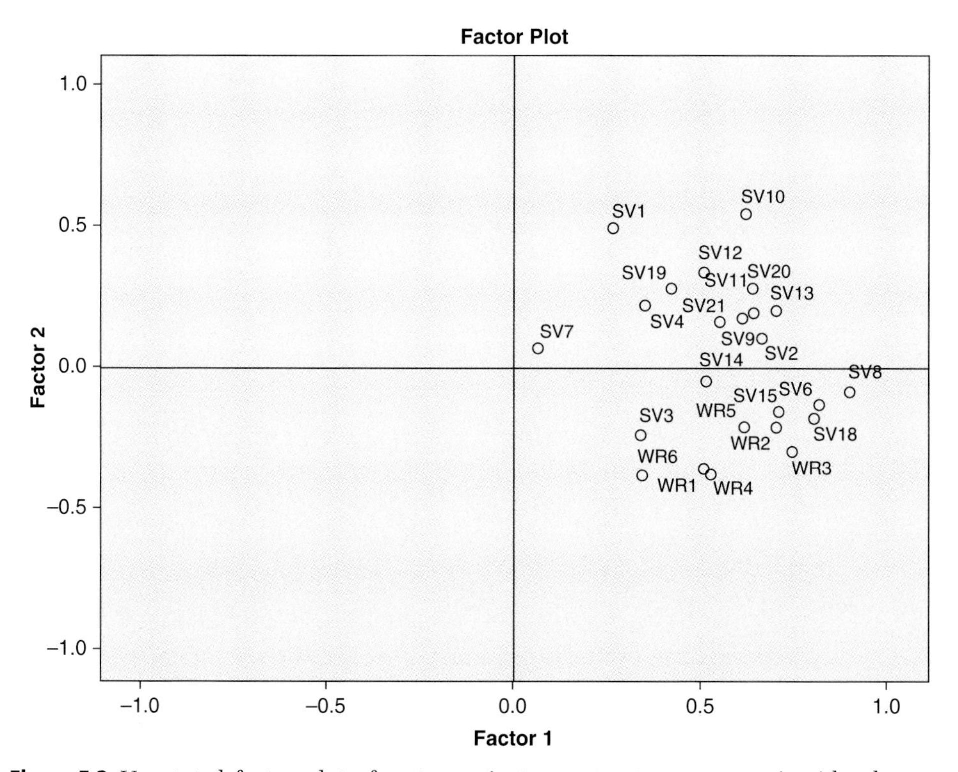

Figure 5.2 Unrotated factor plot of water-project expert raters vs. generic aid-volunteer raters (NOTE: Student Volunteer aid-work raters are SV1 to SV21 and Water-aid expert raters are WR1 to WR6).

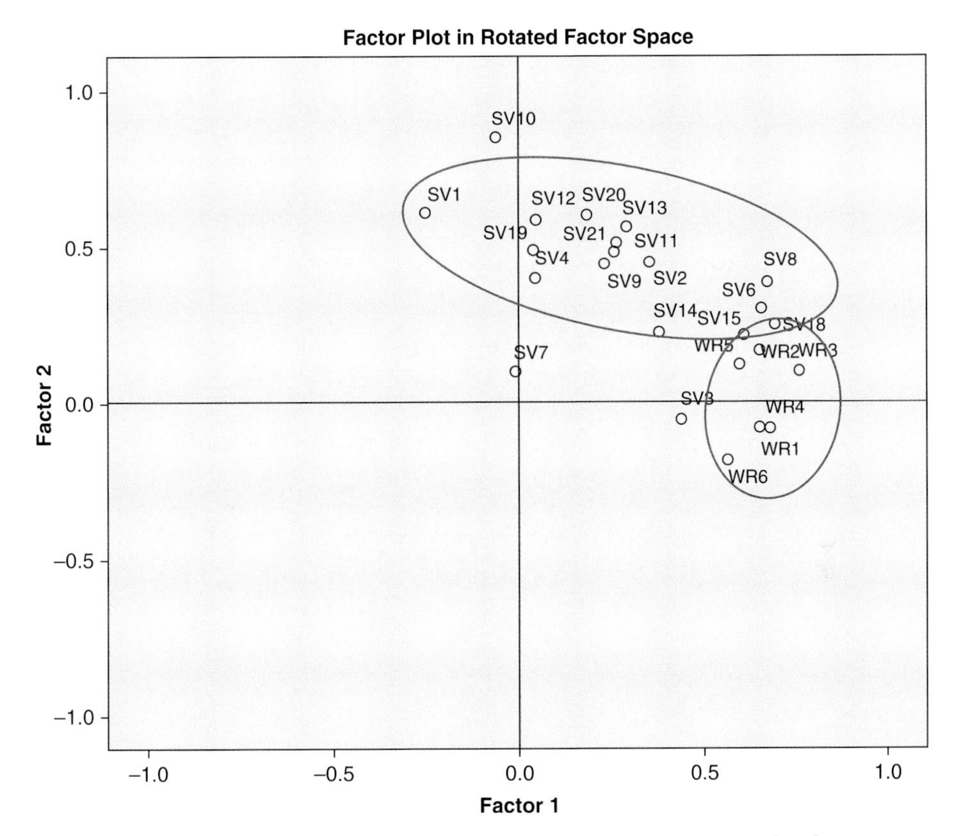

Figure 5.3 Rotated factor plot of water-project expert raters vs. generic aid-volunteer raters (NOTE: Student Volunteer aid-work raters are SV1 to SV21 and Water-aid expert raters are WR1 to WR6).

sufficiently distinct from the one reflecting competencies needed generically in volunteer aid-work. Once factor rotation (routinely undertaken in factor analysis) achieved stable convergence, it was clear that Rotary's experts showed low variance in comparison to the substantial variance in the non-water panel of prior aid-volunteers.

Top 20 water-project competencies suggested by initial application of Q-sorts

As explained earlier, the identification of key water-project competencies is obviously of interest to Rotary water-project talent-recruiters. Given the relatively tight clustering of Rotary's top water-aid experts in their talents-needed ratings (despite the mildly-unwelcomed fit indices in Table 5.1 associated with competing the two models in Figures 5.4 and 5.5), these similar importance ratings, averaged across the six raters, are noteworthy here. Analogous to our approach in interpreting Figure 5.5, we looked for visual evidence of an "A" group distinct

Table 5.1 Goodness of fit indices derived for comparison of theoretical versus empirical model

Model fit indices	CFA derived from visual factor plots (Figure 5.4)	CFA derived from expectation: water experts a distinct panel (Figure 5.5)	Recommended threshold
RMSEA	.098	.110	<.7 as good fit, .8-.10 as adequate fit >.10 as poor fit
CFI	.827	.775	0–1, the closer to 1, the better fitting the model
GFI	.713	.672	0–1, the closer to 1, the better fitting the model

Note: CFA = confirmatory factor analysis.
RMSEA = Root Mean Squared Error of Approximation.
CFI = Comparative Fit Index.
GFI = Goodness of Fit Index.

from a lesser "B" group of water-project related competencies/attributes. Such clear space was evident in the twenty highest rated items listed in Table 5.2 (this clear space separating these top twenty from all items below that cut-point). manifesting as a highly distinct panel.

Averaged ratings on water-project competencies were ranked in a range from 1st to 20th, based on a 1-to-10 rating scale, from a low rating of 8.33 to a peak of 9.50 (mean = 8.63). Averaged ratings on water-project competencies ranked in a range from 21st to 87th, on the same 1-to-10 rating scale, from a low rating of 4.50 to a peak of 8.17 (mean = 6.85). As there was a clean break between the Top 20 water-project competencies and the remaining 67 water-project competencies, it was not surprising to see a statistically-reliable Student's *t*-test outcome when comparing the means ($p < .0001$). In this case, the variance could logically be expected to be smaller in the Top 20, and the mean rating higher. The *t*-test used was for unequal variances, unpaired samples, and with one tail, as in appropriately assuming a directional hypothesis (H_1: $\mu_{\text{Top 20}} > \mu_{\text{non Top 20}}$).

5.3 How modern knowledge-management technologies can make competency tests "time-affordable"

From the preliminary evidence, water-project teams need to recruit team-members in the first instance (again see Table 5.2) with high levels of

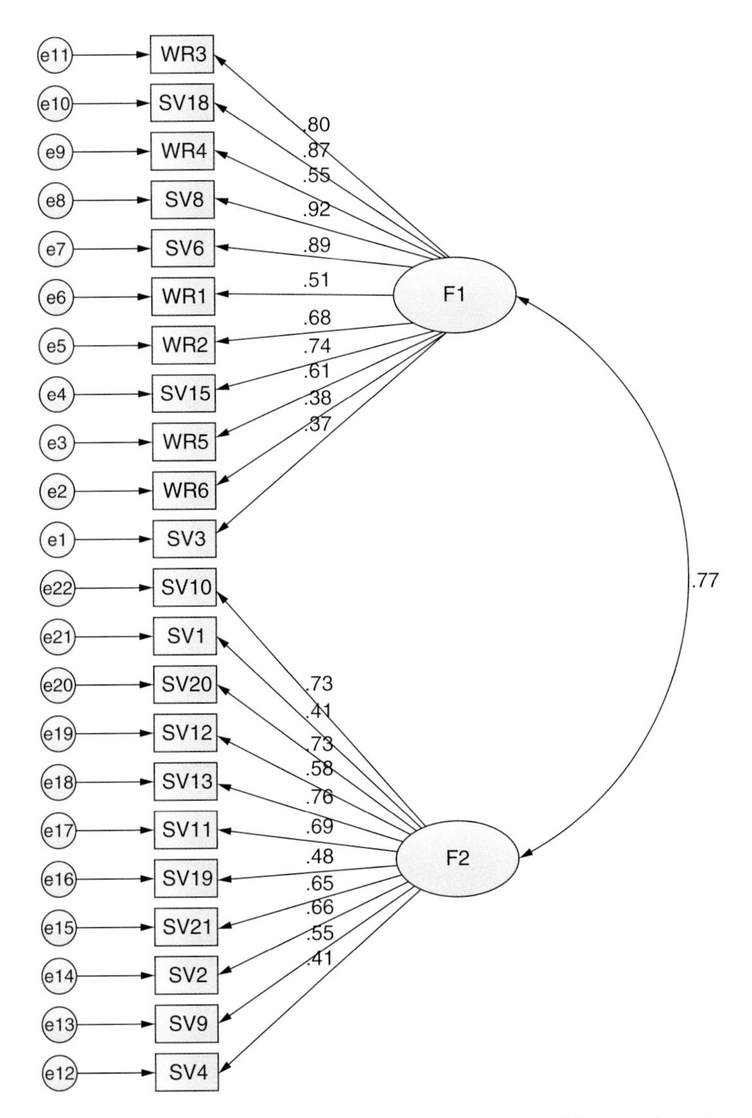

Figure 5.4 Confirmatory factor analysis model based on visual factor plots above.

integrity, emotional-intelligence and team-building skills plus 17 other competencies (noting that necessary water-project technical skills are also assessed via licensing credentials from appropriate professional bodies). Consequently, there are reasonably well-respected reports via reputable science journals, of reliability and validity evidence for mental tests of these crucial teamwork-related attributes.

The same knowledge-management technologies that made the above data collection and analyses possible, also make possible extremely large-set

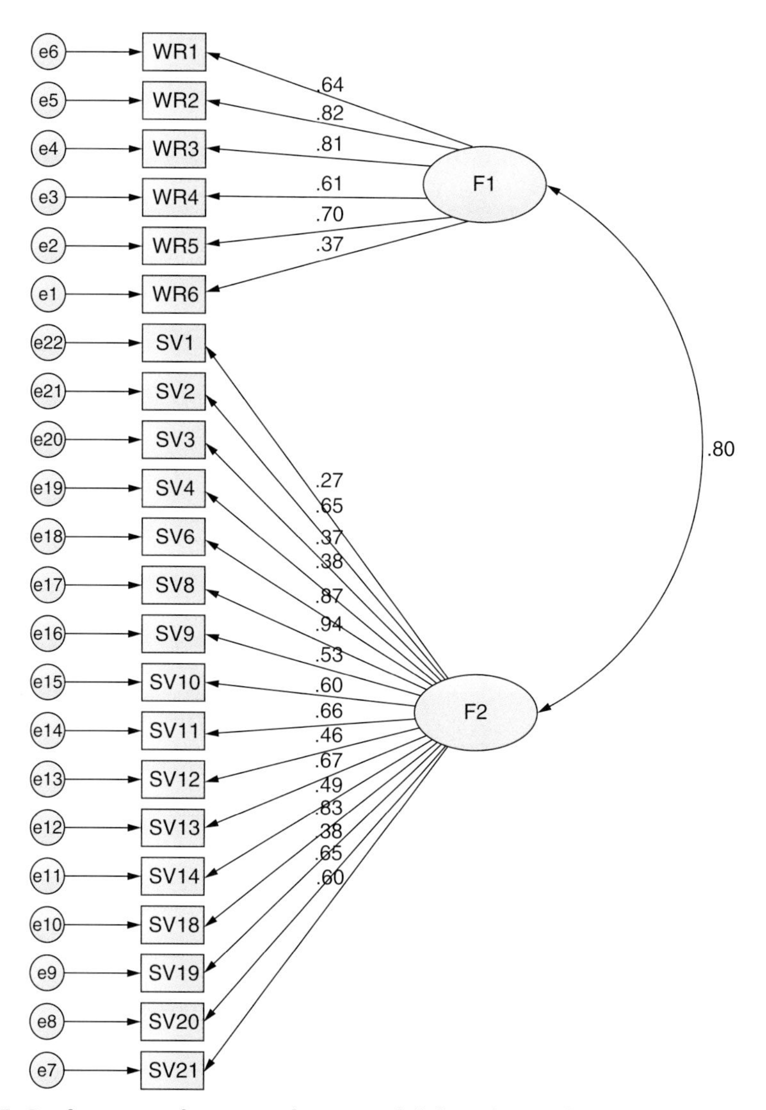

Figure 5.5 Confirmatory factor analysis model based on theoretical expectancy for water-project experts manifesting as highly distinct panel.

multi-dimensional analyses based on very large crowd-sourced samples. Modern calculus-based mental metrics technologies can now measure human attributes, like our Rotary panel's Top Twenty, with fewer questions (although this presumes respondent consistency, in this case, amongst water-project candidates). These modern computer-processing, intensive approaches to multi-dimensional measurement of job-candidate attributes facilitate questionnaire-shrinking, computer-adaptive tests. Anticipated here, if

Table 5.2 Top twenty items rated as most important for water-project field work

Rank	Competency/Human Attribute/Predilection
1st	Integrity
2nd	Establishing maintaining interpersonal relationships
3rd	Developing and building resources
4th	Management of financial resources
5th	Communicating with people outside the organization
6th	Initiative
7th	Making decisions/solving problems
8th	Concern for others
9th	Active listening
10th	Critical thinking
11th	Coordinating the work and activities of others
12th	Guiding, directing, and motivating subordinates
13th	Leadership
14th	Innovation
15th	Social perceptiveness
16th	Communicating with supervisors, peers, or subordinates
17th	Resolving conflicts and negotiating with others
18th	Face-to-face discussions with individuals and teams
19th	Persistence
20th	Cooperativeness

significant increases in water-project human talents are needed in the future, then user-friendly (i.e., relatively brief) online tests of water-project friendly attributes are needed.

5.3.1 *A resurgence to computer-adaptive testing afforded by 21st century crowd-sourcing*

Computer Adaptive Tests (CAT) tests are enjoying revitalized regeneration based on the technological flexibility they afford (Callaghan, 2015). Their adaptability for use on mobile digital devices also helps explain their resurgence (Triantafillou, Georgiadou, & Economides, 2008).

Given greater affordability and uptake, organizations are finding new uses for the tool. For example, the New Zealand Council for Educational Research (NZCER) has opted for a strategic focus on CAT because tests can be adapted "to the test taker, providing harder or easier questions depending on how they are performing" (*NZCER Annual Report*, 2014–2015, p. 5). The ability of the algorithm to select another question based on the previous responses is the crucial strength of CAT.

In a recent large-scale success, across the American state of Georgia, CAT evinced as the best delivery of school graduation exams (for both efficiency and security reasons). It was found to be an "efficient, fair and objective way of [delivering] student assessment" (Bakker, 2014, p. 6).

Further the CAT design has the ability to craft assessments to each test-taker based on iteratively-improving estimates of each test-taker's trait-level, while ensuring the approximate psychometric-equivalence of these varying but idiosyncratically-tailored assessments (Liu, You, Wen-yi; Ding, & Chang, 2013).

Another explanation for the rise in CAT's popularity is its ability to inform decision-making with regard to what to adapt, and how and why, while designing an adaptive e-learning system and course content ("Computer science research; studies from University of Split in the area of computer science research described: Adaptive courseware literature review," 2015).

5.3.2 Why modern Knowledge Management applied to talent management needs CAT

Contemporary Knowledge Management systems relevant to HR can benefit from cutting edge online information sources relevant to talent requirements, to address the variances across vocations. Some *vocational talents-focused* online systems are proprietary and can be very expensive to access. As described above, in the public-domain, a government-developed online vocational exploration-focused system (O*NET) is available at no charge.

Also as mentioned, O*NET databases provide 277 descriptor-dimensions across a 1000+ occupations. Many can very likely be set-aside for our present purposes. However, it is potentially desirable to efficiently assess large numbers of would-be sustainable-water aid-workers on upwards of a hundred dimensions (including, of course, the Top Twenty non-technical talent-dimensions identified in our study). Achieving this reliably would typically require at least a dozen or more pre-employment or candidature test items per talent-dimension, produced via traditional metrics of skills and vocational personality. Such would yield tests with hundreds of questions. Many of these questions would be mentally-taxing and thus fatiguing.

Modern test theories, for example, item response theory (IRT), facilitate the use of large question or test-item pools but ultimately enable shorter online tests via CAT algorithms honing-in on a participant's increasingly-revealed trait level via choosing to display test-questions with difficulty-levels or precision-points just above and below a given test-taker's estimated trait-level. IRT starts this facilitation via the indexing of items by their idiosyncratic points or levels of greatest precision.

Fortunately, modern test theories (e.g., item response theory or IRT) facilitate the use of large question or test-item pools but ultimately shorter online

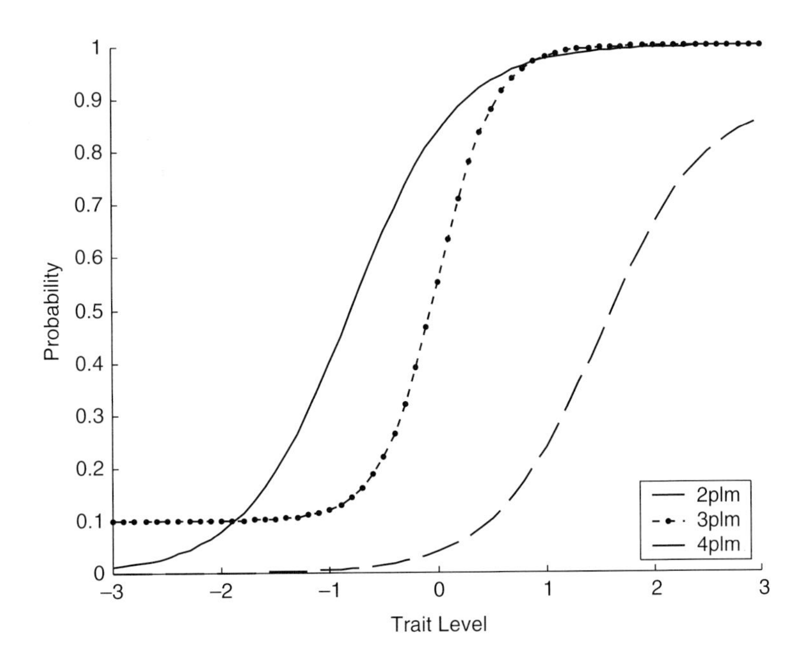

Figure 5.6 Hypothetical example of three item characteristic curves (ICCs).

("CAT'd") tests. IRT starts this facilitation via the indexing of items by their idiosyncratic points or levels of greatest precision.

Precision-points are points of greatest discrimination as evidenced by test-item characteristic curves (ICCs) (Figure 5.6). An ICC is plotted, left-to-right, to show the increasing probability of a correct (or *trait-affirming*) response as test-taker response data is analysed from the lowest-ability (or lowest "traited") project-team candidates. Thus, these curves tend to flatten near 100% probability for the very highest ability (or most "traited") candidates. But it is crucial to remember that only one tightly-defined trait or attribute or ability can be analysed this way. In other words, in the same way that our section 5.2.2 (above) explained that factor analysis allows test-developers to "....empirically group "like-evoking" questions....", modern IRT analyses, and thus CAT, require identification of uni-dimensional sub-tests. These are clusters of questions assessing just one human attribute (e.g., self-deception or *self-award of unlikely-virtues* which like narcissism and Machiavellianism can be theorized as components of integrity tests or their inverse, such as Dark Triad tests – see Jonason & Webster, 2010).

In addition to highly-desirable "water-project" team-member attributes like integrity, emotional-intelligence, etc., should the sustainable-water arena seek an option to test hard-skills (vice using certifications from professional bodies), then CAT online tests can also be developed here (e.g., a sub-test tightly-targeting *Knowledge of Surface Water Features increasing likelihood of disease-conveying contamination* sitting alongside a sub-test targeting *Knowledge of Surface Water Features increasing likelihood of toxic-chemical ingestion*).

Modern IRT analyses can commence once test-developers have used factor analysis to identify "like-evoking" questions, thus forming sub-tests that are, internal to themselves, uni-dimensional (i.e., measuring just one knowledge, skill or attribute). Such will then produce ICCs, for each sub-test item that rise, monotonically, left-to-right. As said above, these normally approach 100% *probability of a trait-affirming response* as the curve flows to the right-most edge. That's because the right-most edge represents the very most "traited" project-team candidates (again, see Figure 5.3).

The first derivative (via differential calculus) of these curves yields item information functions (IIFs) (Figure 5.7), which peak at each test question's point of greatest discrimination or information-provision (NOTE: for readers not trained in differential calculus, one can think of the transforming of an ICC into an IIF as simply the visual plotting of the "instantaneous" slope, angle, or gradient of each point on the ICC as the curve progresses left-to-right, or from the lowest level of the measured human attribute to the highest level).

These item information functions (IIFs) are absolutely crucial to the construction and online operation of CAT exams or project-team candidate assessments. That's because the peaks of these IIFs inform the online CAT algorithm of where a test-taker's trait or ability level falls once a few widely

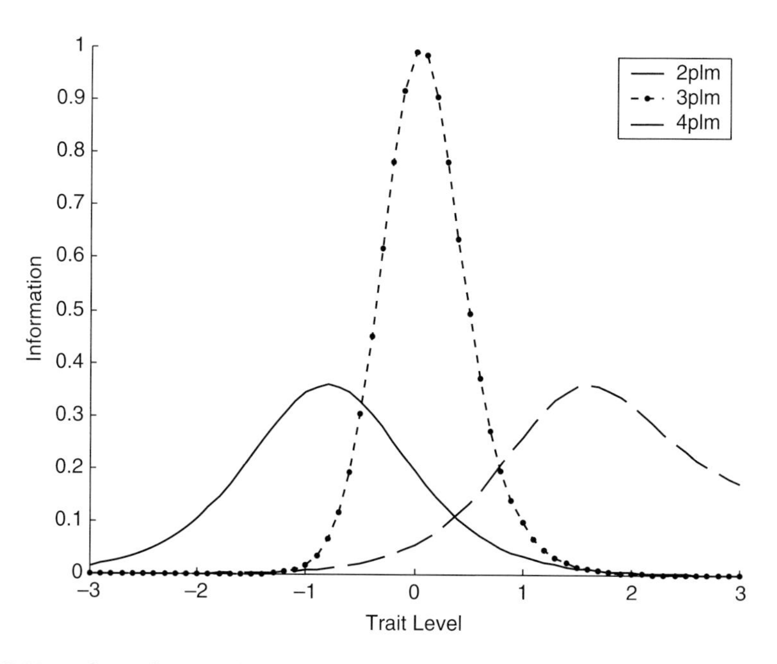

Figure 5.7 Hypothetical example of three item information functions (IIFs).

dispersed test-items have been answered and scored in real-time (see Figure 5.4) (Ainsworth, 2015). This is why test-developers seek to write test-items that, as a set, evince with a diversity of trait-level peaks (looking across a uni-dimensional sub-test's cluster of test-question IIFs).

Arithmetically summing these item information functions, across test questions measuring one human attribute, yields a test information function (TIF) (Figure 5.8). Inversely related to the latter will be a "one-human-attribute" sub-test's Standard Error Function (SEF).

An SEF can be used to assess where a test offers greatest precision (e.g., at what levels of ability it measures most precisely where levels might be ranging from low-skilled to high-skilled as a hydrology-engineer, water-contamination assessor) (Figure 5.9). The SEF can also be used to report standard 95% confidence intervals around an individual water-aid job-candidate score. These would be far better confidence intervals than are typical because IRT (or post-classical test theory) computer algorithms yield intervals specific to a given job-candidate's talent level.

Modern metrics such as these aforementioned, facilitate efficiencies in computer-adaptive testing algorithms, once they can estimate where candidates sit on each attribute's dimension (from low-to-high, or weak-to-strong, or less-talented to more-talented). In sum, online delivery (or administration) of computer-adaptive tests (CATs) allows the test-delivery algorithm to minimize the number of questions asked of each water-project volunteer. Thus, CAT can focus questions based upon the antecedent responses of each online test-taker. Given the large number of talent-related dimensions needing measurement here,

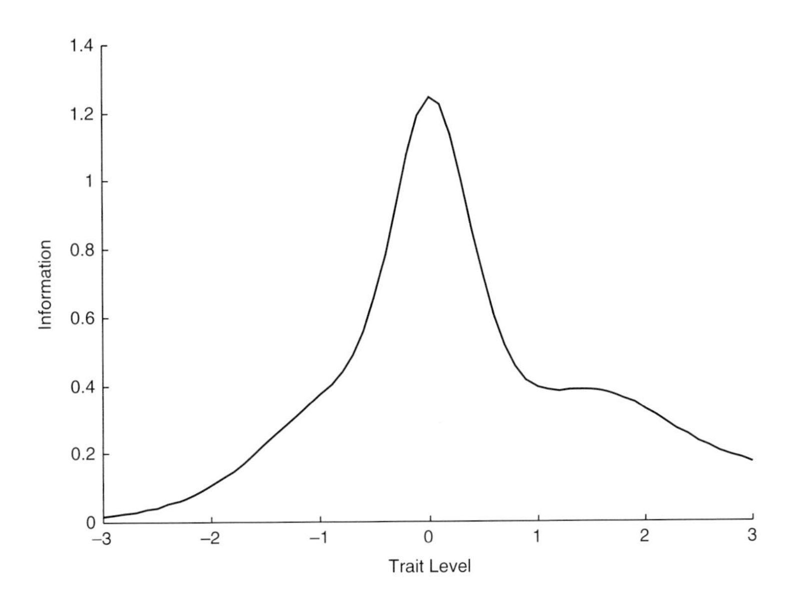

Figure 5.8 Mathematical summation of about three IIFs into a TIF.

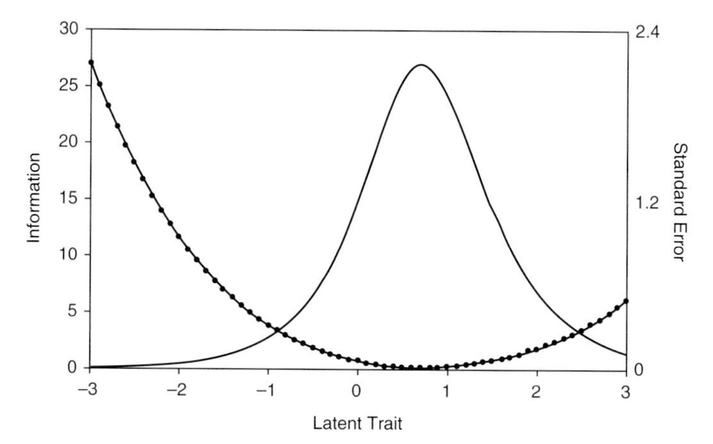

Figure 5.9 Once very large (but still unidimensional) item-pools are created, TIFs can be produced that are intentionally uni-modal, bi-modal or tri-modal, if precision is required in particular and possibly multiple levels of the targeted trait. A test's Standard Error Function (SEF) as shown in the figure, reveals that a test has its greatest precision where its TIF peaks.

it is crucial each sub-test be as brief as possible. Test-brevity is a prime benefit of, and indeed a motive for, CAT use.

5.4 Limitations

In the broadest and most general terms, our empirical research has presumed that traditional mathematical approaches (used in manpower engineering realms) can be productively applied to optimizing the staffing of water project teams. We acknowledge there is a notable difference of perspectives between common HR practices and the mathematical modeling we have detailed in this chapter. However, we are persuaded there is a fruitful middle ground given that water-aid must operate within the realm of human subjectivity and frailty, but also needing, via the huge populations targeted here, the mathematical rigor described above. We have taken this approach because our main funders have highlighted that the world will likely need an enormous increase in the number of rural water projects going forward. To sustain human life in a bleak "water-shortage" future, we need successful water project teams, potentially numbering in the thousands. Those humanist researchers viewing this from a postmodernist or feminist-critical perspective (Wilson, 1998) would likely argue that statistical techniques inherited from the physical and agricultural sciences are unsuitable for our arena. Given the social/emotional complexity of team-member relations and their impacts on team performance, such an argument has credence. However, the sheer magnitude in the numbers of new *human talents needed* calls for team-matching algorithms that can at least minimize the probability of catastrophic *water team-performance* failures.

The humanitarian aspects of future water-aid projects yields an imperative for internal-to-team civility and mutual respect. The harsh climates and cross-cultural tensions frequently encountered in our developing nations' rural locales (Guttentag, 2009; Atkins, 2012) can manifest stressors threatening such civility. The same can be said of in-team/in-situ talent and resource shortages, for instance. We have presumed it prudent to begin development of large-scale team-staffing algorithms (and associated pre-assignment measures or instrumentation). Our hope is that such algorithms, at least when coupled with a humanitarian or a health and well-being ethos, will minimize notable failure risks.

One of our more specific concerns relates to our empirical work's limited generalizability. While the latter is always an empirical question (Hauenstein, 1995), we were specifically looking to see if our panel of Rotary International's top six most-field experienced *water-project expert* raters (Tonny *et al.*, 2016, June) clustered tightly together in a Q-matrix analysis. This clustering is notable when compared to student raters who came with a wide-range of humanitarian-volunteer or aid experiences. Further, as seen in our Q-methods loading-plot, the distinctiveness of our Rotary-provided "world-class experts" was partially supported. In itself, this is not really a limitation. Though, we should note that the distinctiveness of our six experts might be driven by:

1. Possibly greater, or at least a more sustaining, ethos of altruism compared to our student volunteers.
2. Statistically-greater apparent commonality (in rater perceptions of talent needs) via their common long-term membership of Rotary International.
3. Deriving from point 2) above, greater similarity in related workplace values (e.g., Rotary's compulsory 4-way test of its members' ethical behavior, i.e., all *that Rotarians do* must grow goodwill and be truthful, fair and beneficial to all concerned).
4. Unusually high levels of optimism, relatively higher in Rotarians with decades of devoted volunteer service compared to students for whom aid-volunteerism might be seen as more a matter of enhanced-employability post-graduation (Goldweber, 2013). Festinger (2013/1962) explains that giving a harsh or hazardous cause the sustained sacrifice of *your favors to it* will likely generate notable, even stubborn, dedication to your associated beliefs. Thus, if optimism-to-cynicism can be envisioned as a continuum ranging from positive-to-negative, then our massively field-experienced Rotarian aid-workers will likely be far less cynical about water-aid, or aid-service generally, than our student panel-members.

Despite the limitations we have just identified, we see our research program as important to the success of sustainable water-aid project teams.

5.5 Future research

Our initial empirical efforts are largely seen as a methodological proof-of-concept, for example, seen in our novel comparisons of AMOS goodness-of-fit indices across a competing pair of confirmatory factor analysis models applied to a "Q" (as opposed to an "R") data matrix. Our expert panel was not large enough to be fully compatible with U. S. Department of Labor's panel-practices (LaPolice, Carter, & Johnson, 2008; Oswald, Campbell, McCloy, Rivkin, & Lewis, 1999) for rating the differential importance of competencies. In the future, a doubling of our number of expert raters is recommended. It would be prudent to have six water-aid project experienced raters recruited from one of Rotary's competitors (e.g., Lions International, Kiwanis International, etc.) absent the equivalence of the 4-way test. Then a third panel could be recruited comprised of salaried/government or council-employed *rural water-project* professionals, thus increasing generalizability. This is because, thus far we've demonstrated that 21 students with a variety of aid-volunteer experiences have much greater variance in talent-need perceptions than a panel of six Rotarian water-aid experts. But, in case that distinction is flavored by Rotary International (or "Rotarian-peculiar") ethos or zeitgeist, it might be helpful to see how that panel of Rotary's six top water-aid field experts compares to their closest non-Rotary counterparts.

Given our current empirical findings, we hypothesize that these two future clusters of raters would show some internal commonalities, but would likely evince more cross-groups variance in common with our Rotary water project experts. We expect all of these panels' ratings will have less in common with those of our 21 student-raters.

There are several future research needs emerging from the above work, for example, applying these sorts of *talent-requirements* approaches to other competency or worker-attribute domains, including team-roles, work-values and skill-aptitudes. While we intend to pursue these efforts, we also encourage other HR researchers to engage with what is possibly *the* defining challenge to 21st century humanity.

5.6 Conclusion

Better talent or competency detection and integration with online knowledge-sources/big data should lead Knowledge Management to new heights in recruiting needed talents for anticipated drinking-water crises. We conclude that the ill-defined era of big data coupled with the increasing affordability of large data analytics is yielding HR knowledge-managers an unprecedented opportunity to advance the rigor of talent-management practice.

Recruitment and optimal assignment of human talents for *sustainable water projects* should be the highest priority for employing such rigorous analysis advances.

The online talent-management system that Rotary desires currently includes cross-cultural training affordances for the recruited talents. These training materials require input from volunteer experts and this is therefore the first talent-recruitment need. Our research program also seeks short YouTube videos from aid workers with cultural expertise particular to water-project aid-sites. In these short YouTube videos, we ask these aid workers to include common faux pas or social errors that accidentally offend locals. Often such offense is by unintended conveyance of "us-versus-them" hegemonies. However, sometimes it is simply through naiveté or ignorance that a given body language or phrase is interpreted differently across some indigenous cultures. These YouTube videos will act as training materials, organized by geo-regions (internal to this large water-aid application). This app's associated pages also include hyperlinks to recommended sources on local cultural courtesies and common pleasantries (excuse me, thank you, please, welcome, hello, goodbye, farewells, forgive me, etc.). Recommended cross-cultural hygiene-education strategies will also be included along with critical-incidents case study data. The latter will provide examples of likely precursors to water-project successes and failures. Admittedly, what we have described here is an ambitious knowledge-management online-app development program.

Despite the dire circumstance much of the world finds itself in with regard to water-supply, aid-agencies must still accept that there are well-established, but problematic, cultural norms around food, meals, toileting and hygiene. Where these are no longer survivable, eco-sustainable change must occur. However, the way those changes achieve buy-in from locals, particularly tribal seniors or elders, is immensely complex (e.g., diminishing unsafe practices while respecting cultural differences or local cultural norms). Some of this complexity has clearly influenced the human-competencies our experts rated as crucial, but our rating processes and connected CAT technologies are no guarantee that online testing to detect those talent-levels will suffice. This is why this research effort seeks to engage knowledge-managers in a broader fashion.

Earlier, we referred to Davenport and Prusak's (1998) definition of KM. It is judicious to revisit their comment, "Knowledge Management draws from existing resources that your organization may already have in place: good information systems management, organizational *change management* and human resources management practices" (p. 163). We emphasize above, the change management component of KM. While this still connects to KM's HRM component, especially people's well-being as a classic HRM priority, the research programme that yielded the above Q-structured competency model goes further than merely modeling competencies. Talent management is a demanding undertaking when large-scale and complex human endeavors are the context. The global "war on unsafe water" most certainly qualifies, and we hope 21st century knowledge managers will find this war of compelling personal interest. We anticipate such interest will propel their uptake of the challenge to further develop and use these emerging talent-fit technologies (e.g., to enhance efficacy in sustainable water project teams).

Appendix: Example team work-analysis items (Atkins, 2012)

NOTE: Where "Analyzing" appears, in that same space, other attributes or work-activities are inserted, as needed, for adequate domain coverage, by repeating this quartet of questions. For this present example, see definition of "Analyzing" at bottom.

1) How commonly is this attribute (*Analyzing*) required for *acceptable* (i.e., satisfactory) group or team performance? In other words, team or group performance is satisfactory if this attribute has input (or is notable) from:

Not relevant at all	**Any one member**	**Leader alone**	**Leader plus some other members**	**Everyone on the team**
A	**B**	**C**	**D**	**E**

2) How critical is it for this team or group to possess a sufficient amount of this attribute (*Analyzing*)?

Not relevant	**Minimally critical**	**Moderately critical**	**Substantively critical**	**Extremely critical**
A	**B**	**C**	**D**	**E**

3) What proportion of *time* in the group or team's efforts would involve this attribute (*Analyzing*)?

0% to 20%	**21% to 40%**	**41% to 60%**	**61% to 80%**	**81% to 100%**
A	**B**	**C**	**D**	**E**

4) What consequences are likely if the group or team *fails* to possess enough of this attribute (*Analyzing*)?

Not relevant at all	**Diminished team self-confidence**	**Diminished team value or reputation plus possibly that to the left**	**Notable salary, employment, or profit, or program losses plus possibly all to the left**	**Health or life losses and/or corporate demise plus possibly all to the left**
A	**B**	**C**	**D**	**E**

Note: "Analyzing" can be thought of as a combination of deduction, induction, pattern recognition and/or pattern creation (or pattern-ordering). The U.S. Dept. of Labor's O*NET system defines these sub-components as: Deductive Reasoning — The ability to apply general rules to specific problems to produce answers that make sense. Flexibility of Closure — The ability to identify or detect a known pattern (a figure, object, word, or sound) that is hidden in other distracting material. Fluency of Ideas — The ability to come up with a number of ideas about a topic (the number of ideas is important, but to a lesser degree their quality, correctness, or creativity). Inductive Reasoning — The ability to combine pieces of information to form general rules or conclusions (includes finding a relationship among seemingly unrelated events). Information Ordering — The ability to arrange things or actions in a certain order or pattern according to a specific rule or set of rules (e.g., patterns of numbers, letters, words, pictures, mathematical operations).

References

Aguinis, H., Mazurkiewicz, M. D., & Heggestad, E. D. (2009). Using web-based frame-of-reference training to decrease biases in personality-based job analysis: An experimental field study. *Personnel Psychology*, *62*(2), 405–438.

Ainsworth, A. (2015). *Introduction to IRT*. Department of Psychology, California State University/Northridge.

Anwar, A. T., Barends, E., and Briner, R. (2016, June). *Nice model, but what is the evidence? Basic principles of evidence-based management.* Plenary keynote at the annual conference of the European Health Management Association/Porto.

Aritzeta, A., Swailes, S., & Senior, B. (2007). Belbin's team role model: Development, validity and applications for team building. *Journal of Management Studies*, *44*(1), 96–118.

Atkins, S. G. (2013). The role of work analysis in the work of your organization's career development manager, Chapter in *The Handbook of Work Analysis: The Methods, Systems, Applications, & Science of Work Measurement in Organizations* (M. Wilson, W. Bennett, S. Gibson, & G. Alliger, Eds., K. Murphy & J. Cleveland, Series Eds.). Abingdon-on-Thames UK: Routledge Academic Publishers, pp. 491–510.

Atkins, S. G. (2012). Smartening-up Voluntourism: SmartAid's Expansion of the Personality-focused Performance Requirements Form (PPRF). *International Journal of Tourism Research*, *14*(4), 369–390.

Atkins, S. (2009, May). Towards an empirical base to validate and/or improve Belbin team-role formulae. *Poster at European Congress of Work and Organizational Psychology*, Santiago de Compostela, Spain.

Atkins, S., Schaddelee, M., Atkins, E. R., & McConnell, C. (2013, January). *A Belbinesque Team-Diversity Manipulation within a Compulsory Full-Time "Project-Based Learning" year.* Oral paper for XIV Conference of the International Academy of Management & Business, San Antonio, Texas.

Atkins, S. & Foster-Thompson, L. (2009, May). *Work psychology applied to online volunteerism.* Roundtable at European Congress of Work and Organizational Psychology, Santiago de Compostela, Spain.

Belbin, R. M. (2012). *Method, reliability and validity, statistics and research: A comprehensive review of Belbin team roles.*

Bess, T. L., & Harvey, R. J. (2002). Bimodal score distributions and the Myers-Briggs Type Indicator: Fact or artifact? *Journal of Personality Assessment*, *78*(1), 176–186.

Boudreau, J. W., & Ramstad, P. M. (1997). Measuring intellectual capital: Learning from financial history. *Human Resource Management (1986-1998)*, *36*(3), 343–356.

Boxall, P. & Purcell, J. (2008). *Strategy and human resource management.* Basingstoke, Hampshire, UK and New York: Palgrave McMillan.

Boxall, P. & Purcell, J., & Wright, P. M. (2009). Human resource management: Scope, analysis, and significance. In P. Boxall, J. Purcell & P. M. Wright (Eds). *The Oxford handbook of human resource management.* doi: 10.1093/oxfordhb/9780199547029.003.0001.

Brown, R. D., & Harvey, R. J. (1996). *Job-component validation using the Myers-Briggs Type Indicator (MBTI) and the Common-Metric Questionnaire (CMQ).* In 11th Annual Conference of the Society for Industrial and Organizational Psychology, San Diego.

Connell, J. & Teo, S. (Eds) (2010). *Strategic HRM: Contemporary issues in the Asia Pacific region.* Prahran, Victoria, Australia: Tilde University Press.

Dalkir, K. (2011). *Knowledge management in theory and practice.* Cambridge, Massachusetts: The MIT Press.

Daum, J. H. (2003). *Intangible assets and value creation.* Chichester, West Sussex, England: John Wiley & Sons.

Davenport, T. H., & Prusak, L. (1998). *Working knowledge: How organizations manage what they know*. Boston, Mass: Harvard Business School Press.

Dawis, R. V., & Lofquist, L. H. (2000). Work adjustment theory. *Encyclopedia of psychology, 8*, 268–269.

Day, S. X., & Rounds, J. (1998). Universality of vocational interest structure among racial and ethnic minorities. *American Psychologist, 53*(7), 728.

Drewes, D., Tarantino, J., Atkins, S., & Paige, B. (2000). *Development of the O*NET Related Occupations Matrix.* Raleigh: U. S. Dept. of Labour's National O*NET Development Center.

Festinger, L., & Schachter, S. (2013). *When prophecy fails.* NYC: Simon and Schuster.

Festinger, L. (1962). *A theory of cognitive dissonance* (Vol. *2*). Stanford: University Press.

Furnham, A., Steele, H., & Pendleton, D. (1993). A psychometric assessment of the Belbin team-role self-perception inventory. *Journal of Occupational and Organizational Psychology, 66*(3), 245–257.

Furnham, A., Steele, H., & Pendleton, D. (1993). A response to Dr Belbin's reply. *Journal of Occupational and Organizational Psychology, 66*(3), 261–261.

Gill, L. J. (2004) Viewing 'Vocational' through "The Accidental Life by Phoebe Meikle". *University of Otago Management Graduate Review, 2*(1),13–24.

Guttentag, D. A. (2009). The possible negative impacts of volunteer tourism. *International Journal of Tourism Research, 11*(6), 537–551.

Hauenstein, N. (1995). Philosophy of statistical significance testing. Research methods lecture, *Virginia Tech State University Graduate College.* Blacksburg, Virginia, USA.

Harvey, R. J., & Hammer, A. L. (1999). Item response theory. *The Counseling Psychologist, 27*(3), 353–383.

Harvey, R. J., Murry, W. D., & Stamoulis, D. T. (1995). Unresolved issues in the dimensionality of the Myers-Briggs Type Indicator. *Educational and Psychological Measurement, 55*(4), 535–544.

Hesketh, B. (2000). The next millennium of "fit" research: Comments on "The congruence myth: An analysis of the efficacy of the person–environment fit model" by HEA Tinsley. *Journal of Vocational Behavior, 56*(2), 190–196.

Highhouse, S., Zickar, M. J., Brooks, M. E., Reeve, C. L., Sarkar-Barney, S. T., & Guion, R. M. (2016). A Public-Domain Personality Item Bank for se with The Raymark, Schmit, and Guion PPRF. *Personnel Assessment and Decisions, 2*(1), 5.

Jashapara, A. (2011). *Knowledge management: An integrated approach.* Harlow, Essex, England: Pearson Education.

Jonason, P. K., & Webster, G. D. (2010). The dirty dozen: a concise measure of the dark triad. *Psychological Assessment, 22*(2), 420.

Judge, T. A., & Ilies, R. (2002). Relationship of personality to performance motivation: a meta-analytic review. *Journal of Applied Psychology, 87*(4), 797.

Killeen, J. (1996). Career theory. In A. G. Watts, J. Killeen, J. M. Kidd & R. Hawthorn (Eds.) *Rethinking careers education and guidance: Theory, policy and practice.* New York: Routledge.

Macky, K. (2008). *Managing human resources: Contemporary perspectives in New Zealand.* North Ryde, NSW, Australia: McGraw-Hill.

Newell, S., Adams, S., Crary, M., Glidden, P., LaFarge, V., & Nurick, A. (2005). Exploring the variation in project team knowledge integration competency. In K.C. Desouza (ed.) *New frontiers of Knowledge Management.* New York: Palgrave MacMillan.

Osborne, J. W., & Costello, A. B. (2004). Sample size and subject to item ratio in principal components analysis. *Practical Assessment, Research & Evaluation*, *9*(11), 8.

Peterson, N. G., Mumford, M. D., Borman, W. C., Jeanneret, P. R., Fleishman, E. A., Levin, K. Y., & Gowing, M. K. (2001). Understanding work using the Occupational Information Network (O* NET): Implications for practice and research. *Personnel Psychology*, *54*(2), 451–492.

Ployhart, R. E., Van Iddekinge, C. H., & Mackenzie Jr., W. I. (2011). Acquiring and developing human capital in service contexts: The interconnectedness of human capital resources. *Academy of Management Journal, 54*(2), 353–368.

Porfeli, E. J., & Mortimer, J. T. (2010). Intrinsic work value-reward dissonance and work satisfaction during young adulthood. *Journal of Bocational Behavior*, *76*(3), 507–519.

Raymark, P. H., & Tafero, T. L. (2009). Individual differences in the ability to fake on personality measures. *Human Performance*, *22*(1), 86–103.

Ricceri, F. (2008). *Intellectual capital and Knowledge Management: Strategic management of knowledge resources.* New York: Routledge.

Stone, R.J. (2014). *Human resource management.* Milton, Qld, Australia: Wiley.

Sveiby, K .E. (1997). The new wealth: Intangible assets from *The new organizational wealth: Managing and measuring knowledge-based assets* (Chapter 1, pp. 3–18). San Francisco: Berrett-Koehler Publishers Inc.

Tinsley, H. E. (2000). The congruence myth revisited. *Journal of Vocational Behavior*, *56*(3), 405–423.

Tonny, T. & Atkins, S. (2016). *Introducing the Rotary SmartAid research program's first requested application: Field-work talents needed in Water and Sanitation/Hygiene (WASH) aid deliveries.* Presentation to Rotary WASH e-Club Organizing Committee at 8th World Water Summit, Seoul.

Tonny, T., Atkins, S. G., & Howison, S. (2016, May). *Appreciating Methods/Analysis Choices: Humanitarian Aid as a Teaching Context.* In *ECRM2016-Proceedings of the 15th European Conference on Research Methodology for Business Management": ECRM2016* (p. 293). Academic Conferences and Publishing Limited.

Tziner, A., Meir, E. I., & Segal, H. (2002). Occupational congruence and personal task-related attributes: How do they relate to work performance? *Journal of Career Assessment*, *10*(4), 401–412.

Van Iddekinge, C. H., Roth, P. L., Raymark, P. H., & Odle-Dusseau, H. N. (2012). The criterion-related validity of integrity tests: An updated meta-analysis. *Journal of Applied Psychology*, *97*(3), 499.

Vernon, P. A., Villani, V. C., Vickers, L. C., & Harris, J. A. (2008). A behavioral genetic investigation of the Dark Triad and the Big 5. *Personality and Individual Differences*, *44*(2), 445–452.

6 How sustainable innovations win in the fish industry: Theorizing incumbent-entrant dynamics across aquaculture and fisheries

Bilgehan Uzunca[1] and Shuk-Ching Li[2]

[1] *Utrecht University School of Economics, Kriekenpitplein 21–22, 3584, EC Utrecht, The Netherlands*
[2] *University College Utrecht, Maupertuusplein 1–320, Utrecht, The Netherlands*

Introduction

Sustainable innovation,[1] defined as the innovative process, including products, services and technologies that create sustainability through environmental, social and/or economic goals (Charter & Clark, 2007; Boons *et al.*, 2013), is the key to our future. Companies worldwide are exploiting a variety of non-renewable resources (e.g. oil for energy, wood for furniture and paper, and wild fish for food), and these resources will soon be depleted without sustainable innovations.

Knowledge[2] derived from sustainable innovations and the management of such knowledge – i.e., information, communication, human resources,

[1]Please note that sustainable innovation differs from Christensen's (1997) sustaining innovation and Barney's (1991) sustained competitive advantage, the difference and connections will be made clear in later sections.
[2]Knowledge is defined as a justified belief that increases an entity's capacity for effective action (Huber, 1991; Nonaka, 1994). Knowledge has been described as "a state or fact of knowing" with

Handbook of Knowledge Management for Sustainable Water Systems, First Edition. Edited by Meir Russ.

intellectual capital, brands etc. that collectively ensure that knowledge is available when and where needed and can be acquired from external as well an internal sources – prove to be important elements for competitive advantage and firm performance (Russ, Fineman, & Jones, 2010; Russ & Jones, 2011). Properly designed environmental standards can encourage companies to be more innovative to achieve sustainability, which is "the ability to meet the needs of the present without compromising the ability of future generations to meet their own needs" (cf. Hall, Daneke & Lenox, 2010). Achieving sustainability can also maximize companies' performance in a long-run (cf. Bekmezci, 2015). However, sustainability develops in each industry at a different rate. There have been studies looking at diverse aspects of sustainable innovation to explain these differences, using diverse theoretical lenses such as business models (Bohnsack, Pinkse, & Kolk, 2014; Boon *et al.*, 2013), business competitiveness (Tan *et al.*, 2015), market space (Smith *et al.*, 2014), and market dynamics (Hockerts & Wüstenhagen, 2010). But it is still not clear how sustainable innovation changes a particular industry.

To address this gap, we first look at the pursuers of sustainable innovations: new entrants and/or incumbents. Incumbents are defined as established industry leaders in a stable market (Macher & Richman, 2004), and entrants are the first pursuer of sustainable innovations in this study. Entrants are more likely than incumbents to first pursue sustainable opportunities, because they are flexible, opportunistic and focused on dislodging dominant innovations. On the other hand, incumbents tend to be too slow, routinized and uninterested in developing technologies (Obal, 2013; Christensen, 1997), because they prefer to stay close to what they are familiar with and rely on continuing their past successes (Bohnsack, Pinkse, & Kolk, 2014). If entrants with sustainable innovations perform well economically, they can advance the industry's sustainable development (Boon *et al.*, 2013). Incumbents will either be driven out, or force to adapt to the change of the industry by improving their sustainability. One example of an incumbent acknowledging the benefits of sustainability and adjusting to survive the competition is incumbents in the car industry (such as Nissan and Toyota) began to offer electric and hybrid cars around 2007 (Bohnsack, Pinkse, & Kolk, 2014). But if entrants fail to survive, incumbents may have little to no incentive to invest in research and development to boost sustainability. Thus, the survival of entrants pursuing sustainable innovations is essential to increase incumbents' incentives to overcome their inertia and innovate, and the resulting efforts from both incumbents and entrants will lead to the transformation of an industry towards sustainability (Hockerts & Wüstenhagen, 2010).

---

knowing being a condition of "understanding gained through experience or study; the sum or range of what has been perceived, discovered, or learned" (Schubert *et al.*, 1998).

To further investigate an industry's transformation towards sustainability, this study focuses on finding evidence from the emergence of fish farming (i.e. aquaculture)[3] as a sustainable alternative to traditional fisheries. Sustainable innovations in aquaculture can advance aquaculture sustainability and maintain fish supplies while reducing the problem of overfishing. Overall, this chapter aims to answer the questions: "Can sustainable aquaculture innovations survive?" and "How do these innovations affect the sustainability of the fish industry?" Incumbent-Entrant Dynamics are an organizational approach based on incumbent failure and entrant-incumbent models (Macher & Richman, 2004; Christensen, 1997; Christensen & Rosenbloom, 1995; Henderson & Clark, 1990; Tushman & Anderson, 1986). We will build theoretical arguments based on Incumbent-Entrant Dynamics (IED) to support the survival of entrants with sustainable innovations. Empirically, we test the impact of sustainable innovations by looking at incumbents' responses to the challenge in the past nine years. We hypothesize that there is a positive correlation between the increasing concerns over sustainability among large companies (incumbents) and the amount of sustainable innovations developed and applied in their aquaculture segments. Also, we hypothesize that the larger the firms' aquaculture segment, the more sustainable aquaculture innovations are applied there.

This chapter proceeds as follows. Section 6.1 briefly introduces key concepts and provides a review of existing literature, as well as IED in three categories: the industry setting, the incumbent firms, and the entrants. Section 6.2 provides theoretical arguments of the transformation from fishing to farming practice. Then a content analysis and a statistical analysis in sections 6.3 & 6.4 evaluate the transformation towards sustainability of the eight incumbent firms in the industry. Section 6.5 concludes the chapter with limitations and suggestions for future research. This chapter aims to find empirical evidence to support the sustainable transformation model, and to provide a successful example to encourage innovators, entrepreneurs, and policy makers to support and invest in sustainability. Also, it aims to provide a background and suggestions for future research on sustainable innovations.

6.1 Background

6.1.1 Including sustainability in business value

The concept of sustainability was adopted by the United Nations, governments, other institutions and organizations in the 1960s after the realization that some

[3]Fish farming and aquaculture are used interchangeably in this chapter.

incumbents' practices are negatively impacting the environment (Bekmezci, 2015). Environmental issues became an opportunity for new entrants to be the first to pursue sustainable innovations. For example, traditional (gas engine) cars emit pollutants and require fossil fuel to operate, it created an opportunity for entrants to join the market with environmentally-friendly alternatives, such as cars with hybrid and electric engines to compete with incumbents' products. After entrants (such as Tesla) first developed innovations to reduce the production costs and price of electric cars, the competitiveness of these alternatives increased. In response to this challenge, incumbents in the car industry also began to pursue sustainable innovations, and their resourcefulness allows them to develop innovations through multiple and extensive experimentations (Bohnsack, Pinkse, & Kolk, 2014). In 2010, incumbents (e.g. GM, Nissan, Toyota, etc.) were able to develop more new innovations for electric cars as compared to entrants. Once incumbents become aware of sustainability issues, innovations in their products, services, and business models take place almost automatically (cf. Bekmezci, 2015). Tan *et al.* (2015) examine sustainability performance in the construction industry and find the firms with the best sustainability performance also had the highest growth rate in 2 years. Tan *et al.* (2015) state that the effect of sustainability takes time to be reflected on revenue growth and sustainability is becoming a competitive advantage, [4] which is when a firm implements a value creating strategy that is not simultaneously being implemented by any current or potential competitor (Barney, 1991).

However, incumbents' awareness of sustainability is unlikely to increase if entrants with sustainable innovations do not first succeed in the market. Smith *et al.* (2014) examine the "spaces" for sustainable innovation in the UK's solar photovoltaic systems market. Particularly in the energy industry, we can see that entrants with sustainable innovations not only face the problem of challenging powerful incumbents, but they also deal with other barriers, such as the lack of market attractiveness – they often do not match existing production methods, managerial expertise and consumer preferences, and the benefit of potentially resolving environmental degradation does not seem a sufficient condition to generate widespread consumer acceptance (cf. Bohnsack, Pinkse, & Kolk, 2014). Smith *et al.*'s (2014) study finds that sustainable innovations need to be developed initially by entrants, and they also need the support of consumers, suppliers and public policies in order to survive. Sustainable innovations go beyond traditional product and process innovations and are future oriented (Boons *et al.*, 2013). They are not only aimed at improving productivity, but also at allowing businesses to overcome environmental and social challenges. In 2010, around $250 million was invested in developing sustainable innovations globally, and it

---

[4]Note that sustained competitive advantage is defined as when a firm implements a value creating strategy not simultaneously being implemented by any of its current or potential competitors and when these other firms are unable to duplicate the benefits of the new strategy (Barney, 1991). It focuses on sustaining (economic) performance of a firm, rather than sustainability, which focuses on social, environmental, and economic goals. Our study focuses on the latter.

is expected to reach between $500 million to $1.5 million by 2020 (cf. Bekmezci, 2015).

6.1.2 Linking sustainable innovations to Incumbent-Entrant Dynamics (IED)

Nurturing and growing sustainable innovations require companies that are open to radical ideas, experimentation, action and learning (Charter & Clark, 2007). Entrants posing a challenge to incumbents can trigger the adaption of sustainability in incumbents' business value to protect their market power, which lead to more sustainable innovations and practices in mainstream markets (Boons *et al.*, 2013). Large-scale transformation of production and consumption are required to transform the industry towards sustainability (cf. Boons *et al.*, 2013). Thus, the survival of entrants is crucial for this transformation. Incumbent-Entrant Dynamics will be used to explain the survival of entrants pursuing sustainable innovations in three categories: *the industry setting, incumbent firms, and the entrants*, as well as their underlying constructs.

The industry setting explains the outlook of the industry with five propositions, relating to complementary markets, government regulations, supply factors, demand factors, and level of rivalry. Ansari and Krop (2012) define a complementary market as a market that conjuncts with another market, such that greater sales in one market increase demand in the other. Entrants are more likely to survive when their complementary markets are more evolved; when government intervention at the national or industry level is more protective and favorable for entrants; when the suppliers for incumbents are standardized and less diversified; when the existing consumers are less "inert"; and when the rivalry in an industry is weak (Ansari & Krop, 2012).

Then, two key constructs can be identified for *incumbent firms* to explain incumbents' responses to entrants, affecting the likelihood of entrants' survival and the sustainable transformation: entrants are more likely to survive when incumbents engage less in boundary management (i.e. less partnerships with entrants); and when incumbents less effectively build and leverage links between innovations and their complementary assets such as brand familiarity and market power (Ansari & Krop, 2012).

The entrants are subdivided according to three constructs: innovation type, commercialization requirements and incubation period. Entrants are more likely to succeed in an industry when the innovations devalue the incumbent value network, which is the relationship between firms and their suppliers and consumers (Christensen & Rosenbloom, 1995); when the assets required for the profitable commercialization of an innovation are more generic or open; and when the incubation time-horizon – the period between incumbents' awareness of the innovation and its profitable commercialization – is short compared to the industry average (Ansari & Krop, 2012). These propositions will be applied to the fish industry in the following section.

6.2 Theorizing incumbent-entrant dynamics in the fish industry

6.2.1 Industry setting – the global fish industry

Comparing fish (defined as finfish and shellfish) production with meat production shows us that meat producers transitioned from hunting to farming hundreds of years ago, whereas aquaculture production became significant in the fish market and started growing only around the year 1950 (see Figure 6.1). Froese *et al.* (2012) found that the problem of overfishing is more serious than the Food and Agriculture Organization of the United Nations (FAO) predicted (see Figure 6.2). Even though the global fish production has grown steadily for the last five decades with fish supplies increasing at an average annual rate of 3.2% (FAO, 2013), it was mainly due to the growth of aquaculture. While the stagnation of fisheries production is expected to continue, aquaculture production is expected to double by 2050 to satisfy the increasing demand from the estimated 9.6 billion world population (Waite *et al.*, 2014). It is clear that there is limited time for aquaculture to grow. Thus, sustainable innovations in aquaculture are needed to produce efficiently and sustainably. Based on past trends and future expectations, we assume that new entrants in the fish industry are firms in aquaculture, which leaves firms who are active in fisheries as incumbents.

Aquaculture relies on complementary markets such as fishmeal productions and farming technologies, and these complementary markets are very important to the sustainability of the farms. For example, the cost of fishmeal accounts for around 85% of annual running costs of aquaculture in Egypt and 80% in Vietnam (Kleih *et al.*, 2013). Some fish farms use wild pelagic fish to

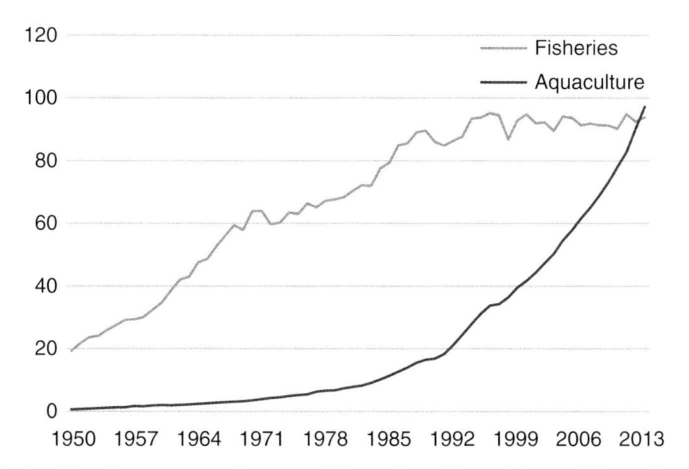

Figure 6.1 Production from aquaculture and fisheries. Note: Adapted from http://www.fao.org. Copyright 2015 by FAO. Reprinted with permission.

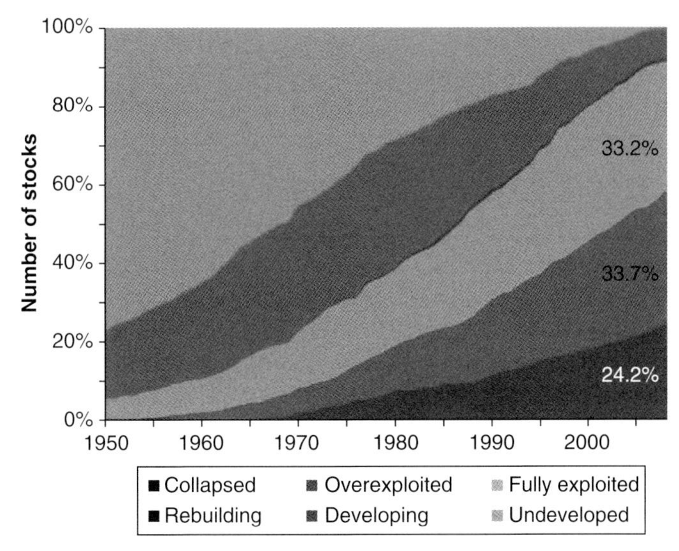

Figure 6.2 The updated algorithm applied to global FAO catch data from 1950 to 2008. Note: Retrieved from Froese *et al.*, 2012. Copyright by Froese *et al.*, 2012. Reprint with permission.

feed their fish, the feed conversion ratio can go up to as high as 15:1 (Li *et al.*, 2011), which means it requires up to 15 kg of wild fish to feed 1 kg of farmed fish, this is considered economically and environmentally unsustainable, as it does not solve the problem of overfishing. Fortunately, there has been progress in sustainable innovations in aquaculture, as well as its complementary markets. Sustainable innovations have been rapidly developing to provide fishmeal alternatives, such as using insects, soybeans, and other ingredients to create suitable food for farmed fish. Aquaculture entrants are likely to survive as their complementary markets are more evolved.

Government interventions such as regulations and subsides are important factors to the survival of businesses in the fish industry. Around $13 billion is paid to fishermen globally in the form of fuel subsidies (Maribus, 2013), and the main reason of that is to avoid market disruption by maintaining stable fish supplies. On the other hand, governments are more supportive of entrants because their sustainable innovations can be a solution to the future rising demand, for example the US government provides direct funding to aquaculture firms, as well as additional supports through educational services and funding for research (Jarvinen, 2008). As aquaculture grows, public funding will gradually be redirected from fisheries to aquaculture. In the short term, fisheries will be able to survive due to high subsidies, but this dependence cannot maintain the business in the long run. Besides, incumbents in fisheries are under more governmental pressure than entrants, such as meeting catch quotas and strict licensing regulations (European Commission, 2014). These legal restrictions are related to the supply factors for fishery incumbents because their main supplier is nature (fish in the ocean). Not only is the "supplier" for the incumbent firms standardized and not diversified, it is also not sustainable. Thus, aquaculture entrants are more likely to survive as it is more protective and favorable by governments, and also the supplier for incumbents are standardized and less diversified.

The question related to demand factors is frequently asked when it comes to fish. Fish is in high demand because it is high in protein and contains essential fatty acids that have critical health benefits to human growth, as well as the growth of other animals and fish (Thurstan & Roberts, 2014). Can farmed fish substitute the demand for wild fish? While some consumers take sustainability into account, most consumers are more sensitive to factors that affect them directly, such as safety, quality and price (Washington & Ababouch, 2011). However, a study by Claret *et al.* (2014) found that consumers assume that the fish they find at the market is safe regardless of where it came from (wild/farmed fish); participants' preference for wild fish was due to their belief that it was higher quality, but three sensory studies (Pohar, 2011; Cahu *et al.*, 2004; Luten *et al.*, 2002) have concluded that most consumers are not able to perceive any difference between the two kinds of fish. In addition, a study by Claret in 2011 even found in a blind test that farmed fish were preferred (cf. Claret *et al.*, 2014). The belief that wild fish are better is perhaps due to the inertness of consumers, or simply because the price is higher (price premium) than that of its farmed counterparts. The effect of price premium is beyond the scope of this chapter but there is no evidence that farmed fish cannot substitute wild fish based on quality and taste. According to a Maribus report in 2011, 47% of fish produced were consumed fresh, 29% were consumed frozen and the rest were used in processed food and non-food purposes. For frozen fish and processed fish products, there is little to no difference in taste between those that use farmed fish and those that use wild fish. The debate between the quality of wild and farmed fish is often actually referring to fresh fish consumption. For the 53% of fish consumers who eat non-fresh fish, the safety and quality of farmed and wild fish are the same. Similar to the case of poultry, most of the 47% of fresh fish consumers will get used to the sensory characteristics of farmed fish and value it in a more highly (Claret *et al.*, 2014). Therefore, we can conclude that wild fish and farmed fish are substitutable.

Lastly, rivalry in the seafood industry is not as strong as industries that are dominated by a few large firms, such as the computer or car industry. For the most part, the production of fish consists of many small businesses. For example in China, the largest seafood producer in the world, most fisheries and aquaculture firms are family-run businesses (NBSO, 2012). Since the rivalry is relatively weak in an industry, entrants face less challenges and incumbents suffer.

6.2.2 The incumbent firms

The history of fishing dates back to hundreds of years ago, yet around 50% of global fish production today still comes from catching wild fish. This section explains how fishery firms' responses affect the likelihood of aquaculture entrants' survival and the sustainability of the industry with boundary management and complementary capabilities.

As mentioned, boundary management is related to the partnership between incumbents and entrants (Ansari & Krop, 2012). Incumbents are well aware of the growth of sustainable aquaculture innovations and some may have established or expanded their aquaculture unit by acquiring entrants and/or developing their own new businesses. However, incumbents would only invest in aquaculture innovations after entrants have successfully survived. This incumbents' adaptation will benefit the industry's transformation towards sustainability. This assumption will be tested in the empirical section.

There is no brand familiarity in the business to consumer segment of the seafood industry, it is rather in the business to business market, in which fishermen and farmers sell their products to distributors such as supermarkets, wholesalers and restaurants. Incumbents' leverage is the consumers' belief that wild fish is better, resulting in distributors' interest in buying incumbent's products. But aquaculture firms are gaining market power because farmed fish is generally cheaper and sustainable. Since it has been previously established that farmed fish and wild fish are substitutable, the leverage of incumbents' is low, resulting the survival of entrants is high.

6.2.3 The entrants

Early aquaculture was established to grow selected species to fulfill specific demand (FAO, 2015), such as low value species to "fill the bottom" of the market. Now, entrants in aquaculture aim to produce fish sustainably. However, fish farms were not sustainable at the beginning. Lebel *et al.* (2010) pointed out the innovation cycles in the seafood sector, which is initiated by traditional fishing, followed by extensive aquaculture, intensive aquaculture, and lastly sustainable aquaculture. In other words, businesses go from fishing to farming, then to intensified farming by using chemicals and putting more fish in the same ponds, and finally they evolve to sustainable fish farming. Some examples of sustainable fish farms are farms that do not use wild fish as feed, use little to no antibiotics, or spend extra resources to minimize water pollution and farm waste. As Hockerts and Wüstenhagen (2010) stated, it takes time for entrants to address a broad range of sustainability related issues given their limited experience and resources. Thus, we believe that fish farms will continue to improve sustainability with new innovations.

When entrants with sustainable innovations enter the fish market, the value network of firms in fisheries is not significantly disrupted at the beginning. However, entrants may gradually show consumers and investors that the environmental value is missing in incumbent firms. To cope with this challenge, incumbent firms may begin to transform their business to protect their value network, perhaps by promoting their business' sustainable actions. Also, the impact of entrants' sustainable innovations may depend on incumbents' business units. If incumbents already have an aquaculture unit, they may invest and develop more sustainable innovations in aquaculture as compare to incumbents who have smaller to no aquaculture unit. These assumptions will be tested in the empirical section.

For commercialization requirements, entrants generally need fishmeal, baby fish, water, land, ponds, and pumps to enter the market. Some governments (such as China) provide training to farmers to increase the efficiency of production ("NBSO", 2012; Jarvinen, 2008). The assets required for profitable commercialization of sustainable aquaculture innovations are generic and open. In addition, aquaculture firms are also able to provide a stable supply all year, as compared to seasonal fishery catch. Thus, aquaculture entrants are likely to succeed.

Lastly, shorter time horizons in an innovation's development is more favorable for entrants (Ansari & Krop, 2012). This construct does not define what is considered short or long. In general, sustainable innovations in aquaculture are developing very rapidly. The following section gives some insights of the quantity of innovations that are developed per company, but it is unclear if this construct is more favorable for incumbents or entrants as we are not discussing specific innovations or companies. While some evidence can be obtained from literatures to support IED theoretical arguments, some sections require further empirical testing. Table 6.1 summarizes the theoretical constructs and hypotheses.

6.3 Data and methods

6.3.1 An analysis of incumbents' sustainability

Prior section discussed that incumbents are only likely to pursue sustainable innovations if entrants successfully survived. Following our theoretical arguments, this section will focus on testing hypotheses and finding evidence of incumbents' increased interest in sustainability. The list of top ten (by revenue in 2012) publicly traded companies in the fish industry is taken from a report by Food for Valuation (2013). Their total revenue is ranging from ranging from $1514 to $9418 million. Excluding irrelevant business segments (such as fish trading and processing), two companies are excluded since they have minimal or no business in aquaculture or fisheries. The remaining four companies are involved in both farming and fishing segments, three are involved only in fish farms, and one only in fisheries. Data collected from their annual reports in the period of 2006–2014 provide a total of 72 observation for content analysis and statistical analysis.

Table 6.1 Summary of the illustration with IED and hypotheses in the seafood industry

Industry setting	
Complementary markets	Fishmeal production and farming technologies are equipped for sustainable fish farming
Government intervention	More restrictions on fishery incumbents and support for sustainable entrants
Supply factors	Standardized and not diversified for incumbents
Demand factors	Farmed fish and wild fish are substitutable
Rivalry	Consisted of many small businesses, the rivalry is relatively weak
Incumbent firms	
Boundary management	**H1:** Incumbents aquaculture units are expanding.
Complementary capabilities	Some market power
Entrants	
Innovation type	Sustainable innovations have impact in incumbents' sustainability. **H2:** Incumbents firms are promoting their business' sustainability. **H3:** The larger the incumbents' aquaculture segment, the more sustainable aquaculture innovations are applied.
Commercialization requirements	Easy access and open
Incubation time-horizon	Not clear

A brief overview of the companies will first be presented. Then, the companies' sustainability will be evaluated. We will compare the firms' claim of sustainability and the amounts of sustainable aquaculture innovations used per firm. One of the ways to promote firms' sustainability is to stress the term "sustainability" more often to promote their sustainability focus. Thus, a keyword search is performed using search terms referring to sustainability in a set of 72 annual reports. The occurrence of "sustainability" in a company's report may not completely reflect the company's actual actions, but it gives some idea, thus a proxy of the development of the company's sustainability focus. Therefore, we carefully interpret companies' production developments and achievements to look at the amount of sustainable aquaculture innovations used in each firm each year. The development of their sustainable aquaculture innovations are not illustrated in detail, but some examples in the following areas are found in their reports: methods and products for breeding, system, feed, fish health, and waste control.

Following that, the difference between the number of occurrences of "sustainability" and the amount of innovations used are observed. This analysis aims to see if incumbents are committed to their sustainability claims and if

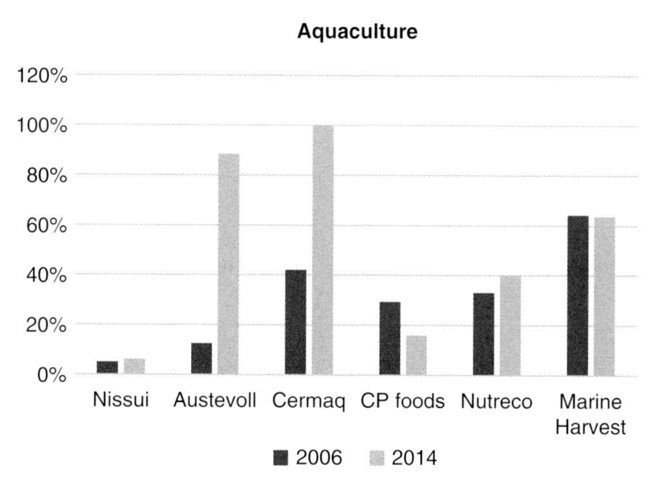

Figure 6.3 Revenue contribution to aquaculture in top 8 seafood companies in 2006 and 2014. Adapted from various annual reports, 2006 and 2014.

sustainable innovations are being utilized increasingly in their business. A positive correlation between the occurrence of "sustainability" and the amounts of innovations used is expected. Next, a regression analysis will be used to look at how the size of aquaculture unit of companies (*Aquaculture*) affects the amount of sustainable aquaculture innovations that they used (*Innovation*). Segment revenue and total revenue from each company are taken from their annual reports and are converted to dollars with the current exchange rate (1 USD = 125.06 YEN = 7.88 NOK = 7.88 HKD = 33.76 THB). The variable *Aquaculture* is calculated by using revenue from company's aquaculture unit divided by their total revenue (see Figures 6.3 & 6.4). While some researchers consider fishmeal production as aquaculture, most fishmeal consists of wild fish and the rest is made using bio-formula (i.e. contains less fish or uses green alternatives). If the firm's fishmeal production came from fishing, it is considered as fisheries in this chapter, otherwise it is considered as sustainable fishmeal and counted as aquaculture business. In addition, total firm size (*total revenue*) and the number of occurrences of "sustainability" may have an effect on sustainable innovations used. A regression model can be represented in the form of equation 6.1 as shown below:

$$Innovation_{it} = \hat{\beta}0 + \hat{\beta}1\ Aquaculture_{it} + \hat{\beta}2\ total\ revenue_{it} + \hat{\beta}3\ sustainability_{it} + \hat{\beta}4\ time_{it} + U_{it} \tag{6.1}$$

where:
"i" represents company; t represents year; U is an error term; $\hat{\beta}$s denote the coefficients.

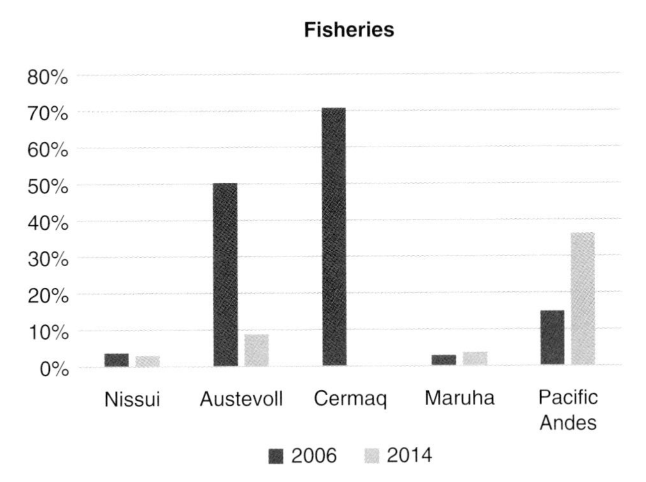

Figure 6.4 Revenue contribution to fisheries in top 8 seafood companies in 2006 and 2014. Adapted from various annual reports, 2006 and 2014.

6.3.2 *Sample*

Nissui, Austevoll, and Cermaq are involved in both fisheries and aquaculture sectors. The Japanese multinational Nissui began as a small fishing business in 1905. As seen from Figures 6.3 and 6.4, live fish production is now only a small part of its business; its major business focus is fish processing and food services. Austevoll is a Norwegian company that focuses on salmon, trout farming and fishmeal production (from fishing). From 2008 to 2009 its revenue contribution from aquaculture drastically increased from 8% to 66%. According to its annual reports, it has been expanding its aquaculture business by acquiring numerous fish farming companies. Cermaq was originally a fisheries business, but it was discontinued completely in 2013. This Norwegian company now has only fish farming business.

Maruha Nichiro and Pacific Andes are the two firms that have fishery units and little to no aquaculture farms. The Japanese multinational Maruha Nichiro focuses largely on fish processing and trading. While Maruha Nichiro's fresh fish production unit has seven subsidiaries, six are fishery companies, only one subsidiary has both fishing and fish farms. Maruha Nichiro's annual reports and financial details do not include a separate revenue from aquaculture like all other firms. Since the revenue contribution from its fish farms is insignificantly small, the size of its aquaculture unit may be even smaller than 0.05% of total revenue, hence we disregard their aquaculture units in terms of revenue, but it will be considered later on in terms of innovations. Moreover, the Hong Kong based company Pacific Andes specializes in food processing and frozen seafood trading. It acquired a fishery subsidiary in 2004 and began harvesting fish from the ocean. Among the top 8, it is the only one with a large fisheries segment in its business.

CP Foods, Nutreco, and Marine Harvest are the major players in the aquaculture segment. The Thai company CP Foods is a leading agro-industrial and food conglomerate. Its business consists of agriculture farming, food processing, aquaculture and sustainable fishmeal. Nutreco is a Dutch company that played an important role in the food value chain between raw material suppliers and farmers. Its main business segment is producing sustainable fishmeal and chicken-meal, and its sub-segments include animal nutrition and meat. Last but not least, Marine Harvest is the world's leading farmer of Atlantic salmon. While salmon farming is the largest part of its business, it has forward integrated to processing into its business since 2013.

6.4 Results

Looking at the total segment revenue in these public companies (Figure 6.5), it is clear that their aquaculture segment is growing while that of fisheries is not. We can conclude that H1 is supported, incumbents' aquaculture units are expanding. Is this rising trend of aquaculture also increasing the sustainability of these companies? We first look at the companies' claims of sustainability in Table 6.2, which shows the occurrence of "sustainability" in their annual reports.

As expected, the occurrence of the term "sustainability" increases from 2006 to 2013 in the annual reports of these eight companies. It appears that incumbents are adapting to the increasing demand of sustainability. H2 is accepted. Although the yearly total of "sustainability" in 2014 is slightly below the number from 2013, the drop is due to more companies began to establish

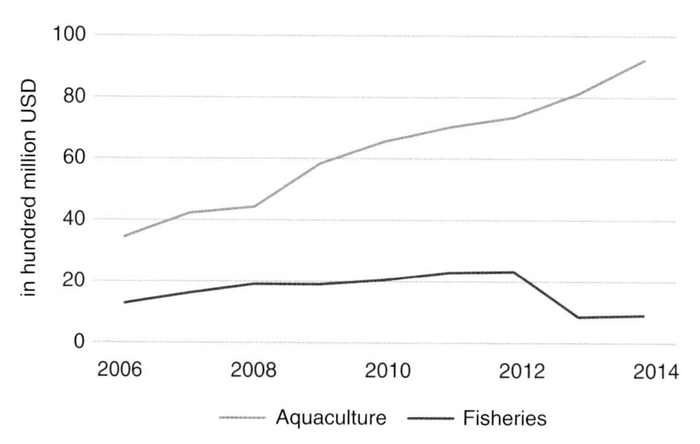

Figure 6.5 The sum of fisheries revenue and the sum of aquaculture revenue from the top eight companies. Adapted from various annual reports.

Table 6.2 The occurrence of "sustainability" in companies' annual reports. Adapted from various annual reports

	2006	2007	2008	2009	2010	2011	2012	2013	2014	Total
Maruha Nichiro	0	0	0	0	0	1	0	0	1	**2**
Nissui	0	8	14	15	11	7	0	0	0	**55**
Pacific Andes	3	11	17	16	21	8	21	29	29	**155**
CP foods	4	7	11	21	34	66	60	88	86	**377**
Austevoll	6	4	23	21	47	63	73	73	80	390
Cermaq	31	25	53	100	163	156	119	47	31	725
Marine Harvest	34	14	19	26	40	142	124	202	182	783
Nutreco	43	64	57	154	193	225	375	402	385	1898
Yearly total	121	133	194	353	509	668	772	841	794	4385

a separate sustainability report. As concerns over sustainability increased, not only did the usage of the term rise in the companies' revenue reports, but most companies (except for Maruha Nichiro and Austevoll) established a separate sustainability report by 2014 and their sustainability goals and achievements are highly visible and accessible on their websites. Another pattern is that European companies seem to have more concerns over sustainability than Asian companies, which are bold in Table 6.2. Now, we look at the amounts of sustainable aquaculture innovations that are being used in these companies to see if the rising pattern continues.

Similar to Table 6.2, there is a rising pattern in the amounts of sustainable aquaculture innovations, which means that in addition to entrants, incumbents are also becoming more sustainable. The numbers in Table 6.3 are accumulative because companies continue feasible aquaculture innovations after initiating them. As expected, some examples of aquaculture innovations found are methods to reduce diseases, methods to shorten the production time, and methods to decrease fish morality rate. When comparing Table 6.3 to Table 6.2, we see the order of the companies based on the occurrence of "sustainability" and the innovations used changes slightly. The two fishing companies have the least sustainable innovations because they had little to no business in aquaculture from 2006 to 2014. In Table 6.2, we can see that Nutreco has a relatively stronger emphasis on sustainability, but Nutreco (and Marine Harvest) are outnumbered by Cermaq. The explanation for this outcome is that Nutreco's innovative focuses are less concentrated on aquaculture than the 100% aquaculture multinational

Table 6.3 Accumulative amounts of sustainable aquaculture innovations from 2006 to 2014. Adapted from various annual reports

	2006	2007	2008	2009	2010	2011	2012	2013	2014
Pacific Andes	0	0	0	0	0	0	0	1	**1**
Maruha	1	1	1	1	1	1	1	1	**1**
Nichiro									
Nissui	5	6	6	7	8	9	12	16	**17**
Austevoll	1	3	6	6	6	9	13	16	19
CP foods	4	4	4	10	10	11	14	17	**20**
Marine Harvest	2	2	5	8	11	15	19	22	24
Nutreco	3	7	11	15	19	23	27	31	33
Cermaq	12	14	17	22	24	27	33	35	38
Yearly total	28	37	50	69	79	95	119	139	153

Cermaq. Even though Nutreco dedicates its business largely to sustainable fishmeal production, it also has other animal feed productions. Table 6.2 shows the number of the word "sustainability" was used by a company as a whole, while Table 6.3 only accounts for the aquaculture segments of the companies. A correlation analysis between the two variables follows.

Since the amount of innovations in Table 6.3 is accumulative, the change in sustainable innovations is used for the correlation analysis. The bold number in Table 6.4 shows that there is a mild positive correlation (smaller than 0.5) between *sustainability* and the change in sustainable innovations. However, the effect of sustainability may take time to reflect on the amount of innovations, thus a correlation between *sustainability* and the change in sustainable innovations with a 1-year time lag is tested, which results in a stronger positive correlation

Table 6.4 Pearson correlation between the occurrence of "sustainability" (Table 6.2) and the accumulative amounts of sustainable innovations in aquaculture (Table 6.3)

	N	Pearson correlation coefficient	Sig (2-tailed)
Sustainability/The change in sustainable innovations	72	**0.4905**	0.00
Sustainability/The change in sustainable innovations with a 1-year time lag	72	0.5600	0.00

(large than 0.5). We can conclude that companies who emphasized more on sustainability invested more on sustainable innovations over time. It also appears as if a company has a larger business aquaculture segment, it is more likely to invest more in research and innovations in aquaculture. The following regression analysis will test the relationship between sustainable innovations on the size of aquaculture unit (H3).

In addition to the significant p-vale ($p = 0.000$) of the variable *time* in Table 6.5, and an additional testing for time-fixed effect is also performed ($p = 0.000$, $a < 0.05$), indicating that time should be controlled in the analysis. Table 6.5 shows both fixed effect and random effects regression. Hausman test is performed to test whether fixed effects or random effects should be used for this model. The p-value of Hausman test is 0.99, which means that the model with random effects is preferred. H3 is confirmed in Table 6.5, there is a significant result between the size of aquaculture unit and sustainable aquaculture innovations. When the size of aquaculture unit of a company increases by 1%, there is a 0.096 increase in the amount of innovations that firm uses when all other variables are constant. *Sustainability* is also founded to be significant ($p < 0.05$), while *total revenue* is not. R-squared shows that this analysis explains 83% of the variance.

Table 6.5 Regression analysis on sustainable aquaculture innovations with the size of aquaculture unit

Variables	(Fixed effects) innovations	(Random effects) innovations
The size of aquaculture unit	0.0961***	0.0980***
	(3.941)	(0.0231)
Total revenue	0.0005	0.000459
	(1.459)	(0.000320)
Sustainability	0.0483***	0.0487***
	(6.630)	(0.00695)
Time	0.967***	0.968***
	(4.215)	(0.218)
Constant	−1,940***	−1,943***
	(−4.220)	(437.4)
R-squared	0.83	0.83
R2-adj	0.80	
Observations	68	68
Number of firms	8	8

Standard errors in parentheses.
*** p<0.01, ** p<0.05, * p<0.1.

6.5 Discussion

Using Incumbent-Entrant Dynamics, we argue that entrants pursuing sustainable innovations in aquaculture are likely to survive as they are beneficial to governments, producers, and consumers. The impact of sustainable aquaculture innovations on incumbents are examined. Results show that the increase of sustainability and sustainable aquaculture innovation are creating economic and environmental impacts in the industry. A regression analysis shows that the size of aquaculture unit has a significant effect on the amount of sustainable aquaculture innovations developed and used. In addition, the social sustainability of these companies is not mentioned in the previous sections, but all eight companies adequately listed their efforts in promoting social goals over the years, such as providing training for farmers, balancing the ratio between women and men working in the industry, creating jobs in developing countries, etc. Our sample shows that the amount of sustainable innovations in aquaculture increases every year. There were 139 innovations in use in 2014, compared to 28 in 2006 in these eight companies.

While some incumbents established sustainable fish farms, others also developed innovations to improve the sustainability of commercial fishing. The term "sustainable fisheries" began to occur in Pacific Andes's and Austevoll's reports in 2011. They pursue sustainable fisheries through meeting catching quotas, reducing CO_2 emissions and waste, and increasing fish traceability and transparency. Even though these measures do not solve the problem of overfishing, they have an impact on the environment. This also confirms that incumbents staying close to their preexisting business models often to lag behind in sustainability. The conceptual model of this sustainability transformation in the fish industry is illustrated in Figure 6.6.

Even though this transformation shows that both entrants and incumbents are moving towards sustainability, it is known that entrants and incumbents take a different approach to achieving their sustainability goals. Entrants see sustainable innovations as an opportunity; sustainability is the core of their business. While incumbents may see developing sustainable innovations as a risk (because consumers might not pay more for eco-friendly products during a recession, or suppliers cannot provide green inputs or transparency), they are likely to see sustainability as part of corporate social responsibility practices, rather than a business objective (Nidumolu, Prahalad, & Rangaswami, 2009). There may also be a misconception that sustainable companies may see lower profits. However, sustainable innovations actually can reduce costs because companies end up reducing the inputs they use and additional revenues can be generated from better products (Nidumolu, Prahalad, & Rangaswami, 2009).

When Incumbent-Entrant Dynamic is introduced in a study, disruptive and sustaining innovations are often also mentioned (Russ *et al.*, 2010). Can sustainable innovations be disruptive or sustaining? According to Christensen (1997), a disruptive innovation appeals initially to fringe consumers, but

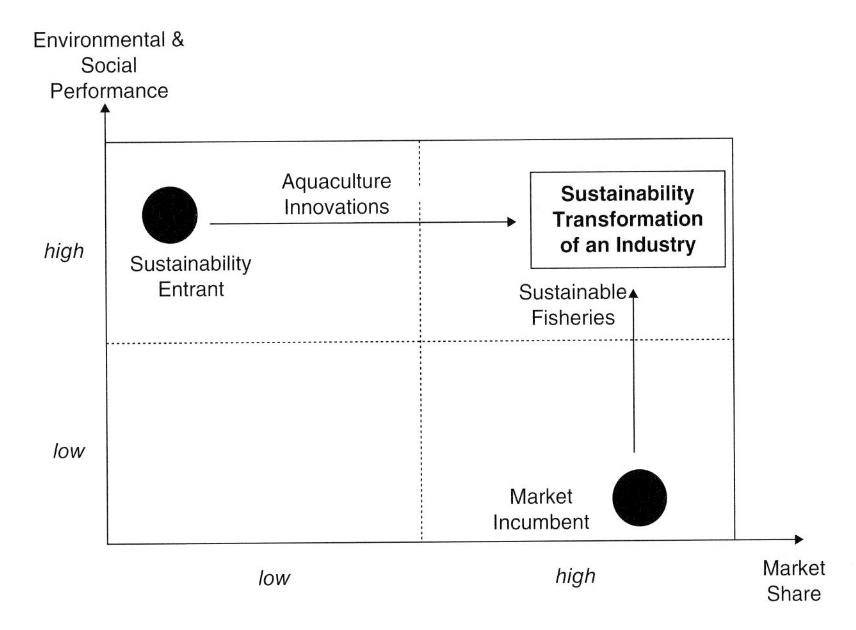

Figure 6.6 Co-evaluation of entrant and incumbent moving toward sustainability. Reprinted with permission. Adapted from Hockerts & Wüstenhagen, 2010.

eventually improves and displaces the dominant technology that had long been used by mainstream consumers. In the fish industry, there are certain correlated characteristics between sustainable innovation and disruptive innovation, such as simpler products (little species diversity), smaller initial target markets, and that the developments are less attractive than existing solutions (the belief that wild fish is better). However, a major point of disruptive innovations is that the innovative technologies develop faster than consumers' need, which is different in the case of sustainability. They could also potentially destroy the value of existing knowledge (Henderson & Clark, 1990; Tushman & Anderson, 1986). If sustainable innovations in aquaculture are able to fully support the global demand for fish, fishermen would not be receiving subsidies, instead they would be forbidden to fish commercially. In contrast to disruptive innovations, sustaining innovations generate sustained improvements in an established industry's performance (Christensen & Bower, 1996). In the future, sustainable innovations in aquaculture may become sustaining innovations as they will improve the industry performance until the next disruption in the industry. As our study focuses on the how sustainable innovations survive and how do they affect the sustainability of an industry, further investigation on these theories are beyond the scope of this chapter. Perhaps new theories building on these three types of innovations can be developed in further research.

A major limitation of this study is data from the companies' annual reports. While some innovations may be undocumented, some companies also exaggerate their innovative achievements. For example, CP Foods stated that a total 527 innovation projects were developed in 2012, 1,601 in 2013, and 3,323 in 2014 in its reports. As shown in Table 6.3, the recorded number of sustainable aquaculture innovations is significantly smaller than the total amount of innovations stated in CP Foods' report. CP Foods focuses on innovations in numerous sectors, and some projects may not be applicable to production. While aquaculture innovations may be relatively smaller, there is also a chance that the numbers in Table 6.3 are underestimated, as undocumented innovations cannot be accounted for. However, the data is still reliable due to companies are motivated to document their innovations to publicly portray their sustainability efforts. Besides, we also assume the currently unsustainable aquaculture will use sustainable innovations to increase their productivity and sustainability. It is difficult to estimate the rate of the implementation of innovations and the impacts that they may have. Little research has been done to show the processes of sustainable aquaculture innovations and their economic and environmental impacts. Some studies focusing on specific species and countries are available, but little to no research has been conducted focusing on the industry as a whole. Further research on this aspect can strengthen our analysis.

In addition, other interesting questions for further research are: What specific sustainable innovations are the most popular among firms? Does the timing of implementation affect the success rate and impact of sustainable innovations? What about the cost and benefit of developing specific innovations? An additional finding from the analysis is that the Asian companies tend to use fewer innovations and focus less on sustainability. Further research is also needed to investigate the sustainable development in Western countries as compared to Asian countries. Is the topic of sustainability more of a concern in Western than Asian countries, and why?

In summary, this chapter uses the fish industry to provide an example of a sustainability transformation within an industry. Sustainable innovations can succeed with a supportive industry setting, the suffering of incumbents, and easy access for entrants. Evidence shows that the number of firms engaging in aquaculture is increasing, and the remaining firms in fisheries are likely becoming more sustainable. Our findings show that incumbents are involved in the development of sustainable innovations to confront the problem of limited marine resources. Based on the sustainable transformation model, we can conclude that the survival of sustainable aquaculture innovations has caused the industry to become more sustainable as a whole.

References

Austevoll Seafood ASA (2006–2014). *Annual Report*. Retrieved from http://www.auss.no/

Bekmezci, M. (2015). Companies' profitable way of fulfilling duties towards humanity and environment by sustainable innovation. *Social and Behavioral Sciences*, *181*(1), 228–240.

Bird, L.A., Wüstenhagen, R., & Aabakken, J., (2002). A review of international green power markets: recent experience, trends, and market drivers. *Renewable and Sustainable Energy Reviews, 6*(6), 513–536.
Bohnsack, R., Pinkse, J., & Kolk, A. (2014). Business models for sustainable technologies: Exploring business model evolution in the case of electric vehicles. *Research Policy, 43*, 284–300.
Boons, F. Montalvo, C., Quist, J., & Wagner, M. (2013). Sustainable innovation, business models and economic performance: an overview. *Journal of Cleaner Production, 45*, 1–8.
Cahu, C., Salen, P., & de Lorgeril, M. (2004). Farmed and wild fish in the prevention of cardiovascular diseases. Assessing possible differences in lipid nutritional values. *Nutrition, Metabolism, and Cardiovascular Diseases, 14*, 34–41.
Cermaq (2006–2014). *Annual report.* Retrieved from http://www.cermaq.com/
Chandy, R., & Tellis, G. (2000). The incumbent's curse? Incumbency, size and radical product innovation. *Journal of Marketing, 64*(3), 1–17.
Charter, M., & Clark, T. (2007.) *Sustainable Innovation.* The Centre for Sustainable Design, Farnham.
Christensen, C.M. (1997). *The Innovator's Dilemma: When New Technologies Cause Great Firms to Fail.* Boston, MA: Harvard Business School Press.
Christensen, C.M., & Rosenbloom, R.S. (1995). Explaining the attacker's advantage: Technological paradigms, organizational dynamics and the value network. *Research Policy, 24*, 233–257.
Christensen, C.M., & Bower, J.L. (2008) Customer power, strategic investment, and the failure of leading firms. *Strategic Management Journal, 17*(3), 197–218.
Claret, A., Guerrero, L., Gines, R., Grau, A., Hernandez, M.D., Aguirre, E., Peleteiro, J.B., Fernandez-Pato, C., & Rodriguez-Rodriguez, C. (2014). Consumer beliefs regarding farmed versus wild fish. *Appetite, 79*(1), 25–31.
Cohen, B., & Winn, M. (2007). Market imperfections, opportunity and sustainable entrepreneurship. *Journal of Business Venturing, 22*(1), 29–49.
CP Foods PCL. (2006–2014). *Annual report.* Retrieved from http://www.cpfworldwide.com/en/about/
Dyllick, T. (1999). Environment and competitiveness of companies. *International Environmental Management Benchmarks.* Springer, Berlin.
European Commission. (2014). *Facts and figures on the Common Fisheries Policy – Basic statistical data.* Publications Office of the European Union: Luxembourg.
FAO. (2000). *The State of World Fisheries and Aquaculture SOFIA 2000.* FAO: Rome.
FAO. (2014). *The State of World Fisheries and Aquaculture SOFIA 2014.* FAO: Rome.
FAO. (2015). *Utilization and Trade.* Retrieved from http://www.fao.org/fishery/topic/2888/en
FAO. (2015). *History of Aquaculture.* Retrieved from http://www.fao.org/docrep/field/009/ag158e/ag158e02.htm
Food for Valuation. (2013). *Investing in Seafood 2013.* IntraFish: Norway.
Froese, R., Zeller, D., Kleisner, K., & Pauly, D. (2012). What catch data can tell us about the status of global fisheries. *Marine Biology, 159*(1), 1283–1292.
Gawer, A., & Henderson, R. (2007) Platform owner entry and innovation in complementary markets: Evidence from Intel. *Journal of Economics and Management Strategy, 16*(1), 1–34.

Hall, J.K., Daneke, G.A., & Lenox, M.J. (2010) Sustainable development and entrepreneurship: Past contribution and future directions. *Journal of Business Venturing*, *25*(1), 439–448.
Henderson, R.M., & Clark, K. (1990). Architectural innovation: The reconfiguration of existing product technologies and the failure of established firms. *Administrative Science Quarterly*, *35*, 9–30.
Hockerts, K., & Wüstenhagen, R. (2010). Greening Goliaths versus emerging Davids – theorizing about the role of incumbents and new entrants in sustainable entrepreneurship. *Journal of Business Venturing*, *25*(1), 481–492.
Huber, G. (1991). Organizational learning: the contributing processes and the literatures. *Organization Science 2*(1), 88–11.
Jarvinen, D. (2008). Federal and state support for aquaculture development in the United States. *Aquaculture Economics & Management*, *4*(3/4), 209–225.
Kleih, U., Linton, J., Marr, A. Mactaggart, M., Naziri, D., & Orchard, J.E. (2013). Financial services for small medium-scale aquaculture and fisheries producers. *Marine Policy*, *37*(1), 106–114.
Li, X., Li, J., Wang, Y., Fu, L., Fu, Y., L.B., & Jiao, B. (2011). Aquaculture industry in china: current state, challenges, and outlook. *Reviews in Fisheries Science, 19*(3), 187–200.
Lebel, L., Mungkung, R., Gheewala, S.H., & Lebel, P. (2010). Innovation cycles, niches and sustainability in the shrimp aquaculture industry in Thailand. *Environmental Science & Policy*, *13*, 291–302.
Luten, J.B., Kole, A., Schelvis, R., Veldman, M., Heide, M., Carlehög, M., *et al.* (2002). Evaluation of wild cod versus wild caught, farmed raised cod from Norway by Dutch consumers. *Økonomisk Fiskeriforskning*, *12*, 44–60.
Macher, J., & Richman, B. (2004). Organisational responses to discontinuous innovation: a case study approach. *International Journal of Innovation Management. 8*(1), 87–114.
Maribus (2013). *The future of fish – the fisheries of the future. World Ocean Review*, Maribus: Hamburg.
Marine Harvest (2006–2014). *Annual report*. Retrieved from http://www.marineharvest.com/
Maruha Nichiro Corporation (2006–2014). *Financial Statements*. Retrieved from http://www.maruha-nichiro.co.jp/english/
NBSO. (2010). *An overview of China's aquaculture*. Netherlands Business Support Office: Dalian.
Nidumolu, R., Prahalad, C.K., & Rangaswami, M.R. (2009). Why sustainability is now the key driver of innovation. *Harvard Business Review*, *87*(9):57–64.
Nissui Susian Kaisha Ltd. (2006). *Annual report 2006*. Retrieved from http://www.nissui.co.jp/english/
Nissui Susian Kaisha Ltd. (2007). *Annual report 2007*. Retrieved from http://www.nissui.co.jp/english/
Nissui Susian Kaisha Ltd. (2008). *Annual report 2008*. Retrieved from http://www.nissui.co.jp/english/
Nissui Susian Kaisha Ltd. (2009). *Annual report 2009*. Retrieved from http://www.nissui.co.jp/english/
Nissui Susian Kaisha Ltd. (2010). *Annual report 2010*. Retrieved from http://www.nissui.co.jp/english/
Nissui Susian Kaisha Ltd. (2011). *Financial Result 2011*. Retrieved from http://www.nissui.co.jp/english/

Nissui Susian Kaisha Ltd. (2012). *Financial Result 2012.* Retrieved from http://www.nissui.co.jp/english/

Nissui Susian Kaisha Ltd. (2013). *Financial Result 2013.* Retrieved from http://www.nissui.co.jp/english/

Nissui Susian Kaisha Ltd. (2014). *Financial Result 2014.* Retrieved from http://www.nissui.co.jp/english/

Nutreco Corporate (2006–2014). *Annual report.* Retrieved from http://www.nutreco.com/

Nonaka, I. (1994). A dynamic theory of organizational knowledge creation. *Organization Science*, *5*(1), 14 pp.

Obal, M. (2013). Why do incumbents sometimes succeed? Investigating the role of interorganizational trust on the adoption of disruptive technology. *Industrial Marketing Mangement. 42*(1), 900–908.

Pacific Anders International Holdings Limited. (2006). *Annual report: 2006.* Retrieved from http://www.pacificandes.com/html/index.php

Pacific Anders International Holdings Limited. (2007). *Annual report: 2007.* Retrieved from http://www.pacificandes.com/html/index.php

Pacific Anders International Holdings Limited. (2008). *Annual report: 2008.* Retrieved from http://www.pacificandes.com/html/index.php

Pacific Anders International Holdings Limited. (2009). *Annual report: 2009.* Retrieved from http://www.pacificandes.com/html/index.php

Pacific Anders International Holdings Limited. (2010). *Annual report: Pioneering Global Traceability.* Retrieved from http://www.pacificandes.com/html/index.php

Pacific Anders International Holdings Limited. (2011). *Annual report: From Ocean to Plate.* Retrieved from http://www.pacificandes.com/html/index.php

Pacific Anders International Holdings Limited. (2012). *Annual report: Traceability System.* Retrieved from http://www.pacificandes.com/html/index.php

Pacific Anders International Holdings Limited. (2013). *Annual report: Structured for Future Growth.* Retrieved from http://www.pacificandes.com/html/index.php

Pacific Anders International Holdings Limited. (2014). *Annual report: Natural Protein from our Oceans.* Retrieved from http://www.pacificandes.com/html/index.php.

Pohar, J. (2011). Detection and comparison of the sensory quality of wild and farmed brown trout (*Salmo trutta*) by consumers. *Acta Agriculturae Slovenica*, *98*, 45–50.

Russ, M., Fineman, R., & Jones, J.K. (2010). C^3EEP taxonomy-knowledge based strategies. In M. Russ, (Ed.) *Knowledge Management Strategies for Business Development*, pp. 133–158. Hershey, PA: Business Science Reference.

Russ, M., & Jones, J.K. (2011). Knowledge Management's strategic dilemmas typology. In D.G. Schwartz and D. Te'eni (Eds.) *Encyclopedia of Knowledge Management,* 2nd Edition: pp. 804-821. Hershey, PA: IGI Reference.

Schubert, P., Lincke, D., & Schmid, B. (1998*) A Global Knowledge Medium as a Virtual Com- munity: The NetAcademy Concept.* In Proceedings of the Fourth Americas Conference on Information Systems, E. Hoadley and I. Benbasat (eds). Baltimore, MD, pp. 618–620.

Smith, A., Kern, F., Raven, R., & Verhees, B. (2014). Spaces for sustainable innovation: Solar photovoltaic electricity in the UK. *Technological Forecasting & Social Chance*, *81*, 115–130.

Tan, Y., Jorge Ochoa, J., Langston, C., & Shen, L. (2015). An empirical study on the relationship between sustainability performance and business competitiveness of international construction contractors. *Journal of Cleaner Production*, *93*, 273–278.

Thurstan, R., & Roberts, C. (2014). The past and future of fish consumption: Can supplies meet healthy eating recommendations? *Marine Pollution Bulletin*, *89*(1), 5.11.

Tushman, M.L., & Anderson, P. (1986). Technological discontinuities and organizational environments. *Administrative Science Quarterly*, *31*, 439–465.

Waite, R., Beveridge, M., Brummett, R., Castine, S., Chaiyawannakarn, N., Kaushik, S., Mungkung, R., Nawapakpilai, S., & Pillips, M. (2014). Improving productivity and environmental performance of aquaculture. *World Resource Report*. World Resource Institute: Washington DC.

Washington S., & Ababouch, L. (2011). Private standards and certification in fisheries and aquaculture-current practice and emerging issues. *FAO Fisheries and Aquaculture Technical Paper 553*. FAO: Rome.

7 Decrease in federal regulations in the U.S.: Preparing for dirty water, can Knowledge Management help?

Breanne Parr

University of Wisconsin–Green Bay, Green Bay, Wisconsin, USA

Introduction

Problem: A decrease in water regulations in the U.S. and a decrease in the budget for water management makes it difficult to maintain current water standards.

In the 1940s, a growing concern for the cleanliness of our water led to the first major law protecting water: the Federal Water Pollution Control Act of 1948. In 1972, the Clean Water Act (CWA) was amended due to increasing awareness against water pollution (EPA, 2016). The Obama administration added additional protection for our streams and wetlands with the expansion of The Water of the United States Rule in 2015 (Eilperin & Phillip, 2017).

Until recently, we have continued to increase regulations and protection of water in the United States. However, in February of 2017, Donald Trump wrote an executive order dismantling that regulation. He instructed the Environmental Protection Agency (EPA) and the Army Corps of Engineers to review the Waters of the United States Rule. Trump's push to reduce regulations on protecting wetlands could contribute to the destruction of critical aquatic habitats and migratory birds (Eilperin & Phillip, 2017).

This decrease in regulation has introduced a potential for a water crisis sooner than expected. It is a change in consideration for certain bodies of water that could trigger a modification in how governments make and apply

Handbook of Knowledge Management for Sustainable Water Systems, First Edition. Edited by Meir Russ.

legislation. Bob Irvin, the president of American Rivers states, "Without the Clean Water Rule's critical protections, innumerable mall streams and wetlands that are essential for drinking water supplies, flood protection, and fish and wildlife habitat will be vulnerable to unregulated pollution, dredging, and filling" (Eilperin & Phillip, 2017, para 7).

As we see a trend of decreasing environmental regulation, it is realistic to be concerned about the reduction of water regulation playing a serious role in advancing the timeline of a water crisis. What type of strategic planning should be done to avoid a water crisis? Is the threat of unpotable water a real concern? *Our objective in this chapter is to identify the risks of an emergency situation due to water contamination; review recent water emergencies and what made the response or recovery successful or unsuccessful; understand steps to prevent a crisis; and finally, understand the role a knowledge management system can play in preparing a plan to reduce the effect of a water crisis.*

7.1 The Clean Water Act of 1972

The Clean Water Act (CWA) of 1972 made it unlawful for people to discharge pollutants into navigable waters without a permit. It essentially established a structure for regulation over discharging pollutants into waters in the United States. In addition, the act provided the EPA with the authority to implement pollution control, such as setting standards for wastewater. It initially funded construction of sewage treatment plants under the construction grant program, however, that was phased out in 1987 and was replaced with the State Water Pollution Control Revolving Fund or Clean Water State Revolving Fund. In 1990, the United States and Canada agreed to reduce certain toxic pollutants in the Great Lakes with an expansion of the Great Lakes Critical Programs Act (EPA, 2016).

The CWA covers major bodies of water like rivers, bays, coastal waters, and the streams and wetlands that flow into larger bodies of water. In 2015, Barack Obama added a clarification to the CWA known as the Clean Water Rule that more clearly defined the regulation of smaller bodies of water. The Clean Water Rule includes more isolated waters that have a less obvious connection to navigable waters, like seasonal streams and wetlands (Sneed, 2017).

7.1.1 Unsafe water

Almost 80% of the world's deadliest diseases are caused by poor sanitation and unsafe drinking water. Contaminated drinking water is more than just a public health concern. Unsafe drinking water has a negative impact that expands beyond being deadly; it also has a negative impact on economic development (Galvin,

2007). Kenneth Behring, the founder of Global Health and Education Foundation, explains, "We've seen the misery in the world that's caused by bad water. When you go to these countries and see the water that people drink, they don't have the background or education to even know that it causes disabilities" (Galvin, 2007 para. 4).

In the U.S., we have the means to educate individuals on safe water and diseases caused from drinking unsafe water. However, due to limited resources and funds, water crises from contaminated water sources still occur regularly. This is one of the reasons that the Clean Water Act progressed slowly, in terms of coverage and applicability. In the 1970s (before federal mandates), only about half of the states had established water quality standards (U.S. House of Representatives, 2014). In 1972 when the Clean Water Act became law, it was the beginning of the federal government and the states working together to ensure safe water standards for everyone. Today, over 65,000 industrial, commercial, municipal, or other sources must obtain discharge permits from the EPA or their state; and over 150,000 sources are required to obtain permits for storm water management. Penalties for violating water restrictions, including wetland protections, can be as much as $25,000 per violation. Criminal violations of negligence are punishable by up to three years in prison and $50,000 per day in fines (U.S. House of Representatives, 2014).

The CWA does not actually monitor drinking water quality. The Safe Drinking Water Act of 1974 provides this protection (U.S. House of Representatives, 2014). However, since our drinking water comes from the same water that fills our lakes and rivers, decisions that affect the CWA deeply affect our drinking water.

7.2 Regulation rollback

The 2015 Clean Water Rule has not yet gone into effect because of an executive order made by Donald Trump to rescind or replace the Clean Water Rule. In February of 2017, the newly appointed EPA administrator, Scott Pruitt, promised serious rollbacks on Obama-era environmental regulations. A few days later, Trump signed an executive order to revoke the 2015 Waters of the United States rule, with Trump calling the rule "destructive" and "horrible" (Eilperin & Phillip, 2017). According to the order, the overall goal in overturning this rule would be to promote economic growth and cut regulations. The control over "ephemeral and intermittent" streams would be lifted. This includes any streams that only flow when it rains or when the snow melts, or any streams that do not flow continuously year-round (Sneed, 2017).

In the 2006 case, Rapanos vs. United States, the Supreme Court determined that any body of water that had a "significant nexus" to a federally protected waterway fell under federal jurisdiction. The Clean Water Rule specifically defines what these waterways are, and the EPA expects that this clarification will only expand federal jurisdiction by 3% (Waskom & Cooper, 2017). Objectors to the rule stated that it was expensive and required excessive overheads to obtain permits to dig up something, or drain a small pond to construct something. On the other hand, environmentalists argue that many of these wetlands and small bodies of water were crucial to aquatic species, migratory birds, and waterfowl, and they were also a source for drinking water (Branham, 2017). If the rule is rejected, then the federal government will nevertheless still need to define which waterways do fall under federal protection.

Scientists have found that these streams still have a major impact on the water quality of other downstream systems. According to the EPA (as cited by Sneed, 2017), around 117 million people, or one in three Americans draw all or some of their water from a system that contains a stream that would be protected by the Clean Water Rule (2017). This rule does not simply protect a few wetlands or small inlet streams; it includes about two-thirds of the total stream miles in the U.S. (Sneed, 2017). While the EPA would have previously covered a wetland that is nearby or connected to an intermittent stream, under the reviewed rule, these wetlands will probably be excluded. Wetlands are a great filtration system. Bacteria in wetlands work to remove nitrates from runoff before it flows downstream, preventing pollution in larger bodies of water. If these wetlands are no longer regulated, they could be destroyed or polluted (Sneed, 2017) thereby impacting other water systems.

In addition to working to eliminate the 2015 rule, Trump has also ordered significant cuts to the EPA's budget. This includes about $300 million in federal funding that was allocated for Great Lakes clean-up. Projects cut include an invasive species project to eliminate Asian carp and other invasive aquatic animals and plants from the lakes and to keep toxic algae out of Lake Erie. The alga that is currently monitored contains a neurotoxin and in 2014, its presence shut down Toledo's water (McEvers, 2017). Along with the budget cut having a negative effect on the environment and water quality, the benefits from the clean-up far exceed the costs. In 2007, projections estimated $6.5 to $11 billion dollars in tourism, recreation, and fishing around Lake Erie; $12 to $19 billion in coastal property value increases; and a $50 to $125 million reduction in costs to municipalities (Austin, Anderson Courant, & Litan, 2007). Overall, for every one dollar invested in sanitation and drinking water, $3 to $34 of economic development is projected in return (Ross, 2013). Protecting water from the addition of contaminates and controlling contaminates at the outset, is not only safer, but more cost efficient than a clean-up.

7.3 CWA offenders

Even with current regulations, the EPA struggles to enforce regulations under their current budget. States continue to fail to meet federal standards. The question therefore is – What will happen when the federal standards are gone and states are left to set their own standards (Duhigg, 2009)?

7.3.1 Arsenic and other chemicals in West Virginia

In West Virginia, Jennifer Hall-Massey and her family are unable to drink the water. Any type of exposure to the water causes a painful rash. Brushing teeth with the water actually ate away the enamel. People in the area had to pretreat their skin with a special lotion to prevent burns. Water tests show arsenic, barium, lead, manganese, and other chemicals at high concentrations, which are known to cause cancer, damage kidneys and affect the nervous system (Duhigg, 2009). The Hall-Massey family along with hundreds of neighbors are currently involved in a lawsuit against a nearby coal company for contaminating the local lakes, rivers, and groundwater. In the last five years, chemical factories and other plants have violated water pollution laws over half a million times. Currently, the EPA often fails to intervene because of limited resources. States often make the same claim (Duhigg, 2009).

7.3.2 Chemical spill in West Virginia

In 2014, about 10,000 gallons of crude MCHM, or 4-methylcyclohexanemethanol, leaked into the Elk River near Charleston, West Virginia. The water supply for over 300,000 people was contaminated. The chemical leaked from a small hole in a storage tank that was dated from the 1930s. The chemical traveled through miles of water distribution lines. The company responsible, Freedom, ended up having to file bankruptcy after two of the company's officials were found responsible for the spill (Fitzgerald, 2015). One Thursday morning, a leak was detected from one of the storage tanks into the Elk River. By the afternoon, West Virginia American Water warned residents not to drink the water, and to use it only for putting out fires or flushing toilets. The governor declared a state of emergency, and the President provided an order for federal assistance. Federal, state, and local agencies started trucking in water for consumption. While the emergency planners were prepared for evacuations and fires, they were unaware of how to handle the leaking of this mysterious chemical into the waterway (Ward, 2014).

The water company was not sure what the dangers of 4-methylcyclohexanemethanol were, and how much of the chemical was too much. When county officials were asked how much planning the county had done in preparation for a leak from Freedom Industries, the Kanawha County Commission President Kent Carper stated, "Not enough." The EPA deployed personnel and were told by the Department of Environmental Protection (DEP) that Freedom Industries did not require permits since they did not manufacture any products at that facility; they were mostly a storage facility. The only permit identified was a storm water permit; however, the last DEP inspection was in 1991 before the building was occupied by Freedom Industries (Ward, 2014).

Multiple businesses were closed due to the contaminated water. Emergency rooms treated about 169 patients for chemical exposure-related symptoms (Davenport & Southall, 2014). The "Do Not Use" order lasted 10 days. West Virginia Americas Water flushed water mains and distribution systems, then began advising residents to flush their systems. While public water sources were tested, home water taps were not. Even after being given the "all clear", many home owners complained that their water still had an odor. Standard residential tap flushing procedures were not a part of the standard emergency recovery plan (Whelton *et al.*, 2015).

To prevent this from happening again, advocates have pushed for increased chemical storage regulation. Around 50,000 tanks were identified for regulation, implementing a statewide regulatory program. Several lawsuits have been filed, including against West Virginia American Water. There have been issues with the clean-up and remediation of the spill, including solid waste dumping that contaminated the air in the nearby city of Hurricane, West Virginia. The West Virginia Water Crisis was one of the worst in the nation, and it was very much preventable. However, one thing that has still not been remedied is the fact that 4-MCHM is not on the Toxic Substance Control Act Inventory and is currently unregulated (Bryson, 2016).

West Virginia has recently had issues regulating the chemical and coal industry. Jennifer Sass, a senior scientist at the Natural Resources Defense Council and a lecturer in environmental health at George Washington University stated, "West Virginia has a pattern of resisting federal oversight and what they consider EPA interference, and that really puts worker and the population at risk" (as cited by Davenport & Southall, 2014, para. 9).

In 2009, the New York Times found that hundreds of companies in West Virginia had violated federal regulations without any fines or penalties. The impression has been that the coal lobby has a strong influence on state regulations, and that they played a significant role in the state's 1990 groundwater protection bill (Davenport & Southall, 2014). West Virginia is a good example of what the country's future could be like if a decrease in environmental regulations and rule enforcements are allowed to affect the country.

7.3.3 Lead in Michigan

A public health crisis occurred in 2014 in Flint, Michigan after it was discovered that lead levels in children's blood was consistently abnormal. It was confirmed that lead from the city's underground pipes was leaching into the water supply after the water source was changed from Lake Huron to a supply that was more corrosive (Laidlaw *et al.*, 2016).

There were two decisions made earlier that led to this disaster. The first was switching the city's water supply from the treated Lake Huron water to the Flint River. The second was the city not replacing the lead pipes despite their negative effects being known. Over 15% of the homes in the city had water with lead levels over the safe limit, some with lead levels that were 900 times the safe limit (Stockton, 2016). The problem started in April of 2014 when a decision was made to change the water supply while a new pipeline was under construction to Lake Huron. The measure was taken to save money. Soon after, residents complained about the water's color, odor, taste, and other health concerns like skin rashes. *Escherichia coli* was detected, a violation of the Safe Drinking Water Act. Then, another violation for trihalomethane levels occurred after measures were take trying to control the bacteria levels (Hanna-Attisha, LaChance, Sadler, & Schnepp, 2016). The Flint River water had a high chloride level and a high chloride-to-sulfate mass ratio. Along with no corrosion inhibitor, these chemicals provided the perfect ingredients for leaching lead from the pipes in the water. Lead poising is irreversible, and it causes declines in intelligence, behavior, and overall life achievement (Hanna-Attisha *et al.*, 2016).

Then, between June 2014 and November 2015, 87 cases of *Legionella* bacteria were confirmed in the Flint area. Ten of those cases were fatal. This was after the water source was switched back to Lake Huron (Hanna-Attisha *et al.*, 2016). In January of 2016, the Governor activated the National Guard and declared a state of emergency. Snyder, Governor of Michigan, asked for help from the Federal Emergency Management Agency to help with emergency planning for the Flint water crisis (Al Hajal, 2016). A federal state of emergency was enacted, authorizing FEMA and the Department of Homeland Security to act and coordinate disaster relief throughout the area.

The clean-up has been very expensive. After the initial bill to help residents and the children affected, an additional bill was signed to replace pipes and help with utilities. By this time, the crisis had cost over $240 million, and residents had been in a state of emergency for over eight months (Eggert, 2016).

7.3.4 Escherichia coli *(*E. coli*) in Ontario*

In May of 2000, an *Escherichia coli* outbreak in the town of Walkerton, ON was detected. *E. coli* is a fecal coliform that is commonly found in human feces and is used as a fecal indicator. Chlorine is a common disinfectant used to kill coliform bacteria. The contamination stemmed from equipment failure that lead to water from a water treatment plant passing through without the proper chlorine disinfection for bacteria (Danon-Schaffer, 2001).

To prevent future contaminations, a thorough investigation was conducted. All of the wells were tested for flow and filtration, as well as water samples from each well were taken. Several improvements were made. Thereafter, the source of contamination was investigated in more depth. Private wells at 13 livestock farms within four kilometers were tested, and all but two tested positive for human bacterial pathogens. Rainfall and runoff patterns were also simulated, concluding that there were several possibilities for a contamination source. To prevent this from occurring again, a remediation was carried out that included water main flushing, distribution water main swabbing, disinfection of building plumbing, replacing water mains, valve and hydrant repairs, and removing private well and cistern connections (Danon-Schaffer, 2001).

7.3.5 Toxin in Ohio

In August of 2014, Toledo's water supply was contaminated with blue green algae growth. Algae blooms in Lake Erie are becoming more common as the amounts of nitrogen and phosphorous are increasing due to additional runoff from agricultural fertilizers or poorly maintained septic systems. Over 500,000 people were affected by the contamination (Mayhew, 2014). The water was contaminated by microcystin, which is believed to have come from farm fertilizer that made its way into Lake Erie. The toxin cannot be killed by boiling. Toledo residents were advised to avoid even brushing their teeth with the water. Restaurants and other businesses were forced to close and bottled water was required for all cooking and drinking purposes. Hospitals had to cancel medical procedures, and food had to be destroyed. A state of emergency was declared, and the National Guard was brought in (Roby & Murphy, 2015).

Part of the quick response and prevention of human death was due to the voluntary testing that Toledo has done since 2011. However, after the incident, several areas for improvement were identified to obtain an accurate warning more quickly, and to improve the communication system to warn residents in the future. Lessons learned included: a need for better early detection tools, additional chemical feed capacity, an improved Emergency Response Standard Operating Procedure, better communication protocol, and a federal microcystin standard based on scientific evidence (Roby & Murphy, 2015). Before the incident, there was no standard for measuring the cyanotoxin in water. Monitoring was done completely on a voluntary basis. This made it more difficult to identify what levels were safe for cooking, bathing, or drinking.

After the incident, many improvements were made. A test buoy and sampling probes now collect data from a water intake crib and from the Low Service Pumping Station. Chlorophyll, pH levels, and the presence of blue-green algae is reviewed and the data is reported to a public website every 20 minutes. Toledo also made $264 million dollars in water infrastructure improvements (Roby & Murphy, 2015). The city worked to improve their emergency preparedness by implementing new operating procedures, a contingency response plan, training and workshops for Toledo and regional emergency response staff, and improvements to the regional emergency communication plan. A nine-member panel was brought in to review the issue, and Ohio colleges and universities have also been included in these efforts. The University of Toledo is helping by studying toxin detection, treatment methods, watershed and nutrient sourcing, and the health effects of microcystin (Roby & Murphy, 2015).

At the forefront, the area has imposed restrictions on fertilizer applications. They have also increased rules on how dredged lake sediment is disposed of, and created a department focused on algae-management. The EPA published health advisories for microcystin in drinking water in May of 2015. Toledo has implemented a water quality dashboard that indicates if there are any risks present in the lake or tap water to help improve communication to residents (Roby & Murphy, 2015).

7.3.6 Case summary

While these are just a few examples of the many water system contaminations that have occurred in the United States, they are but a sample of what can happen anywhere when a failure in a process or system occurs. If standards and regulations are allowed to decline, the chance of a system-failure occurring is greater. Overall, regulating the prevention of a water emergency is going to be more cost-effective than cleaning up a contamination.

Illness from waterborne disease is still a common risk is the United States. Between 1971 and 2002 there were 764 documented waterborne outbreaks from contaminated drinking water, totaling 79 deaths and 575,457 cases of illness. However, these statistics are for a very specific definition of what is classified as an illness. Contaminated drinking water is directly related to the number of pathogens in the source water, the age of the distribution system, the quality of the delivered water, and climatic events. Overall, it is estimated that the true situation is that there are 19.5 million cases of waterborne illnesses in the US alone per year (Reynolds, Mena, & Gerba, 2008).

As far as the financial aspect goes, researchers at Kansas State University estimate that pollution by phosphorus and nitrogen costs Americans over 4.3 billion dollars each year. Increases and improvements in water treatment end up costing the consumer money. This also includes costs due to decreases in property values of waterfront properties, costs of protecting aquatic species which are negatively impacted by human intervention, and the loss of recreation and tourism (Kansas State University, 2008).

7.4 Knowledge Management – dirty water[1]

One thing that needs to be acknowledged about waterways is that they are not exclusive to one farm, city, or state. Streams flow into rivers that expand into lakes and wetlands. The water from these lakes and rivers is part of the hydrologic cycle; the water evaporates into the atmosphere and eventually

[1]The author acknowledges and is thankful to Dr. Vallari Chandna for her helpful and stimulating comments on this part of the chapter.

settles into groundwater. Thus, our groundwater and the water that makes up rivers and wetlands is, at the end of the day, all one and the same closed system.

Establishing and enforcing water regulations that cross state lines is therefore important to ensure that clean water is available to everyone, given these interconnections between all water bodies. If we only seek to protect the larger water bodies such as rivers and not protect the streams and wetlands that connect to those rivers, then we are introducing pollution into our waterways and our drinking water nevertheless. As more people are using the multiple waterways, there are also more people that need improved access to clean and safe drinking water. More people however, means more pollution and more waste. Looking at these factors is reason alone for increasing the restrictions on what type of waste can be introduced into the water system. Without additional enforcement and restrictions, we will certainly see an increase in water pollution and emergencies related to contaminated drinking water.

Water utility companies collect a great deal of data on their water. Wastewater, rivers, streams, wells, and municipal water are all frequently being analyzed. To be useful and beneficial however, the information needs to be turned into knowledge that creates value by using it to predict and prevent a water-related emergency. In Flint, Michigan, Dr. Mona Hann-Attisha reviewed trending patient data to identify abnormal lead amounts in the area's children. The data were collected in a patient's electronic medical record and then compiled to show trends. The problem was that these data were collected and analyzed only after the children were displaying symptoms of lead poisoning.

The Environmental Working Group (EWG) National Drinking Water Database collects information on cities' drinking water quality. While the EWG is pushing for the federal government to assess and maintain drinking water quality, they launched a project to develop a database that stores around 48,000 cities' water quality. The EWG began publishing and ranking water quality from water utilities across the country in 2009 (Environmental Working Group, 2011). The United States Geological Society (USGS) monitors and records data on some waterways across the country. The USGS has several tools to help people review water data nationwide. Now, how can we use this to prevent the next water disaster? We have the data, but it seems that making available the resources to examine data and focus on identifying and eliminating the source of contamination is the real problem. For instance, if the total trihalomethane count is elevated in the drinking water in Western Arkansas, can we identify the source causing the contamination?

Knowledge Management includes multiple knowledge-focused activities, i.e. creation of new knowledge, accessing knowledge from multiple sources, making knowledge accessible to the decision-making processes, ensuring knowledge is embedded throughout the water management process, building a database of the knowledge gathered, facilitating knowledge growth throughout the system, enabling transfer of knowledge to everyone involved and measuring the impact of this Knowledge Management (Ruggles, 1998). Thus, once there is a Knowledge Management plan in place to utilize the data, the next step is to have an emergency response plan in place to fix the issue before a state of emergency is deemed necessary. A Knowledge Management System (KMS) can help with both these crucial aspects. The extremely large amount of data that is gathered from all the different sources needs to be distilled to useable information and then applied and utilized as knowledge (Alavi & Leidner, 2001). An efficient KMS can assist with the decision-making process, which is crucial for dealing with the issues of water supply disruption that could result when there are water issues resulting from regulation reduction (Bertrand-Krajewski, Barraud, & Chocat, 2000).

7.5 Avoiding non-potable water without federal restrictions

When it comes to managing the water supply and preparing for different situations, it is important to have strategies at the ready at multiple levels, i.e. individual, business, and regional while having a proper understanding of all the possibilities and consequences of all the possible decisions (Barbosa, Fernandes, & David, 2012). An efficient KMS would be able to build a knowledge expertise base and allow for personnel changes as well (Grigg, 2006) which could result from regulation changes. Using the KMS, the water supply can be managed at multiple levels and the system can enable all stakeholders to carry out the following efforts proactively:

Know your water quality

It is important to understand what the current quality of your water is, and what the risks are in your region. To begin educating yourself, review your state's Source Water Assessment. Read the assessment, and identify what seems realistic and what might be missing (Frey, 2003). Then, review your local water assessment or have your private well tested. It is recommended that wells be tested annually.

Get active

To avoid a significant decrease in water quality, local and state governments need to take action now. As an individual, call and write to your congress-person and senators. Get active and educate yourself. Volunteer for organizations like the River Alliance or another nearby wetland or waterway management group. Many of these organizations are non-profit and can use the help. Do what you can to avoid polluting waterways. Most importantly, get out and vote for representatives that support protecting our waters, and do not support companies that pollute our waterways or support eliminating EPA regulations.

After reviewing your area's water assessment, reach out to the appropriate agency with any concerns that you have. Ask questions, especially if there are issues with the water quality that seem worrisome. The agency should have a mitigation plan in progress and they should be ready to provide updates on current projects. Do not hesitate to reach out to other community members and get more people involved. If necessary, submit your information to the media and then, pull in your elected officials. These are the people that can help fund or implement necessary changes (Frey, 2003). After reviewing your municipal assessments and carrying out the previous activities, if you still have concerns, continue to collect data. Go out and review local maps and aerial photos, then interview local residents. Walk the watershed and check out the quality of the local streams, rivers, lakes, lakebeds, and wetlands. Take it upon yourself to check out potential pollution sources and collect samples. If you are not getting what you need from your local municipality, find another group from the EPA's database or reach out to the local university or college (Frey, 2003).

Reduce household water pollution

Do not pour grease, oil or fat down the drain. Collect oil-based products and throw them away. Do not dispose of chemicals or cleaning products down the sink or in the toilet. These products not only pollute the waterways, but they can also corrode pipes. Also, try to transition to phosphate-free detergents and limit bleach usage. Do not flush medications or drugs down the toilet. Do not use the toilet as a garbage can; throw used tissues and papers in the garbage. Instead of putting food scraps in the disposal, compost them. Do not drain your sump pump into the sewer. Finally, be careful with fertilizers, pesticides, and herbicides. Try not to use any of these chemicals, but if necessary, use a minimum to prevent runoff into sewers or waterways (Simsbury, 2013). The majority of wastewater is from residential sources, so keeping this discharge clean will help reduce overall contamination in waterways.

7.6 Conclusion

Over two-thirds of U.S. estuaries and bays are severely degraded from phosphorus and nitrogen pollution. Almost half of lakes and streams and one-third of bays are polluted. The situation in places like Michigan is far worse. Forty percent of rivers and lakes are far too polluted for fishing, swimming, or aquatic life. The Mississippi River delivers about 1.5 metric tons of nitrogen to the Gulf of Mexico each year. Worldwide, half of the groundwater is non-potable (Weebly, n.d.). As the population continues to grow, so will the amount of pollution and waste.

Overall, it will become even more important for leadership to regulate corporations and municipalities to decrease contamination of our waterways and our drinking water. When cities and states are competing for economic opportunities, they might be more hesitant to implement environmental regulations that are stricter than their neighbors' fearing that the restrictions might prevent businesses and companies from moving to the area. However, municipalities are responsible for providing safe drinking water. If a state or city is not protecting its residents, who would want to live there? Regardless, if regulations continue to be reduced and with contamination a real concern, individuals need to have their water tested and report irregularities. In addition, regular blood checks to detect abnormal mineral or chemical levels before symptoms or poisoning occurs will be important. Flint, Michigan was a crisis that should serve as an example for other strategic planners to help prepare for a water crisis. Implementing a well-developed Knowledge Management System in a timely manner is important in the preparation of water supply crises that could result from reduced federal regulations. An efficient KMS can ensure that the regulation changes are monitored closely so that cities are prepared for water disruptions. Additionally, it ensures that multiple problem scenarios are prepared for, consequences are mapped, information dissemination to all stakeholders is made possible, and that decision-making at all levels is simplified.

References

Al Hajal, K. (2016). 87 *cases, 10 fatal, of* Legionella *bacteria found in Flint area; connection to water crisis unclear. Michigan Live.* Retrieved from http://www.mlive.com/news/detroit/index.ssf/2016/01/legionaires_disease_spike_disc.html

Alavi, M., & Leidner, D.E. (2001). Review: Knowledge management and knowledge management systems: Conceptual foundations and research issues. *MIS Quarterly*, 107–136.

Austin, J.C., Anderson, S., Courant, P.N., & Litan, R.E. (2007). Healthy waters, strong economy: The benefits of restoring the Great Lakes ecosystem. *The Brookings Institution: Great Lakes Economic Initiative.*

Barbosa, A.E., Fernandes, J.N., & David, L.M. (2012). Key issues for sustainable urban stormwater management. *Water Research*, *46*(20), 6787–6798.

Bertrand-Krajewski, J.L., Barraud, S., & Chocat, B. (2000). Need for improved methodologies and measurements for sustainable management of urban water systems. *Environmental Impact Assessment Review*, *20*(3), 323–331.

Bryson, K.L. (2016). 2-year anniversary. *West Virginia Water Crisis.* Retrieved from https://wvwatercrisis.com/2016/01/06/2-year-anniversary/

Danon-Schaffer, M.N. (2001). Walkerton's contaminated water supply system: A forensic approach to identifying the source. *Environmental Forensics*, *2*, 197–200.

Davenport, C., & Southall, A. (2014). *Critics say spill highlights lax West Virginia regulations. The New York Times.* Retrieved from https://www.nytimes.com/2014/01/13/us/critics-say-chemical-spill-highlights-lax-west-virginia-regulations.html

Duhigg, C. (2009). *Clean water laws are neglected, as a cost in suffering. New York Times.* Retrieved from http://www.nytimes.com/2009/09/13/us/13water.html

Eggert, D. (2016). *Michigan governor signs budget with $165M more for Flint. Associated Press.* Retrieved from http://www.abc12.com/content/news/Michigan-governor-signs-budget-with-165M-more-for-Flint-384904071.html

Eilperin, J., & Phillip, A. (2017). *President Trump: Law is key to protecting drinking water sources, recreation groups say. The Washington Post.* Retrieved form https://www.pressreader.com/usa/the-washington-post/20170301/281526520837281

Environmental Working Group. (2011). *EWG's top-rated and lowest-rated water utilities. National Drinking Water Database.* Retrieved from http://www.ewg.org/tap-water/

EPA. (2016). *History of the Clean Water Act. United States Environmental Protection Agency.* Retrieved from https://www.epa.gov/laws-regulations/history-clean-water-act

Fitzgerald, P. (2015). *Two former Freedom officials to plead guilty in spill case; Chemical leak in West Virginia contaminated water supply for 300,000 people. Wall Street Journal* (online). Retrieved from http://search.proquest.com.ezproxy.uwgb.edu:2048/docview/1654717220?rfr_id=info%3Axri%2Fsid%3Aprimo

Frey, M. (2003). *Source water stewardship: A guide to protecting and restoring your drinking water.* The Clean Water Network, Clean Water Fund, and the Campaign for Safe and Affordable Drinking Water*,* 1–46.

Grigg, N.S. (2006). Workforce development and knowledge management in water utilities. *Journal American Water Works Association, 98*(9), 91–99.

Galvin, M. (2007). New web resource aims to improve drinking water quality worldwide. *The National Academies in Focus, 7*(3), 18–19.

Kansas State University. (2008). *Freshwater pollution costs U.S. at least $4.3 billion a year. ScienceDaily.* Retrieved from https://www.sciencedaily.com/releases/2008/11/081112124418.htm

Laidlaw, M.A.S., Filippelli, G.M., Sadler, R.C., Gonzales, C.R., Ball, A.S., & Mielke, H.W. (2016). Children's blood lead seasonality in Flint, Michigan (USA), and soil-sourced lead hazard risks. *International Journal of Environmental Research and Public Health, 13*(4), 358.

McEvers, K. (2017). *Proposed budget cuts slash funding for Great Lakes clean-up. All Things Considered: National Public Radio.*

Reynolds, K.A., Mena, K.D., & Gerba, C.P. (2008). Risk of waterborne illness via drinking water in the United States. *Reviews of Environmental Contamination Toxicology, 192*, 117–158.

Roby, C., & Murphy, T. (2015). Toledo water crisis and emergency planning for treatment plants. *OWEA Plant Operations and Lab Analysis Workshop.* Retrieved from http://www.ohiowea.org/docs/1345_Toledo_Emergency_Roby_Murphy.pdf

Ross, N. (2013). *World Water Quality Facts and Figures.* Retrieved from http://www.pacinst.org/wp-content/uploads/2013/02/water_quality_facts_and_stats3.pdf

Ruggles, R. (1998). The state of the notion: knowledge management in practice. *California Management Review*, *40*(3), 80–89.

Simsbury. (2013). *Ten things you can do to reduce water pollution.* Retrieved from http://www.simsbury-ct.gov/water-pollution-control/pages/ten-things-you-can-do-to-reduce-water-pollution

Sneed, A. (2017). *Trump's order may foul U.S. drinking water supply: Narrowing the Clean Water Rule could increase pollution in critical waters. Scientific American.* Retrieved from https://www.scientificamerican.com/article/trump-rsquo-s-order-may-foul-u-s-drinking-water-supply/

Stockton, N. (2016). *Here's how hard it will be to unpoison Flint's water. Wired.* Retrieved from http://www.wired.com/2016/01/heres-how-hard-it-will-be-to-unpoison-flints-water/

U.S. House of Representatives. (2014). Navigating the clean water Act: Is water wet? *U.S. House of Representative Committee on Science, Space, and Technology Full Committee: Hearing Charter.* Retrieved from https://www.gpo.gov/fdsys/pkg/CHRG-113hhrg89413/pdf/CHRG-113hhrg89413.pdf

Ward, K. (2014). *Why wasn't there a plan? Charleston Gazette-Mail.* Retrieved from http://www.wvgazettemail.com/News/201401110085

Waskom, R., & Cooper, D.J. (2017). *Why farmers and ranchers think the EPA Clean Water Rule goes too far. The Conversation.*

Weebly. (xxxn.d.) *The oceans are degrading, thanks to water pollution.* Retrieved from http://putanendtowaterpollution.weebly.com/environment.html

Whelton, A.J., McMillan, L., Connell, M., Kelley, K.M., Gill, J.P., White, K.D., Gupta, R., Dey, R., & Novy, C. (2015). Residential tap water contamination following the Freedom Industries chemical spill: Perceptions, water quality, and health impacts. *Environmental Science & Technology, 49,* 813–823.

PART 2

Regional Aspects of Knowledge Management for Sustainable Water Systems

8 Knowledge Management strategies for drinking water protection in mountain forests

Roland Koeck, Eduard Hochbichler and Harald Vacik

Institute of Silviculture, Department of Forest and Soil Sciences, University of Natural Resources and Life Sciences, Vienna, 1190, Vienna, Austria

Introduction

Knowledge about the requirements of sustainable drinking water protection strategies (DWPS) becomes essential, as a safe drinking water supply can be regarded as basic condition for human development. The pressures on freshwater resources are increasing in the modern world, hence water supply systems driven from forested mountainous catchment areas provide in many cases solutions for safeguarding water supply. For securing a safe supply with high quality drinking water, forest management has to be adapted to meet the requirements of the specific situation. Most of the todays foresters were educated in forestry focusing on timber production and the specific requirements of DWPS in forests are new or sometimes even unknown to them. Also in forestry-schools or at universities DWPS for forest ecosystems is actually not the highest priority. Hence knowledge transfer to stakeholders like forest owners and water suppliers regarding DWPS becomes elementary. Applying different Knowledge Management (KM) techniques can therefore become essential in supporting decision-making. This chapter will document the KM processes for the design, evaluation and implementation of DWPS by highlighting several cases within the thematic context.

Handbook of Knowledge Management for Sustainable Water Systems, First Edition. Edited by Meir Russ.

8.1 Knowledge Management basics in forest ecosystems

When considering the spatial dimension, the time horizon of forest management options, differing stand and site factors as well as multiple goals, the possibilities of assessing the consequences of a management option a priori are fairly limited. This often leads to schematic, sub-optimal solutions. Knowledge Management (KM) techniques can facilitate the identification, creation, and sharing of information in order to improve decision-making (Rauscher *et al.*, 2007; Vacik *et al.*, 2013). KM techniques allow managing both explicit and tacit knowledge and allow increasing performance (Plunkett 2001; Heinrichs *et al.*, 2003). Explicit knowledge is knowledge that has been codified in some way, such as scientific journal articles, operating procedures or databases. Tacit knowledge in contrast refers to the knowledge that people carry in their minds. It consists of subjective opinions, intuition, feelings, understanding, or judgments. People often are not explicitly aware of their own knowledge stores ("we know more than we know how to say", Polanyi, 1958). While some methods help decision-makers to exchange knowledge, others make existing explicit knowledge more readily accessible (Hansen *et al.*, 1999). However, KM also concentrates on methods that help to codify tacit knowledge so that it can be converted to explicit knowledge for general use (Heinrichs *et al.*, 2003). Such explicit and implicit knowledge regarding the provision of sustainable drinking water includes facts, propositions, laws and theories that provide general knowledge about the behavior and functioning of ecosystems. It also includes knowledge about events at specific times and places as well as knowledge about how to do things. Plunkett (2001, p. 5) observed that, "Knowledge management consists of three fundamental components: people, processes, and the supporting technology." The scientific literature demonstrates quite impressively how we have progressed in our technical ability to support the management towards the goals of individuals and organizations. However, KM tools do not manage knowledge by themselves, but rather facilitate the implementation of knowledge processes (Vacik *et al.*, 2013). In Figure 8.1 an illustrative example for the purpose of the different KM tools is given. While knowledge maps or Wikis allow a structured identification and generation of knowledge, databases support the storage of explicit knowledge and web-portals allow its transfer. However, it is much less clear that decision-makers have advanced equally far in changing institutional processes to use KM tools efficiently. Additionally it is evident that decision-makers might have different needs for data, information or knowledge depending on their role within an organization. Depending on the decision-making situation and user perspective the selection and application of the right tools and methods will differ (Vacik *et al.*, 2014). It is also clear that people in organizations are participating regularly and effectively as contributors as well as consumers of knowledge, which has to be considered in designing KM strategies for an organization. It is challenging for organizations to properly value and use both tacit and explicit knowledge in the management for the provision of sustainable

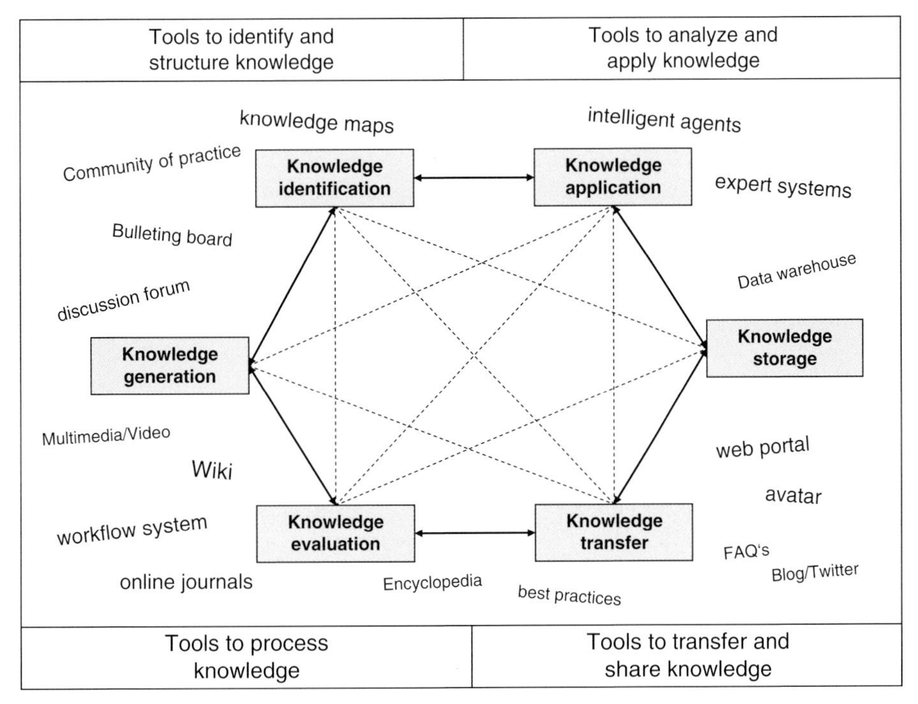

Figure 8.1 Classification of Knowledge Management (KM) tools and techniques according to their potential to support knowledge management processes (adapted from Vacik *et al.*, 2013).

drinking water while promoting a climate of learning that encourages the recognition and sharing of employees' experiences (Boiral, 2002). In this chapter we aim to describe the current challenges of drinking water supply and how different knowledge management techniques can support the forest management of mountainous catchment areas within this context. Examples will be provided on the basic processes for KM: identification, generation, application, transfer, learning and evaluation of knowledge.

8.2 Identify and generate knowledge about DWPS in forested catchments

8.2.1 General outline for knowledge generation

Descriptive KM tools (e.g. tools that support identification and generation of knowledge) in natural resource management focus on the management of data, information, and knowledge. The focus here is on what we know (Thomson *et al.*, 2007) to create a shared, explicit, and accessible understanding of concepts, ideas and relationships. That enables effective communication and understanding of a common societal knowledge base (Heinrichs *et al.*, 2003). It is important that all stakeholders are able to agree on the general requirements of sustainable drinking water protection strategies and have a common descriptive set of knowledge. Such a common understanding of the descriptive, factual knowledge provides a sound basis for reasonable disagreement and discussion concerning interpretations, courses of action, and values. In order to generate and identify knowledge about DWPS in forested mountainous catchments, forest-hydrological processes have to be integrated and studied. For this concern it is important to provide an overview about how water quality, water quantity and water supply systems can be influenced by forest management practices. Literature, expert knowledge and empiric research are seen as a basis for sound recommendations and providing a solid knowledge base. The needs with regard to the processes of identifying, generating and evaluating knowledge are described and the appropriate forest management practices with regard to DWPS are discussed. In this context various tools have been described in scientific literature. Cognitive (or mind) maps are used to generate, visualize, structure, and classify ideas, and are a tool for studying and organizing information and structuring problems. They support a graphical representation of the knowledge mapping procedure used to represent words, ideas, tasks, or other items linked to, and arranged around, a central concept (Tikkanen *et al.*, 2006; Wolfslehner & Vacik, 2011). Ontologies are another important tool for the identification and generation of knowledge in a common domain (Magagna *et al.*, 2006). Within the field of computer- and information sciences ontologies are defined as a formal and explicit specification of a common conceptualization (Gruber, 1993). Ontologies allow the representation of entities, ideas and concepts, along with their properties and relations, according to a system of categories (domain). Especially in the field of DWPS for forest ecosystems it is essential to have a common understanding about terms, objects, functions and processes in order to evaluate the performance of different management strategies, as it can be described as rather new area of research and management applications. Magagna *et al.* (2006) have developed such a concept, where the ontology was interlinked with the application of the Forest Hydrotope Model (FoHyM) for supporting water protection (Koeck *et al.*, 2007).

8.2.2 General knowledge base – the water protection functionality of forest ecosystems

Identification and generation of knowledge on DWPS forms the knowledge base, which will be further explained in this chapter. Forest ecosystems in mountainous areas provide the ecosystem service "water regulation", which contributes to both drinking water protection and flood mitigation or prevention. It is necessary to define, which processes contribute to the water protection functionality of forest ecosystems in order to establish a common understanding. In order to generate the knowledge base the following techniques were used:

A) Literature studies were conducted for defining the state-of-the-art in the field of forest hydrology.
B) Field experiments were implemented to describe the hydrological behavior of different forest stands at comparable sites; done by the set-up of long-term experimental plots in the karstic alpine water protected zone of the city of Vienna in Austria (Koeck, 2008).
C) Integration of the knowledge and discussion with experts in the field of forest hydrology; done in the course of various project activities.

This knowledge generation process resulted in the documentation of the main aspects of the water protection functionality of forest ecosystems (WPF), which is outlined here in concise form:

A) **Adequate infiltration conditions for precipitation water into the soil matrix**
For water protection it is of crucial importance that the precipitation water infiltrates into the soil matrix instead of creating surface flow. Forest soils provide good infiltration conditions for precipitation water, what fosters water storage in soils and also the creation of deep percolation or lateral flow. The forest function for enhancing the infiltration of precipitation water can be explained with the formation of typical forest soils, which provide high macro-pore content created by the roots of both trees and soil vegetation and also by the activities of the soil fauna. In case of forest soils infiltration rates between 6 mm and 206 mm/hour were reported (Harden & Scruggs, 2003). As extreme values, infiltration rates of up to 650 mm/hour were measured in European forests (Eichhorn, 1993). During the summer period the shadowing effect of forest cover on the forest soils provides lower soil temperatures in the upper soil horizons in comparison to open grassland or bare areas (Koeck *et al.*, 2014; Koeck, 2008; Kang *et al.*, 2000). This reduces the tendency for the creation of water-repellent upper soils (hydrophobia of the upper soils) and hence supports better infiltration conditions.
B) **Water storage and retention**
Forests have the capacity of water storage both in the interception storage of the trees and the soil storage. Interception storage takes place on the trunks, branches, leaves and needles of the trees (Cantu-Silva & Okumura, 1996; Carlyle-Moses *et al.*, 2004; Savenije, 2004) and varies within European forests according to Baumgartner and Liebscher (1990) between 0.2 mm and 7.6 mm. The second place of water storage is the forest soil, whose capacity is significantly higher than of the interception storage. The storage capacity of the forest soils is dependent on geology, ecto-humus layers, soil type, soil depth, soil compartments, soil structure etc. The soil storage is an important buffer for balancing the spring or brook discharge. As part of the soil storage, the capacity of ecto-humus layers can sum up to four-times the dry-weight of the ecto-humus substances (Hager & Holzmann, 1997). The mineral soil layers can store water up to 50% of their volume, the measure is called volumetric soil moisture content (cm^3/cm^3 – Ø) and the significance of this storage becomes apparent together with the high infiltration capacity of forest soils.
C) **Snow storage**
For mountainous regions in many areas of the planet a specific form of water storage is given as the snow storage capacity of forest ecosystems. If the forest stands are adequately structured, snow can be trapped by the rough forest canopy and

with its shadowing effect the snow cover can be conserved longer in comparison to open areas. Forest structure is of importance for snow storage as small gaps within forest stands provide the snow trapping effect during strong wind periods, while on huge clear-cut areas or on grassland sites the snow is blown-off (Mayer *et al.*, 1997; Koeck, 2008). Snow storage can contribute to a prolongation of the snow ablation period, what contributes for example, to a more balanced groundwater recharge.

D) **Stabilization of soil- and humus formations**
For providing optimal infiltration conditions for precipitation water (A) and good water storage functionality (B) over space and time, the forest soils together with their humus formations have to be stable. Forest vegetation has the capacity to stabilize the forest soil- and humus layers. The stabilization is given through the dense root network of trees and soil vegetation. The most important aspect is the mechanical stabilization of the soil layers against the gravitational forces effective on steep slopes on mountainous forest sites.

E) **Prevention or mitigation of erosion processes**
Steep slopes of mountains are prone to erosion processes like rock-fall, land-slides or snow avalanches. These erosion processes could jeopardize human traffic routes, settlements and water supply facilities and also could cause contaminations of source waters. Hence these erosion processes have to be prevented or mitigated. A stable and dense forest cover provides in many cases the best protection against these erosive forces. This highlights the prior significance of stable forest ecosystems in steep mountainous terrain all over the world. The protection functionality of the forests against erosive processes gains additional relevance, if the related area is used as watershed for drinking water supply.

F) **Filtration of the precipitation water**
Forest soil- and humus layers have the capacity for filtrating potential contaminants of the precipitation water. This functionality is related to the adsorption of those substances to soil and humus compartments. One part of the precipitation water may always leach directly to the aquifers via preferential flow paths in forest soils; hence pollutants may bypass soil areas where the adsorption of contaminants can occur (Keesstra *et al.*, 2012).

Summarizing, it can be stated that forest ecosystems with a high WPF have to be stable and resilient in order to provide their ecosystem service "water regulation" continuously over space and time. How KM tools can be applied to reach this target is explained in the next section.

8.3 Application of the knowledge-base

8.3.1 The Forest Hydrotope Model – the specific knowledge level

Water protected areas often show major spatial extensions. In mountainous regions there also can be given a high gradient at the level of site conditions.

To establish stable and resilient forest ecosystems with a high level of water protection functionality (WPF), stratification is required within mountainous catchment areas. Forest and water managers need to gain spatial explicit knowledge about their watersheds in order to apply the adequate silvicultural concepts for improving or maintaining the WPF of the forest ecosystems. The Forest Hydrotope Model (FoHyM, Koeck *et al.*, 2007) is a KM tool which provides this spatial explicit stratification hence constituting the specific knowledge level for water protected areas.

FoHyM aggregates knowledge about forest sites into the forest hydrotope type. It is a field survey based model, the data were generated through forest site mapping and geological surveys. Data about geology and soils, terrain information and plant cover information are synthesized into the forest hydrotope type (FHT), which is defined according to the natural forest community (see Table 8.1). Due to the detailed information base the site specific differences are distilled into the FHT, reflecting differences at the level of potential tree species diversity and distribution or applicability of silvicultural concepts and measures.

In the course of several projects FoHyM was applied within the water protected areas of the cities of Vienna (Koeck *et al.*, 2007; Koeck *et al.*, 2011) and Waidhofen/Ybbs (Koeck & Hochbichler, 2012). The forest target definitions for each FHT were elaborated on an ontological base, hence providing a homogeneous frame improving readability and comprehensibility for the FoHyM users (Magagna *et al.*, 2006). The most important information for each FHT is given by the tree species recommendations based on the natural forest community. This ensures the highest level of stability for the forest stands created accordingly. Stability of forest stands through close to nature diversity of tree species is an important issue, as in the past the establishment of homogeneous conifer plantations in Austria, above all Norway spruce (*Picea abies*) was also within the project areas the prevalent forest management focus.

With the establishment of the decreed water protected areas in both municipal watersheds, in the case of Vienna already implemented in the years 1965 and 1973, in the case of Waidhofen/Ybbs work in progress, the overall purpose

Table 8.1 Information classes of the Forest Hydrotope Model (FoHyM)

	Geology and soils	Terrain information	Plant cover information
Forest Hydrotope Type (FHT)	Bedrock type Soil type Soil depth Humus form Humus depth	Site exposition Site inclination Elevation above sea level Water regime class	Soil vegetation type Soil vegetation cover Tree species distribution Crown cover percentage

(Adapted from Koeck *et al.*, 2007).

of integral drinking source water protection and the definition of DWPS became the focus of forest and water management. Hence the demand for a knowledge management tool emerged, which provides for the foresters and water suppliers the capability to know, which tree species distribution should be targeted at the specific forest sites in order to achieve the optimal degree of stability and resiliency for the forest stands as basic condition for an improved WPF.

In the case of the city of Vienna 33.400 ha of the karstic alpine water protected area were mapped (e.g. Mrkvicka, 1996; Koeck *et al.*, 2000). FoHyM was elaborated in the course of several projects for contributing to the integral safeguarding process for the water supply system of the city of Vienna. FoHyM provides now as spatial explicit model the forest target definitions for the whole area, covering more than 60 different FHT, for which tree species diversity and distribution and the related recommended silvicultural regeneration concepts and measures were defined. In the case of Waidhofen/Ybbs 1000 ha were mapped (Figure 8.2), FoHyM encompasses nine different FHT (Koeck & Hochbichler, 2012). The different amount of FHT in the case of Vienna was caused by the higher forest site gradient, where the protected zone ranges from

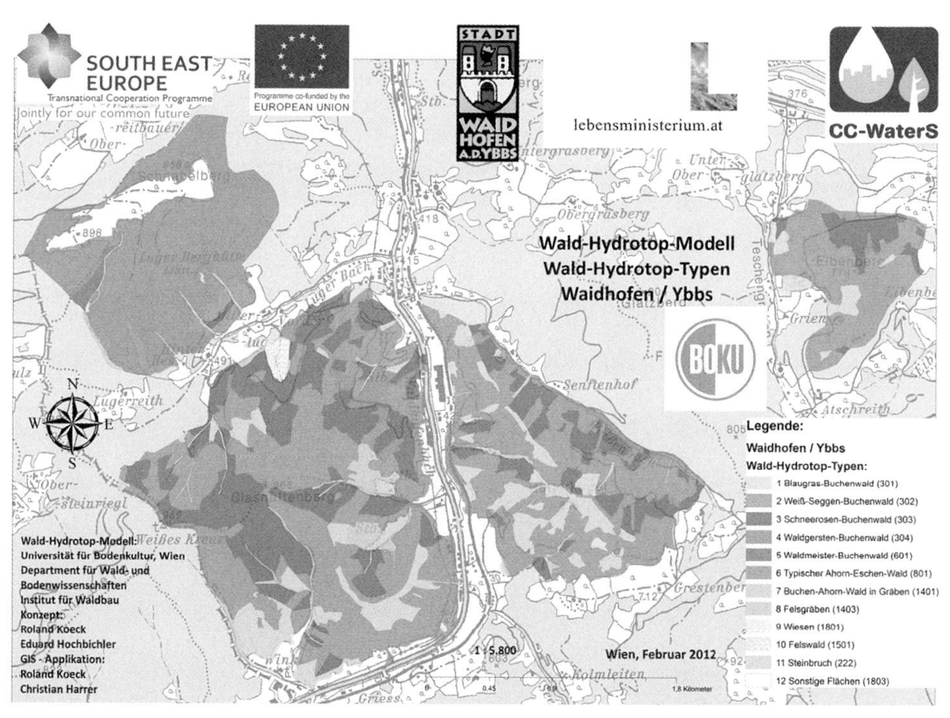

Figure 8.2 Forest Hydrotope Map of Waidhofen/Ybbs, displaying the forest hydrotope types (FHT) within the water protected area. The map was elaborated in German for being useable for the stakeholders. *Source*: Koeck & Hochbichler (2012).

506 m to 2277 m ASL and covers a huge area, in the case of Waidhofen/Ybbs the protected zone ranges from 420 m to 969 m ASL and is significantly smaller.

The foresters and water suppliers have now the GIS-based map (Figure 8.2) and a related detailed report describing FoHyM in their hands, which provide them all necessary information for the application of a forest site-specific adaptive forest management. The tree species distribution was elaborated in the course of the CC-WaterS project (CC-WaterS 2012) for both actual climate and for a climate change scenario, defining the year 2100 as orientation mark. The overall purpose of the FoHyM-definitions is the optimization of the WPF of the forest ecosystems within the water protected areas. FoHyM is a means to support forest managers in strategic and operative decision-making and in monitoring the progress towards the achievement of the desired forest structures in the highly diverse mountain forest area and it covers the specific knowledge level, as all forest-site specific data were translated into forest target definitions as part of the DWPS.

8.3.2 Best Practices – the general knowledge level

In order to keep the WPF of forest ecosystems on an optimal level, the silvicultural concepts and measures have to be adapted to the overall purpose of drinking water protection. As second part of the DWPS and for facilitating an efficient FoHyM application, the general frame had to be elaborated, which is the Best Practice Catalogue for forest management in drinking water protected areas. This became necessary as the implementation of DWPS needs general guidelines for silviculture, which take the specific requirements of water protection into account. Catalogues of Best Practices were already defined as tool to transfer and share knowledge. In the course of the CC-WARE project (CC-WARE 2014) a Best Practice Catalogue (BPC) was elaborated for mountainous forest ecosystems in the south-eastern European region. The BPC as KM tool has the great advantage that the most important recommendations for forest management with the overall purpose of drinking water protection are defined and listed within an operational manual, providing a readable overview for all stakeholders.

The BPC (Koeck & Hochbichler, 2014) was developed in a very detailed form including scientific explanations about the related processes. Here is given a concise summary in order to define and explain the most important parts of the BPC within the context of the DWPS. The discrete Best Practices (BP) are written in italics.

Basic Best Practices

Prevention of clear-cuts is recommended in order to avoid the related negative effects like increased nitrification, mineralization and humus decomposition (Likens & Bormann, 1995; Prescott, 1997) which potentially endanger source water quality. In contrast the *establishment of continuous cover forest systems*

(Thomasius, 1996) is envisaged, which provide the forest functions over space and time and rely on multi-layered, uneven-aged forest stands. For keeping the forest stands stable and resilient towards strong wind-storms (Mosandl & El-Kateb, 1988) the *crown cover percentage of the forest stands* should be given within defined margins, in the montane zone between 70–90% and in the subalpine zone between 60–80%, hence also allowing continuous natural regeneration dynamics. For reaching the above defined goals the *timber extraction percentage should be limited with 10–25% of the stand volume*, what contributes to sustaining the stability of the forest stands and additionally avoids severe mineralization processes (Wang *et al.*, 2006).

The *tree species diversity has to be adapted to the natural forest community* at the specific forest sites in order to achieve the highest level of forest ecosystem stability and resiliency. This BP provides the link to the specific knowledge level given through FoHyM. *Continuous regeneration dynamics* are an additional prerequisite for continuous cover forestry and also contribute to a high level of resilience. The only way to guarantee this is the provision of a *forest ecologically sustainable wild ungulate density*. Unnaturally high densities of wild ungulates are a major threat for the regeneration process of forest trees in Austria and other European countries. The implementation of those BP allows the foresters to *foster stability, vitality and resilience of the forest ecosystems*, what has to be pursued at the level of forest ecosystems, stands and tree individuals. The *improvement of the structural diversity of the forest stands* can also be regarded as basic condition for continuous cover forestry and encompasses the site-specific tree species diversity, uneven-aged trees and multi-layered forests stands. The so created structural diversity of the forest stands should be given both horizontally and vertically and improves forest stand stability (Richards *et al.*, 2012; Schütz *et al.*, 2006). *Protection of the gene-pool of autochthonous tree species* is a crucial management target in times of climate change, because they carry the genetic information, which allowed them the survival of past climate changes in those areas. Also to *foster old, huge and vital tree individuals* is of importance for DWPS, as they can supply younger and smaller tree individuals with nutrients via their common mykorrhizal network (Simard *et al.*, 2002). *Buffer strips along streams, dolines and sinkholes* can protect those from direct infiltration of sediments or nutrient loads (Wallbrink *et al.*, 2002; Boyer 2005).

Best Practices in the field of silvicultural techniques

Also the silvicultural techniques have to be adapted to DWPS. Above all the *small-scale regeneration techniques* have to be mentioned, which can encompass group selection cuts, single tree cuts or small-scale gap cuts. Clear-cuts and also the regular shelter wood cut system should be avoided, as the latter would also involve a clear cut phase as result of its final cut. *Structural thinning operations* follow the purpose of the stabilization of the forest stand and the creation of continuous cover forests. An essential aim of the concept of structural thinning is that the distribution of tree diameters becomes wider (Reininger, 1987). *Artificial recruitment techniques* should be applied if the natural regeneration process of the forest stands does not conform to the defined targets of the related forest hydrotope type (FHT). The *construction of forest roads has to be limited,* as it has to be considered that 4–5 ha of hydrological optimized forests are necessary to compensate surface runoff increase and storage loss of 1 ha forest area used for road construction (Markart *et al.*, 2011). Only if forest roads are necessary for the stabilization of forest areas, their construction could be taken into consideration. As central part for DWPS *adequate timber yield techniques* have to be applied, in the case of mountainous forest areas the application of the cable-crane system is recommended, being the only system which protects the fragile soil- and humus layers. Tractor-skidding should be avoided as far as possible what is due to the soil compaction created by the heavy machinery. The *use and application of chemicals like fertilizers or pesticides should be prohibited*, as their potential entrance in the aquifers endangers source water quality.

Synthesizing planning strategies as Best Practices

The spatial extension of water protected areas implies the necessity of an *integral planning strategy (IPS) for watersheds*, which should encompass all activities in relation to source water protection. Part of IPS would have to be the spatial and temporal dimension of the implementation of silvicultural concepts and measures or the integration of all relevant potential impacts on source water protection. Example for a planning tool which can form part of an IPS on silvicultural level is the application of the spatial explicit Forest Hydrotope Model.

Forest fire prevention is an aspect of synthesizing planning strategies and is of vital interest for source water protection, especially in mountainous regions prone to forest fires, because the spatial extension of forest fires could form a major threat for the WPF of forest ecosystems. Fire ignition is related to the fuel characteristics, the weather conditions, topography and most often human activities, causing a forest fire. The mountain forests in Austria have not been seriously fire-impacted in the past. However, several summer seasons (2003, 2007) have most recently demonstrated how widespread and rapidly forest fires can happen (Vacik *et al.*, 2011). They might become an important issue in the case of the occurrence of certain weather extremes such as a prolonged vegetation period or during periods of drought or heat waves (Müller *et al.*, 2015). It is therefore likely that under climate change the fire danger will increase within Austrian mountain forests (Arpaci *et al.*, 2014). In that context the negative impacts of forest fires on hydrology through the reduced infiltration of precipitation water after the fire event as well as the loss of the humus and vegetation layer have to be highlighted. As a consequence of intensive forest fires the increased soil erosion and probability of mud slides on burned areas have to be considered as well (Papathanasiou *et al.*, 2012, Moody *et al.*, 2013).

8.4 Decision Support System – specific examples

In general terms Decision Support Systems (DSS) are computer-based systems which utilize available knowledge, i.e. facts, expert rules and models which have been found useful in solving specific problems. As DSS are based on formalized knowledge their application in the decision-making process facilitates decisions that are reproducible and as rational as possible. Moreover through the use of DSS the way the decision-maker arrived at a decision is automatically documented and thus the process of decision-making can be evaluated. DSS have been proven to solve ill-structured decision problems by integrating database management systems with analytical and operational research models, thus providing various reporting capabilities. The latest compilation of forest management decision support systems is found in the special issues of Borges & Eriksson (2014) and Vacik *et al.* (2015).

Examples for the application of spatial decision support system (SDSS)

(1) To support silvicultural planning at the stand level in the protection forests of the City of Vienna a spatial decision support system (SDSS) was developed. The SDSS was designed to support forest management for the sustained provision of quality drinking water, timber production, protection against rock fall and avalanches as well as recreation and biodiversity values. The core structure of the SDSS comprised four main components: the information and knowledge base, the tool box, the SDSS generator where the decision model is embedded and a graphical user interface (Vacik & Lexer, 2000). The SDSS supports decision-making about the time when to begin the natural regeneration process, the choice of the future tree species composition (growing stock objective) and selection of the regeneration method. The codification of the relevant knowledge regarding the suitability of tree species and the appropriateness of regeneration methods to achieve the proposed management goals was a prerequisite for the modeling. The application of the SDSS (Figure 8.3) allowed retrieval of expert knowledge and making it available to non-experts.

(2) The Forest Hydrotope Model (FoHyM, Koeck *et al.*, 2007) was introduced as a means to support forest managers in strategic and operative decision-making and in monitoring the progress towards the achievement of the desired forest structures in the highly diverse mountain forest areas. The GIS-based spatial explicit model was elaborated for the drinking water protected areas of the city of Vienna and the city of Waidhofen/Ybbs as landscape-scale spatial decision support system (SDSS). The application of the forest target model is taking place through the use of the FoHyM-map together with the related report, which is at the hands of the forest and water managers in both municipalities. Hence for each site the related forest target definitions are available for the managers. The overall purpose is the selection of an appropriate management system and to assure the water protection functionality of the forest ecosystems in the related watersheds.

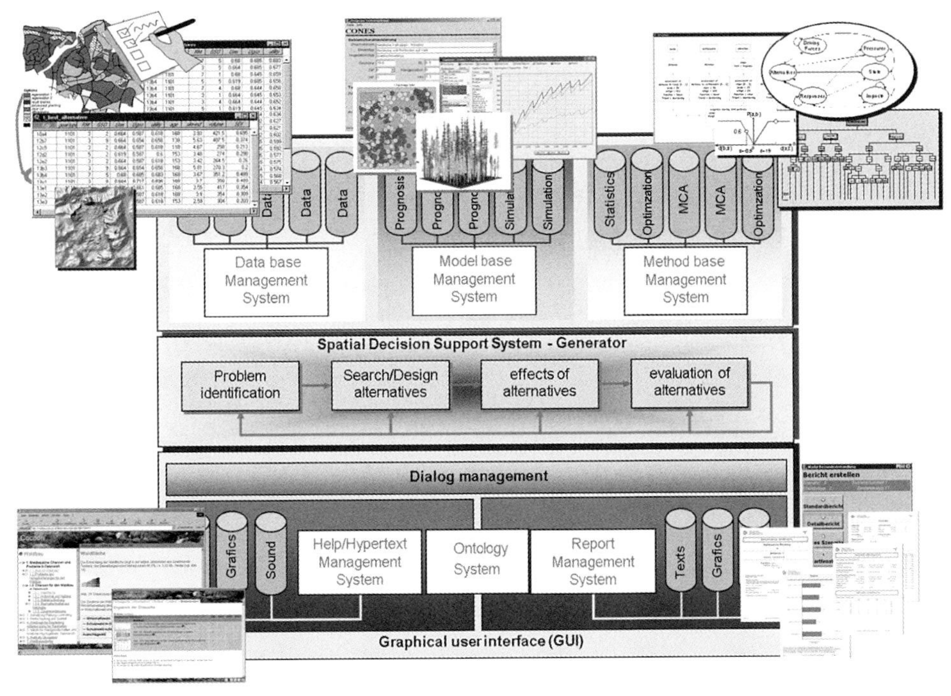

Figure 8.3 Core components of the Spatial Decision Support System (adapted from Vacik *et al.*, 2010).

8.5 Knowledge transfer to stakeholders

The most difficult step in achieving the application of drinking water protection strategies (DWPS) in forested catchment areas is the knowledge transfer to stakeholders. Water suppliers, forest owners and forest managers have to know about the challenges and requirements of DWPS. Predictive KM tools focus on the management of procedural knowledge. They allow prescribing how activities occur, how things are changing in the real world, how specific problems are solved, and how the results of alternative courses of action can be predicted (Heinrichs *et al.*, 2003). The organization, transfer, and sharing of such pieces of procedural knowledge (e.g. focusing on forest-management practices) can be made possible via web sites, web portals, online encyclopedias, Wikis, or communities of practice (Vacik *et al.*, 2013). Another possibility is the direct stakeholder involvement which leads towards a participative process in knowledge transfer. Forest management related to the protection of drinking water resources differs from a solely timber production oriented management. Within water protected areas the management has to follow the principles outlined in the Best Practices section. The challenges and opportunities related to the outcomes of national and international projects dealing with knowledge transfer processes in the context of integral drinking water protection are outlined and explored here.

8.5.1 Participative stakeholder workshops and panel discussions

As a means to inform water suppliers, forest owners and staff from governmental organizations, representatives of those groups were invited to participate in stakeholder workshops. Those were taking place on both local and also on transnational level (CC-WARE 2014). Forest scientists and hydrologists were sharing their knowledge by providing short presentations. The audience was invited to participate through providing feedback on their specific point of view (Figure 8.4). At the panel discussions all participants were confronted with the thematic field of drinking water protection. The workshops and discussions stimulated the consideration of DWPS for all participants, to take a first step towards the implementation.

Figure 8.4 First knowledge transfer workshop for stakeholders in the CC-WARE project in Modena, Italy. *Source*: CC-WARE (2014), meetings and events – photo-gallery.

8.5.2 Field excursions to representative forest stands

For deepening the experiences in relation to the requirements of DWPS, field excursions in forest ecosystems within water protected areas were carried out. Representative forest stands were selected for the analysis and discussion, the respective level of water protection functionality (WPF) was explained by the experts. Possibilities how to improve WPF were jointly discussed among foresters and water suppliers. Especially foresters prefer excursions, where they can discuss in the midst of "the field-lab" about new concepts and ideas, but also water suppliers appreciated the direct experience provided (CC-WaterS, 2012; CC-WARE, 2014).

8.5.3 Application of Best Practices in a pilot case

In the course of the CC-WaterS (2012) project, silvicultural trials were implemented in a pilot case within the water protected area of Waidhofen/Ybbs. Gap-cuts for the induction of natural regeneration were planned and carried out by forest experts in cooperation with the forest owners and forestry workers. The experts facilitated the whole working process, which made the knowledge transfer process very efficient. It was possible to identify the thematic areas where foresters need special support and which strategies could fit best for tackling the challenge of DWPS applications in the future.

As part of the knowledge transfer direct communication towards the forest owners were done. The forest owners were formally informed by the municipality and their homes were visited by water supplier representatives and an expert in the field of DWPS. Aspects of the requirements of DWPS were explained and possible compensation payments (payments for ecosystem services – PES) were highlighted. Through the direct communication it was also possible to discuss with people, who would not be the first speaker in the format of a workshop. It allowed them to express their fears. For most of the forest owners the prospect of PES was motivating enough to accept the requirements of DWPS and for subsequently adapting their silvicultural practices accordingly.

8.5.4 Handbook "Soil Functions for the Water Sector"

For all water suppliers in Austria the handbook "Soil Functions for the Water Sector" (Koeck & Hochbichler, 2016) was published for providing basic knowledge in the field of soil sciences, land use types and hydrology. Two chapters deal with forest ecosystems, as their WPF is strongly interlinked with the status of the forest soils. The WPF of forest ecosystems and the interrelation with soil- and humus layers was explained in precise and concise way. Possibilities to improve WPF through the application of Best Practices (BP) were highlighted. For this purpose the catalogue of BP was explained in a very short and easy to understand version, so that the handbook becomes useable for the target group. The handbook is available for all people, especially for those dealing with the water sector in Austria, hence also for water suppliers, who can use it as additional reference.

8.5.5 *Evaluation*

Despite the multitude of activities already carried out in national and international projects, knowledge transfer to stakeholders remains a challenge which continuously has to be pursued under changing environmental and social conditions in order to guarantee sustainable DWPS. In the course of previous projects the creation, organization and formalization of knowledge was fulfilled in appropriate form, however there still exists the demand of further endeavors in the field of distribution, application and evolution of knowledge, by the way completing the life-cycle of knowledge-based systems (Russ *et al.*, 2008). The evaluation of the available knowledge and of the efficiency of the knowledge transfer strategies is therefore crucial. Quite recently a project funded by the EU was started, where "Efficient Practices of Land Use Management Integrating Water Resources Protection and Non-structural Flood Mitigation Experiences" (PROLINE_CE 2016) will be implemented. One central focus of this project is the evaluation of the processes of providing efficient application and transfer of knowledge. The evaluation of the knowledge base requires the consideration of the latest scientific findings. But evaluation also demands stakeholder feedback, as the success of the implemented concepts and measures depends on their expertise. For achieving this task an intensive stakeholder involvement is planned, focusing on the learning processes within the context of knowledge management. Stakeholders can also provide their feedback about the comprehensibility and clarity of knowledge transfer strategies and hence can help to improve those. Based on the feedback it will be possible to improve the mutual learning process for the engaged scientists and stakeholders. The concept of DWPS will be revised in a transformation process towards a more comprehensible version. It is also planned to support the knowledge transfer and learning processes with an online encyclopedia to meet the demands of the stakeholders.

8.6 Synthesis and lessons learned

Knowledge about drinking water protection strategies (DWPS) can be regarded as crucial, as a safe drinking water supply is a relevant condition for human development. Despite this fact most of today's foresters were educated in timber production forestry, and the managers are often not aware about the specific requirements of DWPS in forest ecosystems. In this field, Knowledge Management becomes a necessity and the processes of identification, generation, application and transfer of knowledge allow facilitating the use of DWPS.

Within the context of knowledge identification and generation it became obvious that the concept of ontologies supports the achievement of a common understanding about terms, objects, functions and processes. In the field of water protection this is necessary in order to evaluate the performance of different management strategies. Fundamental for a common understanding is the knowledge about the water protection functionality (WPF) of forest ecosystems. Forest ecosystems with a high WPF have to be stable and resilient in order to provide their ecosystem service "water regulation" continuously over space and time. Stable forest ecosystems provide good infiltration conditions for precipitation water, they store water in the soil-, humus- and interception-storage, stabilize soil- and humus layers, prevent or mitigate erosion processes and filter the precipitation water.

The Forest Hydrotope Model (FoHyM) provides a spatially explicit stratification of forest site conditions for the whole water protected area. It is a field survey based model, data about geology and soils, terrain information and plant cover information are synthesized into the forest hydrotope type (FHT), which is defined according to the natural forest community. The forest target definitions for each FHT are elaborated on an ontological base. Definitions regarding the tree species diversity guarantee the highest level of forest stability on each site. In order to keep the water protection functionality of forest ecosystems on a high level, the silvicultural concepts and measures have to be adapted to the overall purpose of drinking water protection. The application of the knowledge base can be provided by the definition of the Best Practice Catalogue (BPC), which was elaborated in the course of the CC-WARE project (CC-WARE 2014) for mountainous forest ecosystems in the south-eastern European region. The BPC lists the most relevant recommendations for forest management with the overall purpose of drinking water protection in an operational manual, providing a comprehensive overview for all stakeholders.

As another example for the process of knowledge application, two decision support systems (DSS) were introduced, both applied in the same water protected area. As the DSS are based on formalized knowledge their application in the decision-making process facilitates decisions that are reproducible and as rational as possible. This is often a requirement in the context of the implementation of drinking water protection strategies to safeguard the water quality for the consumers.

The process of knowledge transfer to stakeholders was identified as the most difficult step in achieving the application of DWPS in forested catchment areas. Both predictive KM tools and participative processes in knowledge transfer were mentioned. Amongst the latter various different tools were highlighted, like for example, participative stakeholder workshops and panel discussions, field excursions to representative forest stands, the application of Best Practices in a pilot case, direct conversations with forest owners and the creation of a Handbook "Soil Functions for the Water Sector".

The evaluation of knowledge and within this context especially the knowledge base and the knowledge transfer process are essential for the ongoing work. Through direct stakeholder involvement the evaluation can help to meet the end-users demands and hence facilitates the improvement of learning processes. It can be observed that different tools are used to identify, structure, create, and share knowledge assets, but it is still difficult to locate the relevant information to support collaboration in forest planning with respect to problem identification, problem modeling, and problem solving. The need for further work facilitating the process of KM and knowledge transfer in the field of DWPS can be identified as it requires the selection and application of the right tools and methods for different decision-making situations and user perspectives.

References

Arpaci, A., Malowerschnig, B., Sass, O., & Vacik, H. (2014). Using multi variate data mining techniques for estimating fire susceptibility of Tyrolean forests. *Applied Geography*, *53*, 258–270.

Baumgartner, A., & Liebscher, H.J. (1990). *Allgemeine Hydrologie – Quantitative Hydrologie*. Verlag Gebrüder Borntraeger, Berlin, Stuttgart.

Boiral, O. (2002). Tacit knowledge and environmental management. *Long Range Planning*, *35*, 291–317.

Borges, J.G., & Eriksson, L.O. (2014). Foreword: Special Issue: Decision support systems for sustainable forest management. *Scandinavian Journal of Forestry Research*, *2014*, 29.

Boyer, D.G. (2005). Water quality improvement program effectiveness for carbonate aquifers in grazed land watersheds. *Journal of American Water Resources Association*, *41*(2), 291–300.

Cantu-Silva, I.C., & Okumura, T. (1996). Throughfall, stemflow and interception loss in a mixed white oak forest (*Quercus serrata* Thunb.). *Journal of Forestry Research*, *1*, 123–129.

Carlyle-Moses, D.E., Flores Laureano, J.S., & Price, A.G. (2004). Throughfall and throughfall spatial variability in Madrean oak forest communities of northeastern Mexico. *Journal of Hydrology*, *297*, 124–135.

CC-WaterS (2012). *Climate Change and impacts on Water Supply*. Interreg-SEE project. www.ccwaters.eu

CC-WARE (2014). *Mitigating Vulnerability of Water Resources under Climate Change*. Interreg-SEE project. www.ccware.eu

Eichhorn, K. (1993). Bodenverdichtung durch Forstmaschinen. *Diplomarbeit am Institut für Forsttechnik*, Universität für Bodenkultur, Wien.

Gruber, T.R. (1993). A Translation Approach to Portable Ontology Specifications. *Knowledge Acquisition*, *6*(2), 199–221.

Hager, H., & Holzmann, H. (1997). *Hydrologische Funktionen ausgewählter naturnaher Waldökosysteme in einem alpinen Flusseinzugsgebiet*. Projektendbericht. Hydrologie Österreichs, Österreichische Akademie der Wissenschaften.

Hansen, M.T., Nohria, H., & Tiernex, T. (1999). What's your strategy for managing knowledge? *Harvard Business Review*, *77*(2).

Harden, C.P., & Scruggs, P.D. (2003). Infiltration on mountain slopes: a comparison of three environments. *Geomorphology*, *55*, 5–24.

Heinrichs, J.H., Hudspeth, L.J., & Lim, J.S. (2003). Knowledge management. In: Hossein Bidgoli (Ed.) *Encyclopedia of Information Systems*. Academic Press, Amsterdam, Volume *3*: 13–31.

Kang, S., Kim, S., Oh, S., & Lee, D. (2000). Predicting spatial and temporal patterns of soil temperature based on topography, surface cover and air temperature. *Forest Ecology and Management*, *136*(1–3), 173–184.

Keesstra, S.D., Geissen, V., Piiranen, S., Scudiero, E., Leistra, M., & van Schaik, L. (2012). Soil as filter for groundwater quality. *Current Opinion in Environmental Sustainability*, *4*, 507–516.

Koeck, R., Mrkvicka, A., & Weidinger, H. (2000). *Bericht zur forstlichen Standortskartierung, Revier Gschöder*. Austrian National Library, Vienna.

Koeck, R., Magagna, B., & Hochbichler, E. (2007). *KATER II Handbook – Final report regarding the land use category forestry*. www.ccwaters.eu/downloads – (KATER II).

Koeck, R. (2008). *Waldhydrologische Aspekte und Waldbaukonzepte in karstalpinen Quellenschutzgebieten in den nördlichen Kalkalpen*. Dissertation an der Universität für Bodenkultur, Wien, Department für Wald- und Bodenwissenschaften, Institut für Waldbau (Wien).

Koeck, R., & Hochbichler, E. (2011). Common methodology for land use category forestry. "The Forest Hydrotope Model (FoHyM) and climate change". In: Celico, F., & Petrella, E. (2011). *CC-WaterS, Work Package 5 – Land Uses and Water Resources Safety: Final Report*. www.ccwaters.eu – Output Documentation.

Koeck, R., & Hochbichler, E. (2012). *Das Wald-Hydrotop-Modell als WSMS-Werkzeug im Quellenschongebiet der Stadt Waidhofen/Ybbs. Report in the course of the CC-WaterS project.* https://www.bmlfuw.gv.at – *search for: "ccwaters".*

Koeck, R., Hochbichler. E., Holtermann, C., & Hager, H. (2014). Soil moisture and soil temperature dynamics at the tree line of Mount Rax, 1999–2010. *Austrian Journal of Forest Science*, *13*(1), 45–62.

Koeck, R., & Hochbichler, E. (2014). *Recommendations for Adaptive Management Concepts – Best Practices for Forest Ecosystems in Mountains and Flatlands*. www.ccware.eu – Output Documentation, WP 4.

Koeck, R., & Hochbichler, E. (2016). Waldbewirtschaftung: Trinkwasserschutz in bewaldeten Einzugsgebieten. In: Klaghofer, E. (Ed.) *ÖWAV – Arbeitsbehelf 47, Bodenfunktionen für die Wasserwirtschaft*, pp. 48–55, Austrian Standards Plus Publishing, Wien.

Likens, G.E., & Bormann, F.H. (1995). *Biogeochemistry of a forested ecosystem*, 2nd Edition. Springer Verlag.

Magagna, B., Kollarits, S., Koeck, R., & Hochbichler, E. (2006). Mehr Transparenz in der Entscheidungsfindung durch Ontologien – das Beispiel Quellenschutz. In: Strobl/Blaschke/Griesebner (Hrsg./Eds.) *Angewandte Geoinformatik 2006, 5.7-7-7.2006*. Salzburg, Wichmann Verlag, Heidelberg, 381-388; ISBN: 978-3-87907-437-2, 3.

Markart, G., Perzl, F., Klebinder, K., & Kohl, B. (2011). *Report of work package 5 of the CC-WaterS project, Interreg SEE*. www.ccwaters.eu – Output Documentation.

Mayer, H., Feger, K.H., Ackermann, B., & Armbruster, M. (1997). Schneedeckenentwicklung in Fichtenwäldern im südlichen Hochschwarzwald. *Forstw. Centralblatt*, *116*, 370–380.

Moody, J.A., Shakesby, R.A., Robichaud, P.R., Cannon, S.H., & Martin, D.A. (2013). Current research issues related to post-wildfire runoff and erosion processes. *Earth-Science Reviews*, *122*, 10–37.

Mosandl, R., & El-Kateb, H. (1988). Die Verjüngung gemischter Bergwälder. Praktische Konsequenzen aus 10jähriger Untersuchungsarbeit, *Vortrag. Forstwissenschaftliches Centralblatt*, *107*(1), 2–13.

Mrkvicka, A. (1996). *Bericht zur forstlichen Standortskartierung im Revier Hirschwang-Schneeberg, Forstverwaltung Hirschwang, NÖ.* Austrian National Library, Vienna.

Müller, M.M., Vacik, H., & Valese, E. (2015). Anomalies of the Austrian forest fire regime in comparison with other Alpine countries: A research note. *Forests*, *6*(4), 903–913.

Papathanasiou, C., Alonistioti, D., Kasella, A., Makropoulos, C., & Mimikou, M. (2012). The impact of forest fires on the vulnerability of peri-urban catchments to flood events (The case of the Eastern Attica region). Special Issue Global NEST. *Journal of Hydrology and Water Resources*, *14*(3), 294–302.

Plunkett, P. (2001). *Managing knowledge @ work: an overview of knowledge management.* Knowledge Management Working Group of the Federal Chief Information Officers Council, U.S General Services Administration, Office of Information Technology, 36 pp.

Polanyi, M. (1958). *Personal Knowledge: Towards a Post-Critical Philosophy*. University of Chicago Press. ISBN 0-226-67288-3.

Prescott, C.E. (1997). Effects of clearcutting and alternative silvicultural systems on rates of decomposition and nitrogen mineralization in a coastal montane coniferous forest. *Forest Ecology and Management*, *95*, 253–260.

PROLINE_CE (2016). Efficient practices of land use management integrating water resources protection and non-structural flood mitigation experiences. *Interreg Central Europe Project.* www.interreg-central.eu/proline-ce

Rauscher, H.M., Schmoldt, D.L., & Vacik, H. (2007). Information and Knowledge Management in Support of Sustainable Forestry: A Review. In: Reynolds, K., Rennolls, K., Köhl, M., Thomson, A., Shannon, M., & Ray, D. (Eds.) *Sustainable Forestry: From Monitoring and Modeling to Knowledge Management and Policy Science*, pp. 439–460. CAB International, Cambridge; ISBN 9781845931742.

Reininger, H. (1987). *Zielstärken-Nutzung oder die Plenterung des Altersklassenwaldes.* Österreichischer Agrarverlag, Wien, 163 S.

Richards, W.H., Koeck, R., Gersonde, R., Kuschnig, G., Fleck, W., & Hochbichler, E. (2012). Landscape-scale forest management in the municipal watersheds of Vienna, Austria, and Seattle, USA: commonalities despite disparate ecology and history. *Natural Areas Journal*, *32*(2), 199–207.

Russ, M., Jones, J.G., & Jones, J.K. (2008). Knowledge-based strategies and systems: a systematic review. In: Lytras, M., Russ, M., Meier, R., & Naeve, A. (Eds.) *Knowledge Management Strategies: A Handbook of Applied Technologies*, pp. 1–62. Hershey PA: IGI Publishing.

Savenije, H.H.G. (2004). The importance of interception and why we should delete the term evapotranspiration from our vocabulary. *Hydrological Processes*, *18*, 1507–1511.

Schütz, J.P., Götz, M., Schmod, W., & Mandallaz, D. (2006). Vulnerability of spruce (*Picea abies*) and beech (*Fagus sylvatica*) forest stands to storms and consequences for silviculture. *European Journal of Forest Research*, *125*, 291–302.

Simard, S.W., Jones, M.D., & Durall, D.M. (2002). Carbon and nutrient fluxes within and between mycorrhizal plants. In: *Mycorrhizal Ecology*. Van der Heijden, M.G.A., Sanders, I.R.: Springer Verlag, Berlin – Heidelberg.

Thomasius, H. (1996). *Geschichte, Theorie und Praxis des Dauerwaldes.* Broschüre des Landesforstverein Sachsen-Anhalt e.V., Arbeitsgemeinschaft Naturgemässe Waldwirtschaft, Bücherdienst Ebrach.

Thomson, A.J., Rauscher, H.M., Schmoldt, D.L., & Vacik, H. (2007). Information and knowledge management for sustainable forestry. In: Reynolds, K., Rennolls, K., Köhl, M., Thomson, A., Shannon, M., & Ray, D. (Eds.) *Sustainable Forestry: From Monitoring and Modeling to KM and Policy Science*, pp. 374–392. CAB International, Cambridge.

Tikkanen, J., Isokaanta, T., Pykalainen, J., & Leskinen, P. (2006). Applying cognitive mapping approach to explore the objective-structure of forest owners in a Northern Finnish case area. *Forest Policy and Economics*, *9*(2), 139–152.

Vacik, H., & Lexer, M.J. (2000). Application of a spatial decision support system in managing the protection forests of Vienna for sustained yield of water resources. *Forest Ecology and Management*, *143*(1–3), 65–76.

Vacik, H., Lexer, M.J., Rammer, W., Seidl, R., Hochbichler, E., Strauss, M., & Schachinger, C. (2010). ClimChAlp – a webbased decision support system to explore adaptation options for silviculture in secondary Norway spruce forests in Austria. In: Falcao, A., Rosset.C (Eds.) *Proceedings of the Workshop on Decision Support Systems in Sustainable Forest Management*, Lisbon, April 2010. OP3 http://www.fc.ul.pt/dsfm2010/docs/presentations/paper_15.pdf

Vacik, H., Arndt, N., Arpaci, A., Koch, V., Mueller, M., & Gossow, H. (2011). Characterization of forest fires in Austria. *Austrian Journal of Forest Science*, *128*(1), 1–31.

Vacik, H., Torresan, C., Hujala, T., Khadka, C., & Reynolds, K. (2013). The role of knowledge management tools in supporting sustainable forest management. *Forest Systems*, *22*(3), 442–455.

Vacik, H., Kurttila, M., Hujala, T., Khadka, C., Haara, A., Pykalainen, J., Honkakoski, P., Wolfslehner, B., & Tikkanen, J. (2014). Evaluating collaborative planning methods supporting programme-based planning in natural resource management. *Journal of Environmental Management*, *144*, 304–315.

Vacik, H., Borges, J.G., Garcia-Gonzalo, J., & Eriksson, L.O. (2015). Decision support for the provision of ecosystem services under climate change: An Editorial. *Forests*, *6*(9), 3212–3217.

Wallbrink, P.J., & Croke, J. (2002). A combined rainfall simulator and tracer approach to assess the role of Best Management Practices in minimising sediment redistribution and loss in forests after harvesting. *Forest Ecology and Management*, *170*, 217–232.

Wang, X., Burns, D.A., Yanai, R.D., Briggs, R.D., & Germain, R.H. (2006). Changes in stream chemistry and nutrient export following a partial harvest in the Catskill Mountains, New York, USA. *Forest Ecology and Management*, *223*(1–3), 103–112.

Wolfslehner, B., & Vacik, H. (2011). Mapping Indicator models. From intuitive problem structuring to quantified decision-making in sustainable forest management. *Ecological Indicators*, *11*(2), 274–283.

9 Knowledge Management, openness, and transparency in sustainable water systems: The case of Eau Méditerranée

Chris Kimble[1] and Isabelle Bourdon[2]

[1] *KEDGE Business School, 13009 Marseille, France*
[2] *University of Montpelier, 34090 Montpellier, France*

Introduction

The water crisis in Flint, Michigan, in the United States, began in April 2014 when the city of Flint changed its water source from treated water to water taken from the Flint River. The water from the river was highly corrosive and was not treated properly; this caused lead to leach from the pipes that supplied water to people's homes. Local officials claimed that the water had been tested and the levels of lead were acceptable, but it later emerged that they did not know where the samples had been collected. In September 2015, an independent survey found dangerously high lead levels in samples of tap water taken from people's homes. It was found that the city had disregarded federal rules requiring it to seek out homes with lead plumbing for testing, potentially leading the city and state to underestimate the extent of toxic lead leaching into Flint's tap water (Fonger, 2015). Criminal charges were filed against a number of people and the city eventually returned to using treated water from the Detroit Water and Sewerage Department.

Problems such as those faced by the town of Flint illustrate the importance of openness and transparency when it comes to dealing with the data relating to

Handbook of Knowledge Management for Sustainable Water Systems, First Edition. Edited by Meir Russ.

water treatment systems. In this chapter, we will examine a case study of how a company based in the south of France moved from annual reports based on what had happened in the past to a more open, comprehensive and continuous form of reporting based on an information portal. We analyze their approach to Knowledge Management from three different viewpoints: from the traditional view to Knowledge Management; from the view of the information portal as a Panopticon and from the viewpoint of Michel Foucault the French philosopher and social theorist. We will use this analysis to deepen our understanding of what was occurring in this case in order to find what lessons can be learned and to draw out the practical implications of the study.

9.1 Background/context

9.1.1 Big Data

Today, more and more sources of data have become available to both businesses and their customers. This tsunami of data, often known simply as "Big Data", is produced by machines, sensors and by human activity. For example, YouTube claims to have more than 6 billion hours of video: an hour of video for every person on Earth (YouTube.com, 2015).

Dealing with this amount of data creates both technological and managerial challenges. As long ago as the 1980s Zuboff (1988) noted that as information systems began to be deployed in organizations, they began to produce data on what happened inside the organization, making visible activities and events that were previously unseen. As long as businesses use computers to manage their daily operations, the volume of data they produce will continue to grow.

It is not only the volume of data that presents a challenge. If the volume of data is thought of as a "stock" of data, then the rate at which that stock changes – the speed at which data is generated, the frequency at which it is updated and the rate at which it is delivered – can be thought of as the velocity of data. This also presents a significant challenge; for example, what data should be saved and what could be lost? At present, most businesses are only able to view real time data through a 2 to 10 minute sliding window (ScaleDB, 2015).

Finally, as this data is produced by many different sources, the types of data will also vary, i.e. different formats, structures, and semantics will be associated with the data. This poses perhaps the biggest problem for dealing with Big Data because each different data source needs to be processed in a different way. For a computer to be able to process data in a way that makes it valid and meaningful for human beings, the data first needs to be codified, that is a semantic value – effectively a meaning – needs to be allocated to each item of data. Although this might sound straightforward, in practice it is fraught with both practical and theoretical difficulties (Kimble, 2013a, 2013b).

9.1.2 The regulation of water in France

In France, the water and sanitation sectors have been regulated since the post-war period in order to strengthen the accountability of elected officials and promote transparency. MURCEF Act (2001) defines the regulatory framework between standard public service operating contracts and delegation of public services contracts (Délégation de Service Public or DSP). Standard public service contracts are drawn up between contracting authorities, such as local governments, and private sector organizations. These simple administrative contracts set goals and detail the conditions for remuneration.

DSP contracts, on the other hand, involve a contracting authority that delegates not only the service provision to another organization, but also its management. Remuneration under DSP contracts is not fixed but linked to operating results. Since the public authority retains ultimate responsibility for the service, the accuracy of the data upon which the operating results are judged becomes a matter of crucial importance, to both the delegated service provider and the public authority.

In the 1990s, a series of new laws were introduced, particularly relevant to our case (Guérin-Schneider, Bonnet, & Breuil, 2003). The Barnier Act established an obligation for local authorities to prepare an annual report on the price and quality of public water services. The Sapin Act is concerned with transparency and the prevention of corruption in public life. The Mazeaud Act defines a set of key performance indicators (KPIs) that a delegated service provider must publish in an annual report, known as Rapport Annuel du Délégataire or RAD. It must contain a set of 50 indicators that specify the performance of the activities in technical and financial terms. This not only includes technical characteristics and pricing, but also indicators of performance and financing of investments (Guérin-Schneider, & Nakhla, 2003).

9.1.3 New Public Management

Beginning with the economic reforms of Margaret Thatcher and Ronald Reagan in the 1980s, there has been a steady move towards what is sometimes termed "New Public Management" or NPM. Although the qualifier "new" no longer really applies, Hood (1995) argues that NPM represents a fundamental shift in the way that public services, such as water, are provided – shifting responsibility from the public sector to semi-autonomous agencies or private companies.

Farazmand (2002) notes that although these private-sector corporations initially act on behalf of the public sector, after a while, the public sector entity no longer has the resources to operate independently and effectively becomes an extension of the private sector. He observes that, "One of the key problems of the reliance of NPM on agencification, privatization, and outsourcing … is the loss of accountability to the public. Privatized and outsourced corporations are not accountable to citizens; the government is" (Farazmand, 2002, p. 368).

This creates a dilemma: the goal of NPM is to streamline public services by delegating them to the private sector, but, as the Flint example at the beginning of this chapter shows, the public sector retains the ultimate responsibility for the quality of those services. In practice, this dilemma is resolved by creating a system of audits to ensure that accountability is maintained. However, as Olson, Humphrey, and Guthrie (2001) claim, this approach can contain a potential "evaluatory trap" since these audits tend to encourage the simple pursuit of numerical targets, rather than focusing on the outcomes the targets are supposed to represent. This leads to the need for new controls to be introduced, which runs the risk of becoming a damaging spiral in which new sets of rules are introduced, ostensibly designed to ensure accountability but which, in reality, destroy autonomy and breed resentment.

9.1.4 *Cross transparency requirements*

The introduction of performance indicators in the management and regulation of water and sanitation services in France has led to a new governance in the management of public services in this sector over the last fifteen years (Guérin-Schneider, & Nakhla, 2000, 2003). The issue of local regulation by local authorities is at the heart of the management of performance of these public services. Public authorities have become aware of new needs, including transparency and consumer consideration. In this context, there is a demand for cross transparency between local authority, consumer and delegate. Thus, new regulatory provisions govern these activities that enable public authorities to get tools to control their public service and to ensure transparency vis-à-vis users.

These constraints demand an analysis of the performance of the delegate and a reflection on how best to supply that data to consumers. On the one hand, local institutions need to analyze a delegate's performance, change modes of controls to enhance transparency for users' and regulators' organizations. On the other hand, delegates need to improve transparency in their production and control process and ensure the process of the dissemination of data on water quality is effective. The DSP relationship modifies the autonomy-control dialectic in the relational processes by imposing a new form of control from users to public authorities and from public authorities to delegate.

9.2 The case study – Eau Méditerranée

The case study concerns Eau Méditerranée, a division of Veolia, a multinational water company that makes water fit for drinking, gets it to where it is needed, collects it once used, treats it, and then recycles it for household and business use. In 2015, Veolia had a turnover of almost 25,000,000,000 Euros, supplied

101,000,000 people with drinking water, connected 71,000,000 people to sewerage networks and employed 174,000 people in 43 countries (Veolia, 2015). The division we consider, Eau Méditerranée, provides services and solutions related to water management and sanitation for the south of France. Although it operates in ten administrative departments in the South of France, this case focuses on its activities in Montpellier, the eighth-largest city in France and the administrative center for the Occitanie region.

9.2.1 Methodology

The approach we took in this this research was that of a case study. There is a large body of literature that deals with the use of case studies in research (Glaser & Strauss, 1967; Eisenhardt, 1989; Strauss & Corbin, 1998; Yin, 2009). Below we briefly describe how our work fits into this literature in order to highlight the nature of the analysis we present in section 9.3.

Eisenhardt (1989), views case studies as an inductive approach to theory building. She describes two broad approaches to analysis. The first, within-case analysis, is aimed at getting a better understanding of specific aspects of each case; the second, cross-case analysis, is directed toward exploring the similarities and differences between the cases by searching for cross-case patterns.

Yin (2009) on the other hand views cases as being similar to experimental studies and recommends the use of multiple case studies and replication logic to build theories inductively or to test them deductively. Replication logic implies that cases that confirm emergent relationships enhance confidence in the validity of those relationships, while cases that do not are viewed as opportunities to refine and/or extend the theory. He suggests that targeted, multiple case studies should be used when collecting data in organizations that, a priori, seem to address similar issues but do so in different contexts or with different outcomes.

In this study, we only considered one case, that of Eau Méditerranée. As a result, we were unable to use either Eisenhardt's cross-case analysis or Yin's replication logic, meaning that the opportunity to generalize our findings is somewhat limited. However, our goal was not to generalize but to explore how Knowledge Management could be used in the administration of sustainable water systems and to try to draw out themes and ideas that might be of practical value. Benbasat, Goldstein, and Mead (1987) claim that case study methodologies are particularly useful when the phenomenon under consideration is broad and complex, and when it cannot be studied outside the context in which it occurs, which was the situation we faced in our study.

The bulk of the data for this chapter was collected by a Masters level student during a one-year apprenticeship in the information technology department of Eau Méditerranée. In our work, we combined several qualitative data collection methods such as interviews, documents, an analysis of the company's extranet and intranet, and participant observation of meetings and day-to-day activities. The description of the case study is mainly based on information collected through semi-structured interviews with 12 top managers, middle managers, and operators from September 2013 until July 2014 (Table 9.1).

The topics discussed during interviews were mainly focused on data management methods, data reliability and compliance, the extranet website, big data, quality audits, and DSP management and its constraints. All interviews were tape-recorded and transcribed. The number of interviews undertaken followed the notion of saturation proposed by Glaser and Strauss (1967). The length of the interviews varied from 30 to 150 minutes. Participant observations of project related meetings and day-to-day activities were used to complete the qualitative study.

To code the data, we used the methods suggested by Miles and Huberman (1994). We developed a database containing interviews transcripts, notes, collected documents, a description of the sample, chronological data, the coding scheme, and coded data.

9.2.2 Presentation of the findings from the case study

In the Montpellier area, Eau Méditerranée has more than 600 sites that produce drinking water and 300 sites that treat waste water, and administers more than 350 contracts. It is organized around six centers, each with different operating structures that are served by a variety of systems, which may be centralized at either the local or the regional level. Thus, the vital data on which Eau Méditerranée's performance will be monitored and evaluated was derived from a mixture of sources.

Table 9.1 Interviewees

Job title	Interview
Assistant Director of IT services	Weekly interviews[a)]
Director of IT Services	Weekly interviews[a)]
External Auditor	150 mins
Input operator REPORTER	30 mins
Operations Manager	60 mins
Operator	30 mins
Operator	80 mins
Operator, application C&D	40 mins
Operator, method and planning	90 mins
Quality Manager	140 mins
Quality Manager	30 mins
System Administrator	45 mins

a) Approx. 1 hour per working week.

In addition to the requirement to produce an annual report based on the fifty KPIs specified by the Mazeaud Act, for each contract, some local authorities require the production of monthly or bimonthly reports using a set of different indicators. For example, Montpellier city demands higher visibility of the totality of the business processes and the genesis of the data (i.e. access to the data application's source) as it wants ensure the strategic coherence of the data from different sources. Thus, some of the requirements that shaped Eau Méditerranée's systems lie outside the regulatory framework in which the data is eventually used.

The information technology department of Eau Méditerranée is in charge of processing the data reliably, from its collection to its diffusion. It also creates the data definitions used in the contracts with public authorities, develops tools for construction of the RAD and is responsible for the availability (on-line) of prices and quality of water data for the public authorities (the collectivités). Their initial approach was to look for conventional solution such as master data management to provide a synchronized, consistent, accurate, and timely single version of data. This requires regularly merging records into a single master file, which then provides a complete, consolidated view of the company's key data.

Eau Méditerranée has made considerable investments in improving the quality and reliability of its data, and has been awarded various forms of certification. This required a detailed and costly examination of the organization's internal systems. To deal with the complexity of its data and the diversity of their sources under DSP contracts, it eventually chose to take a different approach than MDM, one closer to the techniques of quality control found in open source software. This involved not only a further investment in technology, but also a change of philosophy toward management transparency: being open and accountable, and ensuring that users could see what happened and when. The most visible manifestation of this approach was the construction of the "extranet des collectivités", an information portal that made data available to the public authorities that had delegated their water services to Eau Méditerranée (Figure 9.1).

The extranet des collectivités provides data in the form of maps, tables, and graphs on a close to real-time basis. The data that is made available is now no longer limited to the 50 indicators specified by the Mazeaud Act, but covers a broad range of data that allows Eau Méditerranée to both meet its statutory obligations and respond flexibly to its clients' demands. In effect, the company has gone from using the 50 KPIs specified by the Mazeaud Act in a single RAD which is produced once a year, to offering a broad range of indicators, which can be created on demand, providing a continual supply of data via a wide selection of reports.

For the generation of RAD (an ISO 14001 and 9001 certified document) and the data for the extranet des collectivités, Eau Méditerranée needed to re-organize its information system and create a new reporting tool called REPORTER, which would connect with more than 20 operational applications

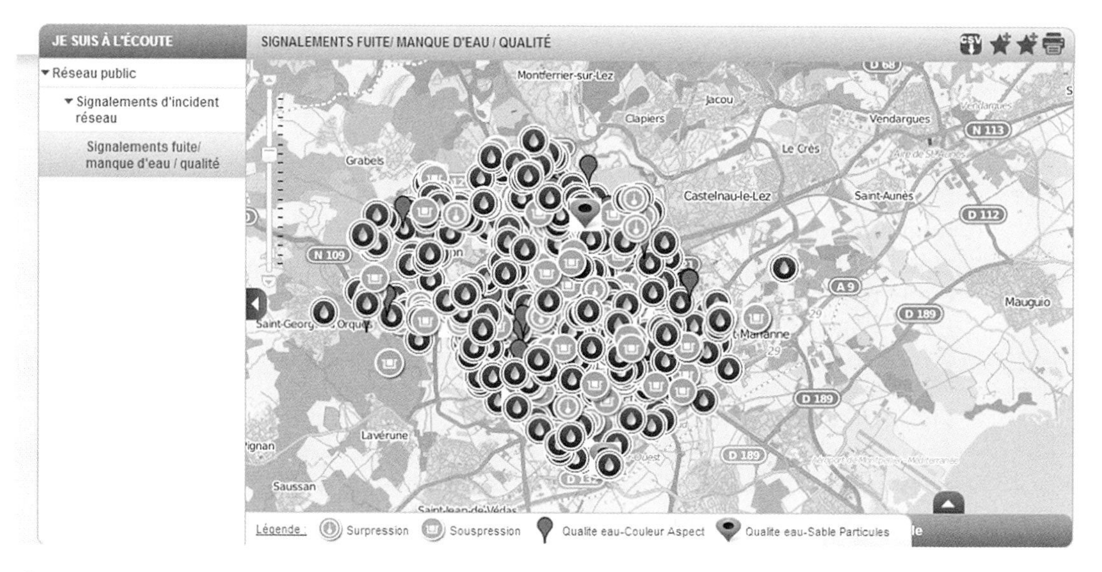

Figure 9.1 A screen shot from the extranet des collectivités.

and external information systems. Three types of data are integrated via a Business Object Data Services application on REPORTER:

- Automatic data, i.e. data whose source applications have a direct link to the REPORTER data warehouse.
- Semi-automatic data, i.e. data that requires some form of manual intervention before it is transmitted to the data warehouse.
- Manual data, i.e. data that is input directly to the REPORTER data warehouse.

The internal process to generate this data is shown in Figure 9.2.

For Eau Méditerranée, data comes from different "métiers" (clients, networks, maintenance reports, etc.) and different places (factories, treatment

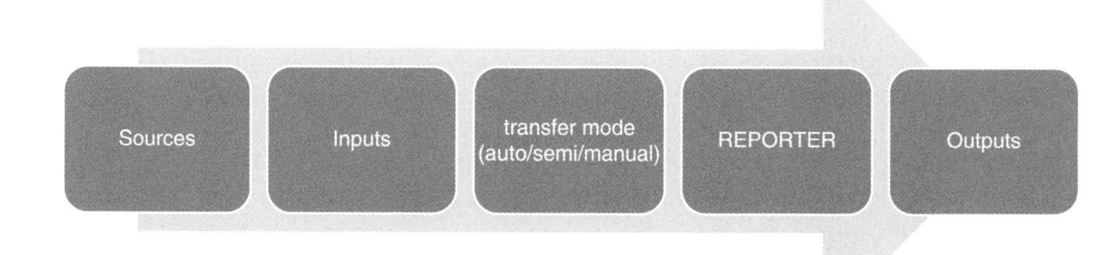

Figure 9.2 The data generation process (internal).

plants, etc.), and some from collectivités; some examples of applications sources are provided in Table 9.2.

For data output, REPORTER collects, transforms and diffuses data to push data in close to real-time to the extranet des collectivités and to present KPI data on service performance as required by the RAD (Table 9.3).

For Eau Méditerranée, the creation of the extranet des collectivités and the related data generation processes meant further investment. It needed to define new business processes to deal with the continuous management of KPIs, construct new monitoring tools, remapping its existing information flows, and create tools to generate the annual RAD on demand. In addition, it also needed to develop technologies for powering the production of near to real-time reports and to invest in training, not only to ensure the quality and reliability of the data, but also to ensure that the new philosophy of management transparency was understood and implemented throughout the organization. Eau Méditerranée deployed methods that it referred to as self-checking to stress the increased importance of the internal data reliability framework in order to ensure that it was transparent to local institutions and to ensure credibility.

The resulting system operates at two levels: Internally, it has helped Eau Méditerranée improve its in-house operations; externally, it has helped Eau Méditerranée build trust with its clients and simplify its interactions with the wider community. This helped the company to define the boundaries between internal and external stakeholders (the employees and the collectivités) more clearly and has helped to clarify the roles and responsibilities of each.

9.2.3 Summary of the case study

The extranet provides greater visibility of the company's affairs, both for its clients and for the public at large, and it can be used to monitor what is

Table 9.2 Some examples of data sources

Data sources	Relation to the REPORTER application
Application A	Application to manage thematic maps in the Extranet
Application B	Contract management framework
Application C	Front office of customer relations
Application D	Application for managing water network maintenance operations
Application E	Application to centralize and provide real-time technical data from controllers, e.g. the water level of all the tanks of a region at some specified time
Application F	Application used in the context of customer complaints
Application G	Application that manages water quality data
Application I	Plant monitoring database

Table 9.3 Details of KPI indicators by category for water and sanitation activities (RAD)

Indicators	Data provider
Customer activity	
Number of municipalities served	Eau Méditerranée
Total population served (estimated)	Collectivités
Total subscribers (customers)	Eau Méditerranée
Number of spill authorizations	Collectivités
Quality of service to user	
Rate of claims	Eau Méditerranée
Rate of delinquencies on previous year's water bills	Eau Méditerranée
Abandonment of credits and payments to a solidarity fund	Collectivités
Sanitation service price	
Price incl. VAT per m^3 for 120˜m^3 (remediation only)	Eau Méditerranée
Collection of wastewater	
Number of network obstructions	Eau Méditerranée
Pipe length	Eau Méditerranée
De-pollution	
Incoming volume (collected)	Eau Méditerranée
Treated volume	Eau Méditerranée
Evacuation of sludge and waste	
Quantity of sludge from sewage works	Eau Méditerranée
Masses of evacuated sands	Eau Méditerranée
Volume of grease removed	Eau Méditerranée

happening at any given time. To summarize, in order to offer a more credible and transparent form of accountability Eau Méditerranée has increased the amount of data available to its clients, both in terms of breadth (by providing access to entire business processes) and in terms of depth (through providing access to the original sources of the data). As a consequence, the collectivités now check the data they receive more frequently and more critically.

9.3 An analysis of the case study

Having seen what happened in the Eau Méditerranée, we now turn to the task of trying to interpreting the case study and drawing some meaningful lessons from it. Knowledge management can be viewed from a number of different perspectives (Russ, Fineman, & Jones, 2010), not all of which are complimentary

(Wilson, 2002). Here we analyze Knowledge Management from three different viewpoints: from the traditional technologically based view to Knowledge Management; from the viewpoint of seeing the extranet des collectivités as a Panopticon and from the viewpoint of Michel Foucault, the French philosopher and social theorist, who is more concerned with how power relations are constructed through discourse.

Our objective is to use this analysis to further our understanding of what occurred in this case, and to attempt to extricate the broader practical implications of the study for Knowledge Management and sustainable water systems. In the discussion that follows, we will illustrate some of our ideas using the problems that occurred in Flint. Our intention in doing this is simply to illustrate a practical application of our ideas – we will not comment on the specifics of the case, we do not wish to imply any particular set of motivations behind the events that occurred, and we do not intend to attribute blame for what happened.

9.3.1 The traditional approach to Knowledge Management

The Sapin, Barnier and Mazeaud acts, as well as the conditions for DSP contracts established in the MURCEF Act, provide a set of regulatory constraints for Eau Méditerranée. If the damaging spiral of new controls described in section 9.1.3 is to be avoided, every aspect of the production, verification, and distribution of this data needs to be accurate, and reliable. As we have seen, the data on which Eau Méditerranée's performance will be evaluated is derived from a wide variety of sources, some of which were created for reasons that lie outside the regulatory framework in which the data will be used. Dealing with issues of data quality and the management of data from diverse sources are not new problems however. The conventional information systems solution is to use such techniques as master data management (MDM) and store the result in a relational database.

MDM is a technique that aims to provide a synchronized, consistent, accurate, and timely single version of data. The objective is to ensure that an organization does not store different versions of data in different locations, but that all data is verified, combined, and reconciled into a "single version of the truth" (Buytendijk, 2008). This requires the organization to regularly merge records into a single master file, which then provides a complete, consolidated view of the company's data which ensures that data is always interpreted in the same way (Otto, Hüner, & Österle, 2012). However, even within the information systems world, MDM is not without its critics. For example, Buffenoir and Bourdon (2013) argue that MDM poses particular problems for organizations such as Eau Méditerranée, which are characterized by multiple relationships with external partners, porous organizational boundaries, and nested control loops. Such concerns are multiplied when this type of approach becomes part of a Knowledge Management initiative.

The traditional information systems approach, often attributed to Ackoff (1989), is to treat data, information and knowledge as a hierarchical progression.

Data originate from datum, phenomena in the physical world that are observable by human agents, information is a collection of data that has been processed in some way so that it becomes useable by, and useful to, that agent, while knowledge is a collection of useful information that has been stored and is available for future use. In essence, both information and knowledge are processed data, the distinction between them rests mainly on how they are used: information is something that is used in the here and now; knowledge is a "stock" of information that has been stored and ready to be used at some later date to create value (Russ *et al.*, 2010).

The approach of treating knowledge as a stock of information underpinned what is sometimes termed the functionalist or first generation approach to Knowledge Management (McElroy, 2002; Edgar Serna, 2012) which focuses on capturing knowledge and storing it in IT-based repositories. This approach is built upon Shannon and Weaver's work on communications theory (Shannon & Weaver, 1949). Briefly, this showed that as long as a suitable "codebook" for encoding and decoding a message existed, then any message could be transmitted without a loss of information. Within Knowledge Management this same approach is applied to a stored item of knowledge, the argument being that as long as a suitable codebook exists then the meaning of the original item can be retrieved (Kimble, 2013a).

In this view, effective Knowledge Management depends on the creation of an adequate representation of the reality that exists outside the computer, within a computer. In other words, the information that is stored inside the computer needs to have a meaning outside of it. However, in order to be meaningful, one needs to know the context in which this information will be used. Cohendet and Steinmueller (2000), suggest this problem can be dealt with simply by encoding the information about how the information should be used along with the information itself; however, this implies that the information must either be self-contained (i.e. its meaning is entirely self-evident) or have access to some other source of information, another "codebook", in order to make its meaning clear. As Duguid (2005) points out this runs the risk of creating circularity (with codebooks explaining themselves) or infinite regress (with codebooks to explain codebooks).

How can this problem be resolved? In the case of human agents the answer is simple, as Miller (2002) suggests, it is knowledge held within the individual that is used to impart meaning rather than codes stored in a machine. Human beings are able to assign meaning to events without the need to refer to a set of pre-programmed instructions. It is this observation that underlies what it called the second-generation or interpretivist approach to Knowledge Management, which places a stronger focus on people and communities as sources of knowledge. The argument here is that although not all members of a community are the same, the fact that they are in the same community implies that they share a set of common beliefs and that this shared worldview will, in effect, act as a proxy for the codebook of first-generation Knowledge Management.

How does this relate to our case study? As all of the data provided comes from computer based information systems, then the approach to Knowledge Management must necessarily involve the method of codification which underlies first generation Knowledge Management (Kimble, 2013a). However, as we have seen, Eau Méditerranée has a diverse set of data sources as well as a diverse, and potentially open-ended, set of uses to which that data could be put. In such circumstances, pursuing a strategy of codification can become both costly and time consuming (Kimble, 2013b). Is it possible to interpret Eau Méditerranée's move towards providing a stream of detailed data directly to its clients as a move away from first generation Knowledge Management towards second generation?

The key issue here is really the level of granularity, i.e. the level of detail at which the data is collected and presented, at which codification takes place. The move away from highly aggregated data (a single report based on 50 KPIs) towards providing a wider range of more detailed data, more regularly. The raw data is now codified at a relatively low level which, taken together with the policy of openness/transparency, means that this could be viewed as a move towards a social/collaborative second generation approach to managing knowledge where, to some extent, the client community becomes responsible for imparting meaning to the data.

If this were so, what would be the practical implications? Clearly one implication is for this strategy to be effective the client community would need to have the skills and resources to interpret the data. If this were the case, and if the linkage between Eau Méditerranée and its clients was strong, then this could be an effective way to manage complex activities in a business setting. Taking the Flint case as an example, if the level of granularity of the data had been low enough (e.g. if it recorded where the samples were collected), and if the community was able to interpret the data and was sufficiently organized, it might have prevented the problem escalating the way it did.

9.3.2 Zuboff's Information Panopticon/Open Source Model

The analysis above takes what might be termed a technological or operational view of Knowledge Management and how to realize it in practice, however there are other ways to look at this case study. The next view we take has its roots in the work of an 18th-century British philosopher and social reformer, Jeremy Bentham, and his innovative design for a prison building. We term this view an autopoietic (self-referential/self-productive) view of Knowledge Management.

Bentham is most often associated with the doctrine of Utilitarianism – the principle of proving the greatest happiness of the greatest number; however, like many present day reformers he was also concerned with the shortcomings in public administration. In particular, he was dismayed by the practice of transporting offenders to penal colonies and with the wastefulness and inefficiency of Britain's prison system. He developed the idea of the Panopticon penitentiary as a substitute. Bentham's Panopticon is a design for a prison consisting of a circular building divided into cells; each cell holds a single prisoner so that the occupants are isolated. There is an observation tower at the center of the building and the cells have both inner and outer windows so that the occupants are backlit and subject to the constant scrutiny of an observer in the central tower. His idea was that, in time, the prisoner would internalize the mechanism of surveillance established by the building.

He laid down the principle that surveillance should be both visible and unverifiable: visible in that the inmate would always be able to see the central tower and unverifiable in that the inmate must never know whether or not he is being observed. Towards this end, Bentham envisioned a system of blinds to avoid any indication that might betray the actual presence of an observer. The actual and imagined inspections would reinforce each other in the minds of the prisoner and, thanks to the continuous exercise of the disciplinary power by the Panopticon, the prisoners would internalize the established rules and norms of behavior and self-regulation would replace direct control. Although Bentham's prison was never built, his idea of a Panopticon has been taken up by several authors such as Foucault (see below), and Zuboff (1988).

In her prescient book In the Age of the Smart Machine, Shoshana Zuboff explores the relationship between information technology, work, and organizations through a series of detailed ethnographic studies. The core of her argument is that, in contrast to much of the rather deterministic literature at the time, computer technology presents us with a choice about how work is structured: computer technology can automate or "informate" work. Automating processes builds on the speed and consistency of computers but deskills those who use them, whilst "informating" builds on the transparency provided by computer based information systems to create skills and increase the effectiveness of both production and administrative processes (Kimble & McLoughlin, 1995).

Zuboff's work was mainly concerned with the effect of computer technology on managers in general and middle managers in particular. She argued that networked information systems were able to collect data about business processes, which could then be stored and analyzed to make the way in which the work was performed more visible. She called this an Information Panopticon. She believed that this new technology had the potential for allowing lower level employees to develop new knowledge and skills that would allow them to take part in decision making and to exercise discretionary judgement, areas which had formerly been the preserve of managers; simultaneously, the same systems would make the work of middle level managers more visible to the layers of management above. Consequently, "the informated environment means opening the information base of the organization to members at every level, assuring that each has the knowledge, skills and authority to engage with the information productively" (Zuboff, 1995, p. 203). Here Knowledge Management is viewed as a self-referential or self-reflecting process. People use their existing knowledge both to understand what is happening in their environment and in order to select what is relevant (Hildreth & Kimble, 2002), which makes autopoietic systems both enabling and constraining at the same time (Leflaive, 1996).

Again, we ask, how does this relate to our case study? Although her book was both prescient and influential, it is not really her views on how the emerging "information revolution" would affect managerial work that is of interest here but her concept of an Information Panopticon. Why should a company like Eau Méditerranée subject itself to such a regime where, via the extranet, the collectivités could assume the role of unseen observers and the company the role of prisoner? Furthermore, how would this help the organization avoid Olson *et al.*'s (2001) evaluatory traps?

The autopoietic nature of the Information Panopticon, creating both horizontal and vertical visibility, means that the company and its clients are not really in the position of prisoners and observers but are closer to partners in the shared custodianship of the data the system contains. The situation is in some way similar to the self-referential model used in open source software. The model for open source development relies on a core-and-periphery structure: core programmers produce most of the code and rely on a periphery of users to review it, identify bugs, and suggest improvements. It is claimed that a direct consequence of this is that "given enough eyeballs, all bugs are shallow," meaning that any problems that exist will eventually be identified and resolved by the wider community (Raymond, 2001). The visibility provided by the Information Panopticon also plays a part in avoiding potential evaluatory traps (Olson *et al.*, 2001) as both Eau Méditerranée and the collectivités have a stake in ensuring that the data remains accurate.

If we were to accept that Eau Méditerranée had effectively created an Information Panopticon, what would be the practical implications? Zuboff's work highlighted the effect of visibility in information systems and opened the possibility of choice in the potential outcomes, without giving much guidance as to how those outcomes might be reached. What is clear in the case of Eau Méditerranée is that for such a system to work, Eau Méditerranée and the collectivités must trust the data they rely on and to trust the way it is produced: both parties need to remain vigilant. However, it is not enough to know that the data is being monitored, there also needs to be some way of taking action if things go wrong. For example, in the Flint case the idea of an invisible and unverifiable watcher would not have prevented something that might have begun as carelessness or a simple error. Mistakes need to be detected and, of equal importance, there needs to be some way of taking action when they are – effectively "closing the loop" by adding external feedback to the internal processes shown in Figure 9.2.

9.3.3 Foucault's perspective

Michel Foucault, a French philosopher, developed a theory about discourse and discipline in societies (Foucault, 1975). He argued that discourse embodies the exercise of power and acts as a form of social control. The Foucauldian theory of power has been used to understand organizational and political situations where discourse is constituted to control and discipline conduct and to convey representations. Foucault's ideas have also been used to explain what happens during the implementation and use of information systems (Orlikowski & Robey, 1991) and to show how surveillance is facilitated through the increased visibility that information systems provide (Doolin, 2009).

Foucault believes that technologies and practices govern, supervise, and discipline actors in society and is concerned with the process of normalization, that is, the way in which societal norms develop, that occurs in societies. He built on the metaphor of Jeremy Bentham's Panopticon assuming that actors, when placed in a permanent and omnipresent area of visibility, are driven to act as if they were being constantly watched, and consequently are led to integrate a set of norms and disciplines into their everyday lives. However, unlike Bentham, the core of his argument is that power does not need to manifest itself physically to be effective.

The interpretation of Bentham's Panopticon given by Foucault is that the Panopticon is the formula of liberal governmentality and that this new governmental rationality is solely concerned by interests and aims at manipulating them (Foucault, 1979). Foucault's concept of governmentality is a type of power that controls and manages whole populations with the aim of trying to regulate social behavior. When considering the problem of dissemination of norms and production of normalized subjectivity, Foucault explains that normalization is a phenomenon that results from the process of disciplining; that force and create habits, rituals, thereby creating norms of behavior. For him, a norm is something that can be applied to both a body one wishes to discipline and to a population one wishes to regularize (Foucault, 1975).

The use of Foucauldian analysis for the case study of Eau Méditerranée requires a shift of the standard viewpoint of information systems. The technologies that are implemented can be seen as objects of disciplinary power, while simultaneously, the development of discourses relating to transparency and data reliability legitimizes its implementation beyond the specified regulatory framework. In this perspective, Eau Méditerranée's information system can be considered as a focal point for the concentration of control and the production of norms, which are then immediately generalized and circulate through a network woven between the company and its partners. The conceptual framework offered by Foucault is particularly effective in analyzing the role played by visibility, transparency and accountability of actors in the deployment of new forms of control mechanisms (Willcocks, 2004) such as those created by the information systems within Eau Méditerranée and through its relationship to the delegators. It shows how information systems promote a disciplinary power over individuals both within Eau Méditerranée and between Eau Méditerranée and collectivités.

The concepts of "Electronic Panopticon" or "Electronic Eye" (Lyon, 1994; Willcocks, 2004) express the role of IT in this surveillance and normalized visibility. For Deleuze (1992), the abstract formula of Foucault's Panoptism is no longer "seeing without being seen," but "imposing any conduct on any human multiplicity". It seems that the contractual and legal framework of the DSP refers to a panoptic device where the delegator wants to be able to see the activities of the delegatee permanently, who has an obligation of results under penalty of sanctions. Deleuze (1992) had already confirmed the omnipresence of control in the information society, defining it as a control society.

Foucauldian theoretical framework can also help interpret the paradoxes and contradictions in the new processes for organizing data in Eau Méditerranée that make strategic data available for collectivités in real time. Leclercq-Vandelannoitte (2011, 2013) note that the tensions between servitude and freedom, dependency and independency, control and flexibility emerge as conflicting consequences of the use of new IT in organizations and offer contradictory results and paradoxical effects. Foucault provides an interpretative framework that clarifies the paradoxical consequences of Eau Méditerranée new information system deployment in an organizational setting (Leclercq-Vandelannoitte, 2013).

What are the practical implications of the Foucauldian perspective? Firstly, it enables us to understand how regulatory discourses on transparency and prevention of corruption in public life in the Sapin Act lead to the systems of control in the DSP and new cross transparency requirements. It also explains the intrinsic contradictory and paradoxical nature of organizational change through a disciplinary context (Leclercq-Vandelannoitte, 2013). Taking the example of the Flint case, Foucault's perspective reinforces the importance of embracing a wide diversity of viewpoints during the construction of power relations through discourse so that differing views and opinions can be taken into account. Indeed, the whole notion of power being exercised through discourse, and that individuals are in a position to both submit to and exercise this power, provides some ideas for how purposive action can be taken that are missing from Zuboff's analysis.

9.4 Lessons to be learned/practical implications

We began this chapter by saying that we wanted to analyze the approach to Knowledge Management in a case study from different viewpoints to deepen our understanding of what was happening in the case and to try draw out some of the practical implications of the study. However, before we do this, we should point out that this was one case study of one part of a large multinational enterprise. We did not have access to top-level managers at Eau Méditerranée, nor did we have access to senior managers at the parent company, Veolia; consequently, our results are not generalizable as such. Nevertheless, as we noted in the section on methodology, our intention was not to produce generalizable findings but to explore a specific instance of the use of Knowledge Management in the administration and control of sustainable water systems with the aim of exploring themes and ideas.

Perhaps the first point to address is what might be called the cynics view, that all Eau Méditerranée was doing was trying to hide bad news in plain sight. The argument would be that by providing more information, more often, in greater detail, to a wider audience, the company was simply attempting to shift responsibility for identifying problems from itself to its clients, who may well lack the expertise to be able to process large volumes of data. There may be some force in this argument, indeed, Robert Glass, one of the founding fathers of software engineering, leveled exactly this type of criticism against the open source model, claiming that the rate at which bugs are detected actually falls as the number of reviewers increases (Glass, 2003). Similarly, the fact that a data set contains billions of items does not mean that it is either accurate nor representative (Kimble & Milolidakis, 2015). Without access to the strategic decisions taken by senior managers in Veolia and/or Eau Méditerranée we are not able to resolve this question, however, there was never any suggestion by our interviewees that this was the case, and the cost in time and effort to introduce the extranet des collectivités would appear to be a poor return on investment when considered purely in public relations terms.

What then are the practical implications? We gave some indication of this in the individual parts of the analysis in section 9.3; we return to these earlier observations to try to place them in a broader context.

9.4.1 *Granularity*

When looking at the traditional, IT based, approach to Knowledge Management (section 9.3.1) we observed that this approach was founded on codification, the ability to build an accurate and complete representation of the world outside the computer within the computer. As discussed elsewhere, although widely used, this approach has its limits, both conceptual (Kimble, 2013a) and practical (Kimble, 2013b). In this case, the level of granularity needs to be appropriate for this approach is to work. If the data is too highly aggregated one loses the advantages of transparency, both in terms of traceability and in terms of customer relations. If it is aggregated at too low a level, then the data becomes too complex, too costly to process, and is effectively unmanageable. Knowledge management is clearly a sociotechnical phenomenon (Hildreth & Kimble, 2002) and when balancing economic, business, social and environmental concerns it is unlikely that investment in technology alone will ever provide the necessary returns.

9.4.2 *A diversity of viewpoints*

When looking at Knowledge Management from the viewpoint of Zuboff's Information Panopticon (Zuboff, 1988), we saw that the horizontal and vertical visibility provided by information systems such as the extranet des collectivités created the possibility of transforming Bentham's design for a prison designed to direct and control inmates into something altogether more empowering. In terms of this case study, it helps to illustrate the importance of allowing a diversity of viewpoints. From the social/collaborative viewpoint it is important that the data can be interpreted in a number of different ways so that differing interpretations and opinions can be taken into account; the extranet des collectivités facilitates by providing its users with a variety of different views of the data and by giving them close to real time access to the data. However, although Zuboff's work provided a useful contrast to the more deterministic views of technology that were current at the time of publication (Kimble & McLoughlin, 1995), its autopoietic standpoint did not provide much in the way of practical guidance about how the promise of empowerment it contained might be brought about. For that we need to turn to the work of Foucault (1975).

9.4.3 Closing the loop

We ended our analysis in Section 9.3 by noting that, for the approach of openness and transparency embodied in Eau Méditerranée's extranet des collectivités to be effective, it was not enough to simply make internal data available, there needed to be some external means of transforming that data into action – closing the loop. Being able to turn abstract data into concrete political action is particularly important in cases such as Flint, where it appeared that the consequences of earlier decisions were deliberately covered up in order to avoid their repercussions (Fonger, 2015). This view of Knowledge Management moves well beyond the confines of a specific type of technology or the management of a particular firm into the realm of the social and the political. A common interpretation of the notion of sustainability is that the economic needs of one generation should be able to be fulfilled without sacrificing the ability of future generations to meet their own needs (FDSD, Undated). As we shall see below, it is in the tradeoffs between the needs and desires of different groups where Foucault's ideas on how power is exercised through discourse come into play.

9.5 Knowledge Management and sustainability

In this chapter we have analyzed the approach to Knowledge Management used by Eau Méditerranée, a company providing water management and sanitation services in the south of France, from three different viewpoints. Our aim was to explore how Knowledge Management could be used in the administration of sustainable water systems; in this final section, we discuss the underlying complexity of this issue.

The topic of sustainability and sustainable development is fraught with difficulties. Although appealing, the definition of sustainable development used above - the needs of one generation not putting at risk the development of future generations - masks a host of complex issues. For example, does sustainability imply the ability to grow indefinitely?

The question of how the world's constantly expanding population can continue to live on a planet that has only a finite amount of resources was first brought to the public eye in 1972 by the Club of Rome report "The Limits to Growth" (Meadows, Meadows, Randers, & Behrens, 1972). Since then it has become clear that debates about sustainability need to include discussions of economics and (geo)politics, for example, questions about the viability of assumptions of continuing economic growth, and issues such as the sharing of resources (and responsibilities) between the industrialized nations and emerging economies. When dealing with sustainable water systems, particularly in a time of global warming, the division between the relatively water-rich nations of the temperate North and the water-poor nations of equatorial regions adds a further layer of complexity to these questions. Is it possible to talk about sustainability simply in terms of making better (i.e. more efficient) use of resources?

There is no doubt that computer technology and developments such as Knowledge Management can help to make better use of resources but, as we have indicated above, truly sustainable water systems imply not only questions about how best to manage a finite resource but also questions about the equitable distribution of that resource.

Nation states are built around clearly defined political boundaries and economic growth is usually given primacy over environmental issues. Furthermore, in the industrialized west, individual choice and economic freedom tends to be the dominant ethos. However, sustainable development, and notion of inter-generational stewardship, carry with them certain long-term implications that run counter to this. Particularly in the case of water, the impact of decisions in one area can cross geographical and political boundaries to effect those who live in another. Sustainable development requires tradeoffs between economic development, environmental protection and social justice; it implies the existence of a set of shared values that are oriented towards the future and concern for the natural world, rather than the short-termism and reliance on abstract economic models that characterizes much of the current approach.

While our case study, based on one division of one company operating in one area in the South of France, cannot provide the answers to these questions, it does show that Knowledge Management has a part to play. While the traditional IT based approach to Knowledge Management clearly has a role to play in making the most efficient use of existing resources, viewing Knowledge Management from the perspectives of Zuboff or Foucault help to show that it can also play a role, at least at the local level, in building trust and facilitating communication within communities. Above all our case shows that, to be effective, Knowledge Management initiatives for sustainable water systems need to go beyond technological and the managerial issues and embrace the social and political aspects of the environment in which such systems are embedded.

References

Ackoff, R. L. (1989). From data to wisdom. *Journal of Applied Systems Analysis, 16*, 3–9.

Benbasat, I., Goldstein, D. K., & Mead, M. (1987). The case research strategy in studies of information systems. *MIS quarterly*, 369–386.

Buffenoir, E., & Bourdon, I. (2013). Managing Extended Organizations and Data Governance. In: P.-J. Benghozi, D. Krob & F. Rowe (Eds.) *Digital Enterprise Design and Management 2013: Proceedings of the First International Conference on Digital Enterprise Design and Management* (pp. 135–145). Berlin: Springer.

Buytendijk, F. (2008). *The Myth of One Version of the Truth*. Redwood Shores, CA: Oracle.

Cohendet, P., & Steinmueller, E. W. (2000). The codification of knowledge: a conceptual and empirical exploration. *Industrial and Corporate Change, 9*(2), 195–209.

Deleuze, G. (1992). Postscript on the Societies of Control. *October, 59*, 3–7.

Doolin, B. (2009). Information systems and power: A Foucauldian perspective. In: C. Brooke (Ed.) *Critical management perspectives on information systems* (pp. 211–230). Oxford: Butterworth Heineman.

Duguid, P. (2005). The art of knowing: social and tacit dimensions of knowledge and the limits of the community of practice. *The Information Society: An International Journal, 21*(2), 109–118.

Edgar Serna, M. (2012). Maturity model of Knowledge Management in the interpretativist perspective. *International Journal of Information Management, 32*(4), 365–371.

Eisenhardt, K. M. (1989). Building theories from case study research. *Academy of Management Review, 14*(4), 532–550.

Farazmand, A. (2002). Privatization and globalization: a critical analysis with implications for public management education and training. *International Review of Administrative Sciences, 68*(3), 355–371.

FDSD. (Undated). *What is sustainable development?* Retrieved February, 2017, from http://www.fdsd.org/the-challenge/what-is-sustainable-development/

Fonger, R. (2015). *Documents show Flint filed false reports about testing for lead in water [Electronic Version]. Michigan Live*. Retrieved November 12, 2016 from http://www.mlive.com/news/flint/index.ssf/2015/11/documents_show_city_filed_fals.html.

Foucault, M. (1975). *Surveiller et punir*. Paris: Gallimard.

Foucault, M. (1979). *Naissance de la biopolitique*. Paris: Gallimard-Seuil.

Glaser, B. S., & Strauss, A. (1967). *The discovery of grounded theory: Strategies for qualitative research*. London: Weidenfeld and Nicolson.

Glass, R. L. (2003). *Facts and fallacies of software engineering*. Boston, MA: Addison-Wesley.

Guérin-Schneider, L., & Nakhla, M. (2000). Le service public d'eau délégué: du contrôle local des moyens au suivi de la performance. *Politiques et management public, 18*(1), 105–123.

Guérin-Schneider, L., Bonnet, F., & Breuil, L. (2003). Dix ans de loi Sapin dans les services d'eau et d'assainissement: évolutions et perspectives du modèle de délégation à la française, *Responsabilité & Environnement* (pp. 44–57): Annales des Mines

Guérin-Schneider, L., & Nakhla, M. (2003). Les indicateurs de performance: une évolution clef dans la gestion et la régulation des services d'eau et d'assainissement. *Flux* (*2*), 55–68.

Hildreth, P., & Kimble, C. (2002). The duality of knowledge [Electronic Version]. *Information Research*, *8*(1). Retrieved 12 October 2012 from http://informationr.net/ir/8-1/paper142.html.

Hood, C. (1995). The "New Public Management" in the 1980s: variations on a theme. *Accounting, organizations and society*, *20*(2), 93–109.

Kimble, C., & McLoughlin, K. (1995). Computer Based Information Systems and Managers Work. *New Technology, Work and Employment, 10*(1), 56–67.

Kimble, C. (2013a). Knowledge management, codification and tacit knowledge [Electronic Version]. *Information Research*, *18*(2). Retrieved 19 June 2013 from http://informationr.net/ir/18-2/paper577.html.

Kimble, C. (2013b). What cost Knowledge Management? The example of Infosys. *Global Business and Organizational Excellence, 32*(3), 6–14.

Kimble, C., & Milolidakis, G. (2015). Big Data and Business Intelligence: Debunking the Myths. *Global Business and Organizational Excellence, 35*(1), 23–34.

Leclercq-Vandelannoitte, A. (2011). Organizations as discursive constructions: A Foucauldian approach. *Organization Studies, 32*(9), 1247–1271.

Leclercq-Vandelannoitte, A. (2013). Contradiction as a medium and outcome of organizational change: a Foucauldian reading. *Journal of Organizational Change Management, 26*(3), 556–572.

Leflaive, X. (1996). Organizations as structures of domination. *Organization Studies, 17*(1), 23–47.

Lyon, D. (1994). *The electronic eye: The rise of surveillance society*. Cambridge: Polity Press.

McElroy, M. W. (2002). *The New Knowledge Management: Complexity, Learning, and Sustainable Innovation.* Burlington, MA: Butterworth-Heinemann/KMCI.

Meadows, D. H., Meadows, D. L., Randers, J., & Behrens, W. W. (1972). *The limits to growth: a report to the club of Rome.* New York: Universe Books.

Miles, M. B., & Huberman, M. (1994). *Qualitative Data Analysis: an expanded sourcebook* (2nd revised edn). London: Sage.

Miller, F. J. (2002). I = 0 (Information has no intrinsic meaning) [Electronic Version]. *Information Research*, *8*(1). Retrieved 12 October 2012 from http://informationr.net/ir/8-1/paper140.html.

Olson, O., Humphrey, C., & Guthrie, J. (2001). Caught in an evaluatory trap: a dilemma for public services under NPFM. *European Accounting Review*, *10*(3), 505–522.

Orlikowski, W. J., & Robey, D. (1991). Information technology and the structuring of organizations. *Information Systems Research, 2*(2), 143–169.

Otto, B., Hüner, K. M., & Österle, H. (2012). Toward a functional reference model for master data quality management. *Information Systems and e-Business Management, 10*(3), 395–425.

Raymond, E. S. (2001). *The Cathedral & the Bazaar: Musings on Linux and open source by an accidental revolutionary*. Sebastopol, CA: O'Reilly Media, Inc.

Russ, M., Fineman, R., & Jones, J. K. (2010). Conceptual theory: what do you know? In M. Russ (Ed.) *Knowledge Management Strategies for Business Development* (pp. 1–22). Hershey, PA, USA: IGI Global.

ScaleDB. (2015). *High-Velocity Data – The Data Fire Hose.* Retrieved July, 2015, from http://www.scaledb.com/high-velocity-data.php

Shannon, C. E., & Weaver, W. (1949). *The mathematical theory of communication.* Urbana, IL: The University of Illinois Press.
Strauss, A., & Corbin, J. (1998). *Basics of qualitative research – Techniques and procedures for developing grounded theory* (2nd edn): Sage.
Veolia . (2015). *Veolia en bref: rétrospective 2015.* Retrieved November, 2016, from http://www.veolia.com/fr/groupe/profil
Willcocks, L. (2004). Foucault, power/knowledge and information systems: reconstructing the present. In J. Mingers & L. Willcocks (Eds). *Social Theory and Philosophy for Information Systems* (pp. 238-296). Chichester: Wiley.
Wilson, T. D. (2002). The nonsense of 'Knowledge Management' [Electronic Version]. *Information Research, 8*(1). Retrieved 16 October 2012 from http://informationr.net/ir/8-1/paper144.html.
Yin, R. K. (2009). *Case Study Research: Design and Methods* (4th edn). Thousand Oaks, CA: Sage.
YouTube.com. (2015). Statistics - YouTube. Retrieved July, 2015, from http://www.youtube.com/yt/press/statistics.html
Zuboff, S. (1988). *In the age of the smart machine.* New York: Basic Books.
Zuboff, S. (1995). The emperor's new workplace. *Scientific American, 273*(3), 202–203.

10 Complexity, collective action, and water management: The case of Bilbao *ria*

Laura Albareda[1,2] and Jose Antonio Campos[3]

[1] *School of Business and Management, Lappeenranta University of Technology, Finland*
[2] *Department of Strategy, Deusto Business School, Deusto University, Avenidad de las Universidades, Bilbao, Spain*
[3] *Department of Marketing, Deusto Business School, Department of Industrial Technologies, Faculty of Engineering, Deusto University, Bilbao, Spain*

Introduction

Water is a fundamental natural resource for life on our planet. It is a common resource shared by many users and local communities, cities and nations. Clean and fresh water is needed to guarantee life on Earth, not only for humans but for all forms of life. Over the last century, human over-consumption and pollution has accelerated water cycles on Earth, though these will accelerate even further over the next few years due to climate change (Oki & Kanae, 2006). Currently, two billion people live in highly water-stressed geographical areas arising from the unequal distribution of renewable and fresh water resources. In order to understand how much renewable fresh water is available on the planet, Oki & Kanae (2006, p. 1068) measured the water cycle. The water hydrological cycle combines all the processes by which water circulates across the Earth ecosystems, including the oceans and sees, the atmosphere or the land. Therefore it involves all the water physical processes such as evaporation, condensation, precipitation, infiltration, surface run of and subsurface flow. According to their estimates (Oki & Kanae, 2006), humans currently extract approximately 3,800 km^3/year of renewable fresh water: this represents less than 10% of the maximum available

Handbook of Knowledge Management for Sustainable Water Systems, First Edition. Edited by Meir Russ.

renewable fresh water in the world. Evapotranspiration, that is, the amount of water that evaporates and transpires, is estimated to be 7,600 km^3/year from croplands and 14,400 km^3/year from grazing lands. Both represent about 33% of total terrestrial evapotranspiration (Oki & Kanae, 2006).

Transnational water resources are naturally circulating and constantly recharging across the Earth. The total mass of water is fairly constant over time; however, its state and localization vary depending on changes in climate conditions. Analyzing and understanding water cycle processes and stress are key to see the current stress on water resources in Earth. Water transforms to and from liquid, solid and vapor states and is continuously moving. Water cycles include different physical processes such as evaporation, condensation, precipitation, infiltration, surface run-off, and subsurface flow. However, these water cycles have suffered many changes since the beginning of the Industrial Revolution, especially due to pollution stemming from toxic waste, including chemicals, metals, pesticides, and oil and derivatives in water ecosystems. As a result of human activity and industrial development, especially in the last 200 years, the world's water cycles have changed considerably. An important change is the acceleration of how water circulates, creating significant risks and uncertainties regarding the long-term sustainability of clean water ecosystems, especially due to the planet's growing urbanization and the increasing impact of climate change (Oki & Kanae, 2006). Consequently, water over-use and scarcity have become key issues in the global governance agenda, recognized as a fundamental human right. To this end, the United Nations General Assembly (July, 28th 2010)[i] passed the Human Right to Water and Sanitation (HRWS) resolution, declaring water a limited resource shared by human and natural ecosystems and indicating that it is a fundamental human right.[ii]

Although there is a lot of water on our planet, only about 2.5% is fresh and potable (Oki & Kanae, 2006). Most of this fresh water is stored in glaciers or deep groundwater, providing us limited or difficult access to these reserves. Rivers are key circulating clean water resources on our planet, especially in their upper course. Rivers are also important ecosystems in the fresh water cycles with circulating resources. Rivers' upper stages include the river's source, a spring, a lake, or a glacier. Many rivers are born in the mountains, taking fresh run-off from melting ice. They also collect run-off water from precipitation and rain and move this water across the middle and lower stages toward the seas and oceans. According to Oki & Kanae (2006, p. 1068), the amount of water stored in all the rivers in the world is only 2,000 km^3, much less than the amount of water extracted: 3,800 km^3/year. The same authors have also studied the measure of water availability in rivers, consisting of 45,500 km^3/year of annual discharge and flowing mainly through rivers, from continents to the sea (Oki & Kanae, 2006, p. 1068). Rivers are used to supply water to cities, farms, and factories. Rivers' water resources are also common resources shared by the populations of cities and towns located on the river banks and estuaries.

In this chapter we analyze a 50-year collective water management experience in the Biscay-Bilbao city region in the Basque Country (Spain)

Figure 10.1 Map of the Bilbao-Biscay Region with the Estuary (2017). *Source*: https://goo.gl/maps/16o9TBy7BWq, viewed on 15 January, 2017.

(see Figure 10.1). The research analyzes the management of a river estuary, Bilbao *ria*. We apply a collective action and adaptive water management analysis to the co-management process. The Bilbao estuary or *ria* begins at the Nervion River delta, encompassing approximately seven municipalities along the lower part of the river where it meets the ocean. The estuary is actually located in a valley formed by the Nervion and Ibaizabal rivers. With high tide, the ocean penetrates the valley because the basin is below sea level, creating the geographic formation known as an estuary. The Nervion estuary encompasses the city of Bilbao, the capital of the region, but also includes neighboring cities and some of the most important towns in Biscay such as Barakaldo, Erandio, Getxo, Portugalete, Santurzi, Sestao and Zierbena.

Until recently, the *ria* also housed the region's most important infrastructure, Bilbao Port.[iii] The latter was moved from its centuries-old location in downtown Bilbao, 15 km upstream from the mouth, to Abra Bay, near the Nervion River delta. Currently, the port's facilities encompass 3.13 km^2 of land and 16.94 km^2 of water with 17 km of waterfront. It is fourth largest ports in Spain in terms of port traffic (tons) and among the six largest ports in Spain in terms of total goods processed after the ports in Algeciras, Valencia, Barcelona, Tarragona and Cartagena, respectively (Puertos del Estado, 2016: statistics 3.1: Goods traffic).

In the 1960s, this estuary was among the dirtiest in Europe due to the region's industrialization and urban development (Eizaguirre, 1996). For more than a century, the estuary received industrial wastewaters with a bed full of heavily polluted sediments and a high concentration of metals. Industrial activities and urban development led to the continuous deposit of hazardous and solid industrial and urban waste in the estuary (e.g., from mining and heavy metals, blast furnace and high toxic chemical materials, and urban solid waste). In addition to these, the urbanization and the growing of Bilbao and neighboring towns and cities also affected the pollution of the estuary. From these cities and towns along the estuary, water use, discharge and extraction covered a large geographical area of the Biscay region, including other towns and cities and from different Nervion and Ibaizabal river watersheds and tributaries. Pollution was so extreme that in the 1950s and 1960s the estuary's water and the area's water sources were not suitable for human-use and primary-sector activities. In the early 1970s the estuary's extreme water pollution overlapped with the severe economic crisis that gripped the metallurgical and shipyard industries, forcing many companies to close or redefine themselves.

In the 1960s, the estuary municipalities together with the Provincial Government of Biscay and local industry had already begun debating about the need to clean the estuary and to cooperate to generate the region's water system distribution. Specifically, our case analyzes five decades of collective commitment to water management in the estuary, creating an institution for collective action to sanitize the estuary and co-manage the water and hydric resources, adopting a long-term approach and building a set of complex infrastructures and Knowledge Management to re-distribute and keep fresh and clean water for the population and the natural ecosystem. The *Bilbao-Biscay Water Consortium* was created in 1967 as a collective action institution which included participation by 19 municipalities across the estuary banks and the region, together with regional and national governments. It represents an advanced and successful case of collective and adaptive water management reacting to high levels of water pollution that was seriously affecting human life and the region's economic and social well-being. The consortium designed and implemented a *Plan for the Integral Sanitation and Clean-up of the Estuary* which led to a series of infrastructures of water treatment and distribution (e.g. Galindo waste water treatment plant) to supply and distribute clean water and channel and clean up the waste from 1979 to 2006 (Barreiro & Aguirre, 2005). Furthermore, we analyze the connections between the region's water management policies and its social and economic development. While the estuary's water quality was being improved, there was a long-term parallel process to launch and implement a *Strategic Plan for the Revitalization of Metropolitan Bilbao.* The latter included re-locating and modernizing Bilbao Port and building the new Guggenheim Museum located on a city river bank. Cleaning the estuary was a key driver for the revitalization of the region, helping rethink the region's economic and urban development, as well as its identity. This collective project has lasted five decades thus far.

The structure of this chapter is as follows: first, we present a discussion on the conceptual analysis including the discussion on common resources and complexity, commons' governance and collective action, and the analysis of control and adaptive water management. Second, we examine the case study. Third, we analyze the main dimensions of the case from the inquiry of adaptive water management perspective and Knowledge Management. Finally, we present our main conclusions.

10.1 Conceptual analysis

10.1.1 *Common resources and complexity*

Water resources such as rivers, estuaries, lakes, oceans are common goods. A common good is a resource or a set of resources shared by a group of users that mutually depend on its use and long-term self-management (Ostrom, 1990). Neoclassical economists (Samuelson, 1954) defined common goods as rivalrous and non-excludable, i.e., they have a competitive nature, that is, the consumption of a given good by one person precludes its consumption by another, although, at the same time people cannot be excluded from using them because they are shared resources. Common goods are vulnerable resources affected by over-use, exploitation, and social dilemmas (e.g., free-riders) (Hardin, 1968); consequently, the over-appropriation by some users destroys the capacity of other users to use them. Therefore, managing common resources has been studied as a tragedy (Hardin, 1968). In contrast, later research based on Ostrom (1990) showed how common resources require adopting a collective action approach among users.

Due to the rise of global markets and globalization, the analysis of commons resources today focuses on the expansion of commons resources across and beyond regions and nations. They are seen as new regional or global commons that are affected by extended networks of resources (Hess, 2008) (e.g. the atmosphere, the water cycle, the rivers, the sea, and the oceans). These modern and global commons have attracted the attention of scientists and researchers who are keen on studying the impact of the current transformation of common resources into more complex and broad, interdependent ecosystems as part of Earth system resources (Hess, 2008). This debate has appeared as a consequence of globalization and global market growth and the over-use of these resource systems linked to global production and transnational supply chains. Given the impact of globalization and industrialization, water resources – rivers, lakes or oceans – cannot be studied as separate units, but as interconnected and complex set of water cycle as common resources. The water cycle encompasses many interconnected water resource units which include different ecosystems: a river and its tributaries, deep groundwater resources, estuaries, lakes or oceans. The over-use or over-pollution of any single one affects the others due to the nature of water cycle, leading to global over-exploitation and precluding its future use and appropriation by other users and even future generations across different regions in the world. Governance refers to the long-term use of these natural ecosystems that are exposed to overexploitation and degradation (Keohane & Ostrom, 1995).

In order to set up an adequate long-term governance of common resources, the analysis of common resource complexity becomes a core factor. In the case of a river estuary, these ecosystem include a complex set of hydric resources that involve multiple biological and biogeochemical, physical and also human components (networks of heterogeneous users and appropriators) affected by complex natural, social and economic interconnected resources (Pahl-Wostl, 2007). An estuary and the adjacent hydric resources are open-access resources spreading across region with many set of individual users and city beneficiaries. Therefore, it is important to understand the mutual interdependences of users and appropriators regarding the causes, effects, and responsibilities of commons degradation.

Thus, to understand the complexity of new commons, Ostrom (2009) proposes studying them as "Social-Ecological Systems" (SES). She bases her idea on complexity theory and the notion of "complex adaptive systems" (CAS) (Gell-Mann, 1994). A CAS is understood as "partially connected agents or components whose interactions give rise to the complex behavior that characterizes the systems" (Eisenhardt & Piezunka, 2011, p. 508). SES is framed by an analysis of a multilevel, nested framework of resources, and not the simple traditional system of resource units (e.g., lakes or rivers) (Ostrom, 2009). An estuary and its adjacent hydric resource can be analyzed as an SES, including a multiple set of variables organized into four, first-level subsystem core dimensions and their interconnections (Ostrom, 2010):

1 **Resource units:** the number of resource units contained in the estuary as a resource system (e.g., the number of rivers, groundwater sources, lakes, channeling systems, and small dams). Resource systems "are stock of variables that are capable under favorable conditions, of producing a maximum quantity of the flow variable without harming the stock or the resource system itself" (Ostrom, 1990, p. 30).
2 **Users:** the number of individuals who use the water resources connected to the river estuary. They can be expanded across different local communities, towns and cities with a complex set of use and goals, from industry to households, recreation, and commercial purposes. Users might be distributed across different types, individual, families, farmers, companies, public administration. Multiple users or appropriators can appropriate these units, and it is costly or infeasible to exclude one appropriator or group of users from a resource system (1990, p. 30).
3 **Resource socio-ecosystems:** the different resource systems that are based in the region, the estuary and protected areas connected by different ecosystem flows.
4 **Governance systems:** the government of the hydric resources based at the estuary, including the organization(s) (Bilbao-Biscay Water Consortium) that manage the

water system, with specific rules related to the use of these resources, and how the rules are made.

In addition, each of these four core subsystem variables includes multiple second-level variables (e.g., the size of a resource system, the mobility of a resource unit, the level of governance, users' knowledge of the resource system, the ease of measurement, the time scale for its regeneration, and their the spatial extent); these, in turn, consist of other, deeper-level variables (Ostrom, 2010). So the analysis of these variables could be validated by the analysis of data regarding the system-one place, including Knowledge Management data.

For the purposes of this study, we consider the river estuary to be a water socio-ecosystem. We frame our discussion on the analysis of one such regional water ecosystem, the Bilbao *ria*, including all its connected hydric resources. First, water resources have a greater and more dispersed regional geographic scale (two rivers, tributaries, small dams, and wastewater treatment plants). Second, we also consider society at large and industry as a key water system component that acts as an important agent of change within the system (Pahl-Wostl, 2007). Consequently, our study encompasses a wide variety of users (households, local communities, cities, and industry), often in conflict due to the multi-level nature of the system and the multiple types of actors. The river estuary is affected by large groups of users across different cities, all searching for their long-term sustainability, and economic, social and environmental value creation. In contrast to the local, common-pool resource approach arguing that resources (e.g., a lake or forest) only affect a small group of users, the regional estuary ecosystem (Bilbao *ria*) requires the effective management, preservation and long-term sustainability of that resource based on its fair distribution for millions of users and the area's industrial, economic, social, and cultural development. Third, collaboration and collective action are particularly important to manage the long-term use of the estuary across many different groups of users, economic and social appropriators, and other stakeholders. This is clear when we analyze how the estuary is affected by water use and discharge, water distribution, and pollution.

10.1.2 Commons' governance and collective action

Managing and protecting the regional estuary from over-use and over-exploitation as a new commons require a long-term vision and collective action between different users and stakeholders (Ostrom *et al.*, 1999). The concept of collective action could be understood as the action of a given group working towards a shared common goal, by which individuals take advantage of the strength of the group's resources, knowledge, and efforts (Ostrom, 1990). Sociologists, political scientists, and economists have been studying how collective action works. In sociology, collective action can be defined as a voluntary action taken by a group to achieve a common interest. In economic theory, the study of collective action is related to the provision of public and common goods. Economists have also studied the impact of externalities on group behavior. In the last few decades, different approaches have attempted to explain the difficulties of assuming collective action to manage commons and public goods. A key approach is based on the neoclassic rational economy perspective, consisting of individual, rational self-interest as proposed by Hardin (1968) with the notion of "Tragedy of the Commons". The latter is based on the hypothesis that common goods shared by a collectivity of users will always be depleted by over-use or degradation when a free-access regime exists for those resources. According to Hardin (1968), individuals acting rationally will always act contrarily to the group's best long-term interests, depleting common resources. Hardin (1968) argued that there are two possible (imperfect) solutions: full private property rights and privatization or, contrarily, public management and legislation/regulation. As part of his analysis on how to avoid depleting commons, he proposed increasing privatization and ensuring full property rights and sometimes strong centralized management, arguing that common-pool resources should be progressively regulated by the principles of economic efficiency and scientific management. He also suggested that certain global commons (e.g., the environment and water) couldn't be privatized and had to be protected in different ways, including via administrative law (regulations) and agreed-on mutual coercion.

By contrast, Olson's research (1965) explains the logic of collective action based on rational self-interest that often leads to inaction depending on the size of the group or organizations. He argued that, in large organizations, a significant part of individuals will act in keeping with their own interests, adopting free-rider attitudes, instead of adopting collective actions. In this view, individuals do not act moved by common interest. The costs of participating in collective action are relatively high for them, and they do not take any actions in the end. Olson (1965) argued that, unless the number of individuals in a group was quite small or there were some coercive mechanisms or some other special arrangements to make individuals act in the common interest, rational, self-interested individuals would not act together to achieve those common interests.

Aiming to discuss how common and collective action works, Ostrom (1990) examined Hardin's (1968) and Olson's (1965) approaches. Her research demonstrates the existence of long-lasting common-pool resource regimes as institutions of collective action. She studied how different local common-pool resources (e.g., water basins, fisheries, pastures, and forests) across the planet had been self-managed by users for centuries, generating common rules. The key question Ostrom (1990, p. 27) pursued was understanding how collective action works coping with free-riding, arranging for the supply of new institutions, solving commitment problems, and monitoring individual compliance with a set of rules: a group of principals (users) in an interdependent situation (related to a common-pool and affected by free-riders and other social dilemmas) have been able to self-organize and self-govern themselves for a long period of time to obtain ongoing, joint benefits when they all faced temptations to free-ride and deplete the resource in question. Her main conclusion was that, traditionally, commons users have been able to self-organize themselves, adopting rules and collective institutional arrangements, setting up a different space between the market and free enterprise and the public economy. Users are able to define principles and rules regarding: access to the group (membership), the use and distribution of the resources and services the parties own collectively, and the introduction of monitoring and sanctioning functions. Ostrom (1990, reviewed by Cox *et al.*, 2009) studied the following principles: (1) users set up clear defined boundaries regarding the local common-pool resource; (2) they set up rules that are well matched to local needs and conditions; (3) individuals affected by these rules can usually participate in modifying the rules; (4) the right of community members to devise their own rules is respected by external authorities; (5) a system for self-monitoring members' behavior is established; (6) a graduated system of sanctions is available; (7) community members have access to low-cost conflict-resolution mechanisms; and (8) nested enterprises exist, that is, the appropriation, provision, monitoring and sanctioning, conflict resolution, and other governance activities are organized in a nested structure with multiple layers of activities.

This is the case with the Bilbao-Biscay Water Consortium that can be studied as a collective action institution, with its rules, governance structure, and broad complexity. We study how collective action principles have been developed for water management in the Bilbao estuary while transforming the old approach to control water management towards a more adaptive and flexible novel approach to water resources.

10.1.3 Water management: From control to adaptive water management

In the last few years, with the increase of water cycle pollution, overstress and scarcity, as well as the growing impact of climate change and the emergence of water cycle disruptions, a new approach to water management has emerged, attracting the attention of researchers and practitioners (Tilmant, van der Zaag, & Fortemps, 2007). This new approach has been named *adaptive water management* (Pahl-Wostl, 2007, Pahl-Wost *et al.*, 2010) and has been framed in contrast to traditional control water management.

The traditional management paradigm is based on prediction and control, with a goal-oriented, centralized and hierarchical governance structure. In that case, water management aimed to plan, develop, distribute, and manage water resources optimally, adopting a control and top–down managerial approach, balancing the competing demands for clean water from cities, regions, and countries as a centralized and hierarchical governance system (Pahl-Wostl, 2007, p. 573). The control management approach focused on the important role of technical, oriented water management, promoting highly centralized technological infrastructures and Knowledge Management, which attempt to distinguish between water cycle units, though without understanding the interconnections between water units and the main challenges that might be related to environmental problems in the affected areas. Regarding the multi-actor interaction process, it is important to understand how control management systems imply a weak participation by stakeholders in the governance structure. At the same time, sectorial or industry integration is based only on policy conflicts and chronic problems. The traditional water system also includes single and independent design sources for each unit or infrastructure.

In contrast, Pahl-Wostl (2007) and Pahl-Wostl *et al.* (2010) propose a paradigm shift in water management, from the hierarchical and controlled water management regimes (based on technological and centralized predictions, technological solutions, and the control of public policies) towards a more adaptive and integrated focus that includes system and complexity analysis and a learning approach to adapt to continuous changes in the water cycle. Adaptive water management is based on an evolutionary approach with a process-oriented strategy. A main principle consists of accepting risks and water cycle uncertainties of water cycle, while also enhancing risk dialogue and multi-actor interactions. This implies taking into account the interconnected changes in environmental and socio-economic systems, and adopting a social learning process (Pahl-Wostl *et al.*, 2010). In contrast to traditional water management techniques, adaptive water management forces organizations, managers, and users to understand the complexity of the water cycle and socio-ecosystems and the interconnections among water users, units, resources, and ecosystems. In order to manage this interconnection, and the systems of information, complex Knowledge Management plays an important role. Water socio-ecosystem analysis thus involves a system-based approach to understand the complex adaptive ecosystems of the estuary and its adjacent hydric resources. As mentioned above, adaptive water management implies an integrated perspective that takes into account environmental and physical components, biological and biogeochemical components, human, and physical and technological factors and, in particular, the mutual interdependences established across these systems (Pahl-Wostl, 2007). However, it is very difficult to predict future key drivers, as many different components can influence future trends.

Thus, adaptive water management approach has to be holistic, systemic, reflective, and adaptive (Pahl-Wostl, 2007) requiring different strategies. First it requires embedding the analysis of the region's economic, social, cultural, and technological characteristics, including human and natural life. Second, adaptive water management involves a process of analysis, reflection, and collective management of natural and environmental water ecosystems, including the water cycle, biodiversity, meteorology, and climate change, but also the economic development of the broad geographical area, agriculture, the industrial model, its economic development, and innovation system. Third, it is based on the transformation of technological and control mechanisms and infrastructures, broadening the components studied, including human and social interaction, urban routines, and economic impacts to understand, analyze, predict, and adapt to continuous changes and future interconnections with water ecosystems challenges. Fourth, it is framed under the "learn and adapt" principle. New environmental challenges worsened by climate change and other environmental ecosystem transformation have increased our awareness of how climate change will affect the water cycle and water ecosystems. All of these continuous environmental changes affect the risks and uncertainties associated with water ecosystems, and it is even more difficult to predict probabilities to manage these risks. Therefore, it is important to introduce systems-based analyses and predictions, including the complexity of water systems at different scales.

In the case of the Bilbao estuary and its ecosystems, the goal is to improve the river basin's adaptive capacity and interconnect adjacent water resources with economic, social, and cultural factors at different scales (Pahl-Wostl, 2007).

This implies three main changes:

1. Move from technical and technological solutions for specific water units to more holistic, systemic, and system-based flows.
2. Move from control management to more adaptive and flexible practices to make it more adaptive and operational in response to fast-changing socio-economic boundaries, climate change transformation, and environmental challenges.
3. The main goal of adaptive water management is to analyze water resources and interconnected ecosystems along with other natural ecosystems (forestry and biodiversity), and social and economic ecosystems (agriculture and manufacturing). When adopting the adaptive water system concept, the different components cannot be understood in isolation, thus requiring a search for technological, environmental, and socio-economic solutions in interconnected water ecosystems.

10.2 Case study: Water management and collective action in the Bilbao estuary

This case focuses on a collective initiative undertaken by the Bilbao-Biscay Water Consortium together with public governments, non-governmental networks and industry to clean up the Bilbao *ria* and create a modern system for the region's water distribution. The estuary was extremely polluted with heavy metals and waste from the city's uncontrolled urbanization during the 19th and 20th centuries. First, we study the natural ecosystems that make up this estuary, the area's industrial development, and the resulting pollution. Second, we study the different stages of the process to create and implement different initiatives to increase the entire water ecosystem's ability to respond to and repair the consequent damage, remove the existing waste, address the ecosystem's vulnerability, and improve the long-term sustainability of the river basin and the adjacent water resources and water sources.

For the analysis of the case study we adopt the adaptive water management approach (Pahl-Wostl, 2007; Pahl-Wostl *et al.*, 2010) to analyze collective actions and the institutional perspective. In the case of adaptive water management, this framework includes different components, linkages, and processes that constitute all the water management resources and those related to river basin and adjacent water resource governance (Pahl-Wostl *et al.*, 2010). It includes five key system dimensions or components that users need to understand:

1 The entire water cycle and water ecosystem with its environmental and geographical components: the river basin's geographical location and relevant indicators regarding the water system, including Knowledge Management.
2 The ecological systems regarding water pollution and water cycle stress (abiotic and biotic water system components, environmental services, water quality, natural storage capacity, and environmental hazards).
3 The technical infrastructure of water management systems, including water canalizations, small dams, reservoirs, waste water treatment plants, flood management systems, maintenance, and ownership.
4 The key economic and industrial systems, including the analysis of how the economic system and challenges as well as industrial development affected the entire process of cleaning up the estuary and transforming the water management process.
5 The societal and cultural systems, including the analysis of how cultural and societal attributes (social needs, urban development, inequality, economic crisis and growth, etc.) affected the process.

The case analyzes the period ranging from 1967 to 2016. Data was gathered mostly from secondary data. We review different documents published regarding the process to renew the Bilbao estuary, including books, public reports, local and regional media, and scientific documents. We also gathered primary data, including: (1) interviews with executives and professionals involved in the process; and (2) direct experience. One of the authors was an active participant in different local organizations. Among other roles, he was a member of the Bilbao Port Authority's Board of Directors from 1995 to 2005, a member of the Board of Directors of the public/private think-tank, Bilbao Metropoli 30, from 2000 to 2005, and a member of the Board of Directors of Bilbao Ria 2000, from 1995 to 2005, thus providing direct information that we contrasted with secondary data.

10.2.1 The estuary's natural ecosystem as a pole for economic growth: Industrial development and pollution

The Bilbao *ria* spans 20 km from the city of Bilbao at the top of the estuary to its delta in the ocean. Within these two extremes we find the most populated cities in the Biscay region, representing approximately 60% of the area's population with nearly 670,000 inhabitants.[iv] The left bank of the *ria*, including the mountains, are the source behind the Biscay region's industrialization: iron mines. The left bank includes five key cities: Barakaldo, Sestao, Portugalete, Santurtzi and Zierbena. In contrast, the right bank has two: Erandio (the most industrial) and the residential municipality of Getxo (the preferred residence of the emerging bourgeoisie and business class since the 19th century). There are other towns and cities not directly located on the banks of the river that use and discharge their waters. Currently, the estuary receives discharges from a total population of approximately one million inhabitants, 91.26% of the Biscay region's population and 48.20% of the population in the Basque Country.[v]

Since the middle ages, Bilbao's *ria* has served as shelter for boats attempting to reach the top of the estuary where the old village of Bilbao was located. Bilbao's *Casco Viejo* ("old quarter") held its market in Plaza Vieja, around San Anton Church. The village of Bilbao was also the key port for the Crown of Castile, exporting Castilian raw materials (e.g., wool and timber) to the rest of Europe. It was also the main port importing products to the Iberian Peninsula from Atlantic and European countries and part of the colonies in the Americas.[vi] With time, more docks were built from the estuary neck to the left and right banks, furthering the growth of Bilbao and the emergence of new, port-based towns such as Barakaldo, Sestao, and Portugalete.

In the first half of the 19th century, Bilbao *ria* experimented an important process of economic growth based on the exploitation of the iron mines located on the left bank of the estuary. The iron was the source of Bilbao Port's growing naval industry and steel and blast furnace industry. In this sense, the estuary is responsible for the emergence of a modern, entrepreneurial, and well-established mercantile industrial class that encouraged the establishment of an advanced naval industry and blast furnace industry. In the 20th century, Bilbao Port and the estuary that connected it to the ocean became a center of economic and industrial development in Spain and all over Europe (García Merino, 1987).

Until the middle of the 19th century, the quality of the water in the estuary still permitted fishing and recreational activities. People would swim on the beaches of Portugalete and Getxo. However, mining and industrialization, especially the iron and the blast furnace industry, along with shipbuilding led to the construction of a significant number of factories located along the estuary, dumping hazardous waste into the estuary without being treated. A second main factor behind the water's growing pollution was the increasing number of urban settlements along the estuary, especially in the first part of the 20th century and during the reindustrialization period in the 1950s. Most of the old towns along the estuary were transformed into big cities that grew disorderly to provide shelter to the huge influx of people from other areas of the Basque Country and Spain heading to Bilbao to work. Between 1900 and 1975, the region's population quadrupled,[vii] and all the waste generated by the new construction and new urban infrastructures were dumped indiscriminately into the estuary.

As a consequence, Bilbao *ria* became extremely polluted. In the 1960s, the estuary received approximately 2,000 tons of waste daily (Barreiro & Aguirre, 2005). According to the acclaimed philosopher and educator, Miguel de Unamuno (Bilbao-Salamanca, 1864–1936), the estuary was the dump for industrial and domestic waste, without any type of prior treatment (Eizaguirre, 1996). It was a giant navigable sewer (Eizaguirre, 1996) that, according to some informants, was so polluted, inhabitants of the area used to joke about the level of waste and dirtiness. In this sense, Dr. Javier Franco, a biologist and researcher at the Azti Technological Center, and one of the professionals responsible for analyzing the

Figure 10.2 Panoramic view of Bilbao *ria* as it passed through the University of Deusto at the beginning of the 20th century. *Source*: © University of Deusto.

estuary, indicated that in 1989, "this turbid yellowish-brown fluid presented a very serious oxygenation problem" (Elejabeitia, 2015), especially in the urban stretch of the estuary and on the bed of the river where there were highly polluted sediments with a high concentration of heavy metals. In addition, as a result of rapid population growth, all the cities along the estuary experienced serious shortages of drinking water, especially in left-bank cities. At different times during the 1960s, only two hours of drinking water could be supplied to most of the municipalities (e.g., in Bilbao, Barakaldo, Portugalete, Sestao, & Santurtzi). At that time, many city governments worked individually to find solutions to their water supply problems.

Public discontent began to grow in the 1960s during the Spanish dictatorship. In March, 1967, 19 municipalities across the estuary along with Bilbao's local government and the former Administrative Corporation of Metropolitan Bilbao created the *Bilbao-Biscay Water Consortium* with two main goals: first, install and manage water supply systems for the estuary's towns and cities; and, second, implement a sanitation plan for the estuary's water and improve the discharge and extraction processes. The Bilbao-Biscay Water Consortium has been spearheading the promotion of collective action since then.

During the second half of the 1970s an important economic crisis affected the entire blast furnace and naval industry in Europe, forcing the closure of steel mills, shipyards, and chemical industries in the Bilbao-Biscay region. All the cities along the *ria*'s left bank and in Erandio were seriously affected by the severe economic recession, leading to high unemployment and extreme poverty among the working class population. At the time, Spain was in the last stages of the Franco dictatorship (1939–1975), and most of the region's social and political forces were fighting for democracy. General Francisco Franco died in 1975, and the country established a new democratic government. During this transformation, the Basque Country recovered its self-government, thus beginning a new stage in its autonomy in 1978 after the Referendum approved the bilateral negotiations between the Basque Country and the Spanish Government. To address the economic crisis, the new Basque Government clearly visualized the goal of regenerating the Biscay region's and Metropolitan Bilbao's economic and industrial development, promoting a new economic and urban development model. This endeavor was enacted together by the Basque Government, the Provincial Government of Biscay and local governments. They aimed to clean up the estuary and build new water channelization and distribution systems.

In August 1983, the estuary was affected by heavy rains that the existing infrastructures could not drain properly. This led to dangerous floods in Bilbao's old quarter and in left and right-bank municipalities. Those cities were already badly affected by the economic crisis, with high levels of poverty, unemployment, and a lack of urban facilities. More than 100 cities and towns in Biscay were considerably damaged as a result of the floods, with 48 human victims (dead and missing) and approximately 1.2 billion euros in property damage. This natural disaster also become a catalyst to stimulate public, private and social awareness about the need to regenerate the entire metropolitan Bilbao area and the Biscay region, specially addressing the state of urban development and its infrastructures.[viii]

10.2.2 Collective action: Bilbao-Biscay Water Consortium

In fact, Bilbao City Council had been attempting to eliminate wastewater in the estuary for the last two centuries. In 1885, due to a cholera epidemic that struck the region of Biscay and Bilbao, in particular, Bilbao City Council issued a call for proposals to develop a sanitation system for the estuary. At that time, Bilbao residents still extracted fresh water from the estuary (Cárcamo, 2012). As a result of this call for proposals, Bilbao had an exceptional sanitation system in place by 1900, only allowing wastewater to be discharged directly into the sea and not the estuary. However, after the Spanish Civil War (1936–1939), these rules were no longer respected. During the 1960s and 1970s, all the estuary cities, blast furnaces, and shipyards dumped their wastewater into the *ria* without any sort of prior treatment. Fernández Pérez (2005) describes this as a "situation of extreme environmental degradation".

In this context, in the 1960s, public authorities and also private businesses and industry associations joined together to discuss the need to sanitize the estuary and design and create a collective solution to channel fresh water and control wastewater dumping in the *ria*. The estuary was increasingly seen as a common resource shared by multiple municipalities along the riverbanks for use by both citizens and industry. They felt that the estuary had to be preserved over the long-term; otherwise, it would be impossible to live in the region. Consequently, in March 1967, 19 municipalities located along the banks of the estuary and adjacent areas launched the Bilbao-Biscay Water Consortium, a supra-municipal organization that would allow city councils to work together to solve the problems with the water supply and sanitation services (Barreiro & Aguirre, 2005). The Bilbao-Biscay Water Consortium was founded to meet two clear needs: first, ensure the supply of fresh water to the different municipalities in the Biscay region, in particular to the most important cities along both banks, and, second, clean and rehabilitate the estuary, properly channeling fresh water and wastewater discharges. Later on, with the return of democracy in Spain, the Bilbao-Biscay Water Consortium was enlarged to encompass 80 municipalities in the Biscay region. Today, the consortium serves 91.26% of the population in the region and 48.2% of the population in the entire Autonomous Community of the Basque Country (Figure 10.3).[ix]

Two people played a leading role in creating the Bilbao-Biscay Water Consortium. Angel Galindez was the Deputy Councilor of Water Management for Bilbao City Council.[x] He was an agricultural engineer with great experience in

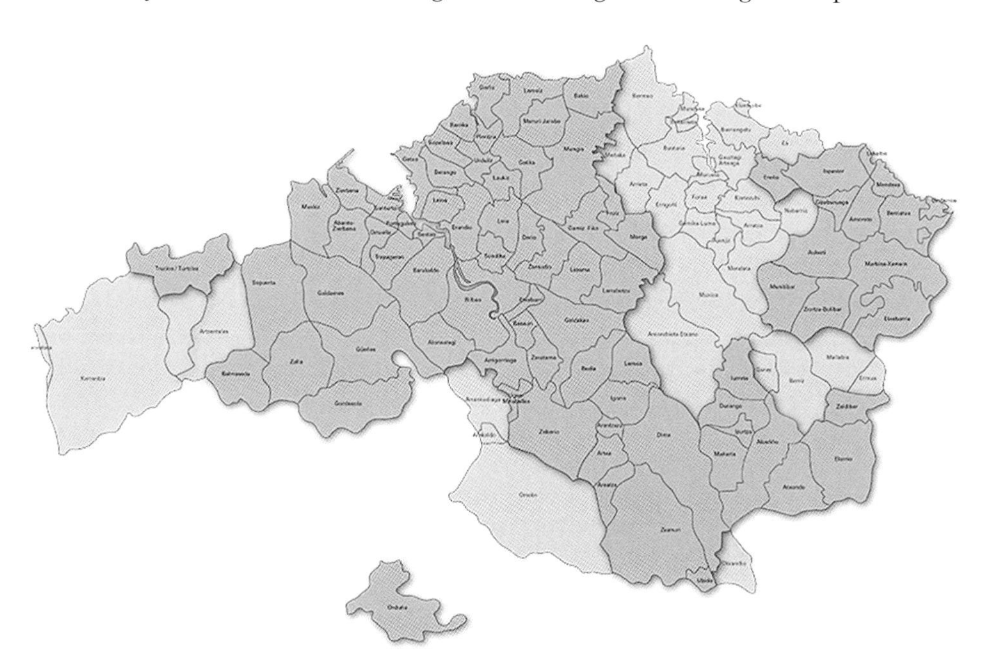

Figure 10.3 Map of the towns and cities member of Bilbao-Biscay Water Consortium (2017). *Source*: Bilbao-Biscay Water Consortium. http://www.consorciodeaguas.com/Web/QuienesSomos/Entidades.aspx (viewed on February 21st, 2017).

the construction of dams for the local firm, Iberduero. In 1960 he visualized and proposed launching a supra-municipal initiative. He first created a small office for Bilbao's local government though later championing the creation of a collective initiative that would become the Bilbao-Biscay Water Consortium in 1967 (Lakunza, 1997). The second person was José Miguel Eizaguirre, an engineer who joined Galindez in the local Bilbao office in 1960. He became the first Technical Director and Manager of the Bilbao-Biscay Water Consortium in 1967, leading the design and implementation of the plan for more than 35 years until his retirement. Technically, he was the main designer of the various water infrastructures and the sanitation and clean-up plan, as well as the chief promoter of collective awareness and reflection.

The consortium developed along different processes and stages: (1) from 1967 to 1975, the consortium focused primarily on solving water supply needs by capturing water from Zadorra River, building small dams in the mountains around the estuary, building the Venta Alta (Arrigorriaga) water purification station and several pipelines to distribute the extracted and treated water; and (2) from 1975 on, the consortium began studying how to clean and regenerate the *ria*'s healthy water cycle, until finally launching the *Plan for the Integral Sanitation and Clean-up of the Estuary* in 1979 (Figure 10.4).

10.2.3 Water supply, collection and distribution

With the goal of collecting and distributing clean water to cities, households, industry, and others, the Bilbao-Biscay Water Consortium works throughout the water management cycle, building an integrated value chain and forming

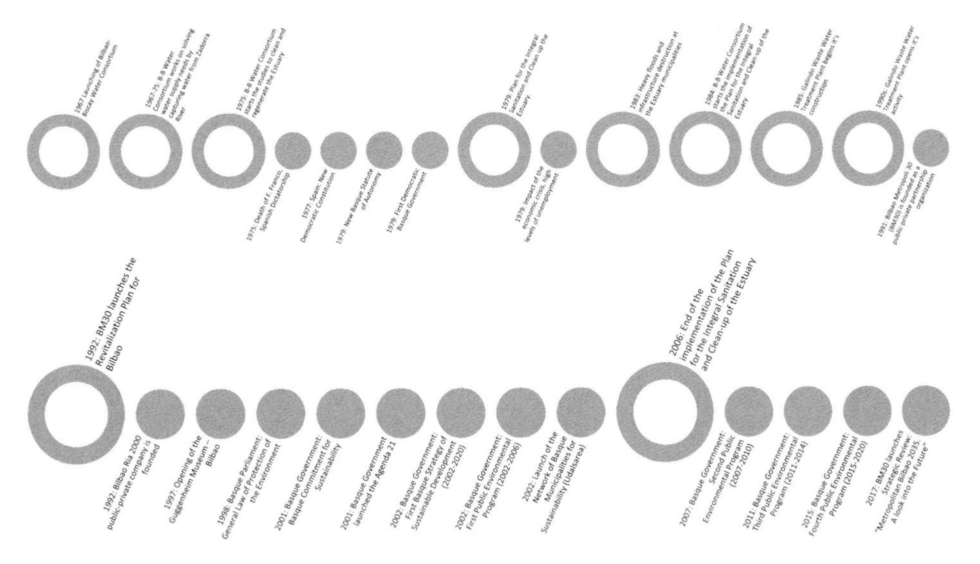

Figure 10.4 Timeline with main events on the process of cleaning Bilbao *ria* (1960–2017).

a virtuous water management circle in the region.[xi] The whole process begins by capturing water primarily from the Ulibarri Gamboa and Santa Engracia reservoirs (located in the basin of Zadorra River in Alava, a province to the east of Biscay). This collection accounts for 90% of the water distributed by the consortium. In addition, there are other secondary contributions from various marshes and underground sources. Also, in case of emergency, the consortium has provisions in place to capture water from other rivers in the region.

After its capture, water purification occurs in five plants, treating up to 111 million cubic liters.[xii] The main plant is Venta Alta (Arrigorriaga), treating the water collected from the Zadorra reservoirs, which, despite its high quality, contains some impurities (solid particles and some metals). Water quality parameters are controlled throughout the entire process: from its springs, rivers, and reservoirs to the water treatment stations and distribution networks. This rigorous monitoring includes quality certifications for both water supply and its treatment.[xiii]

Once in the plant, the water undergoes an initial chlorination process to disinfect it from pathogens. Later, it goes to the decantation deposits where it rests and helps impurities settle to the bottom, forming sludge. This operation is accelerated by adding chemical reagents to the water to allow the aggregation of dissolved particles. In this way, heavy clots are formed that settle to the bottom of the decanters, leaving clean water on the surface. The sludge is then removed, dried, and taken to a controlled landfill. The decanted water is distributed through various filters consisting of successive layers of sand of different thickness, which retain the smaller particles. At this moment only the chlorine is added to ensure its potability in the distribution process and adjust the pH balance. Once the water has been treated, it is distributed through a network of more than 400 kilometers of pipes, 40 regulating tanks, and 37 pumps that allow water to be supplied to homes, industries, and shops.

The consortium also collects wastewater through independent systems. Its infrastructures include collectors and sewage treatment plants located in the region's different watersheds. Once collected, several wastewater treatment plants clean the water and incinerate the sludge. The purified water, in addition to being reused in the purification process, is discharged to the Ballonti River and ends in the ocean, thus restarting the water cycle again when it becomes rainwater.

10.2.4 The plan for the integral sanitation and clean-up of the estuary

In 1975, after the Bilbao-Biscay Water Consortium had successfully resolved the water supply problem for all the municipalities in the region, it turned to the serious problem of the estuary's sanitation. In the 1970s, there was little environmental legislation. Thus, in addition to diagnosing the problem, the consortium's technicians had to:

Prospect the future development of the region in order to establish achievable quality objectives and to make the domestic and industrial uses of the water

> compatible, determine the purification system to be implemented and establish regulations for the discharge and treatment of wastewater. During these years, the technicians undertook several complementary studies necessary to implement the plan: sociological studies, environmental impact, reuse of the water from the purification processes and analysis of the incidence of industrial discharges in the biological purification processes. (Eizaguirre, 1996, p. 74)

During the following two years, the consortium carried out different environmental and sociological studies to analyze the environmental impact of the plan: an oceanographic impact, how to reuse treated water, and how to reduce the negative impacts of future discharges on the natural channels (Eizaguirre, 1994, 1996). After the diagnosis, engineers also developed mathematical models of the *ria* (vertical flows) and the bay (horizontal flows) to model water flows for the holistic sanitation and clean-up of the estuary. An international call for projects was also proposed (Barrero & Aguirre, 2005). All of these reports and discussions served as the foundation for the final plan drafted by the consortium in 1979.

Results from water quality analyses were worrying. Arranz (2012) notes that these studies estimated that:

> The *ria* received 900 tons of solid waste daily, primarily from mining, 400 tons of acid spills, 80 tons of metals, one ton of cyanide compounds and 600 tons of DOC, 20 tons of nitrogen compounds, etc. The estuary worked like a chemical reactor, consuming the oxygen in the water, flocculating large amounts of suspended matter, transferring metals and toxic organic substances to sediments, inhibiting the emission of smelly gases from the decomposition of organic matter that would have created an unbearable atmosphere, ultimately leaving it with no possibility of life. (Arranz, 2012, p. 275)

Therefore, Bilbao's estuary was an extremely polluted water ecosystem. According to Eizaguirre (1996, p. 73), the "wastewater of almost 1 million inhabitants and a formidable industrial load were dumped, which, measured in one of the pollution indexes, was equivalent to 3.5 million inhabitants."

Cearreta (1992) prepared the following table detailing the presence of heavy metals in the *ria* according to analyses carried out by various public institutions and NGOs (Basque Government, Provincial Government of Biscay, Bilbao-Biscay Water Consortium, and Greenpeace) published between 1985 and 1991 (Table 10.1).

After analyzing the different studies, consortium technicians together with all the members of the Bilbao-Biscay Water Consortium proceeded to discuss and diagnose the situation of the river and the estuary basin in order to find the best solution. They even analyzed different scenarios, from non-intervention to holistic and integral sanitation. Consortium members, Basque Government and local public administration officials, boards of directors, and technicians discussed different possibilities to carry out realistic interventions to clean up the estuary, while also promoting economic and urban development and controlling economic budgets. During the discussion on the plan's implementation, two different scenarios and alternatives were faced (Eizaguirre, 1996): first, dumping wastewater directly into the sea through submarine outfalls, and, second, constructing treatment plants along the estuary and then pouring wastewater

Table 10.1 List of heavy metals found in the Bilbao estuary between 1985–1991

Station	Cd	Pb	Cu	Zn	Cr	Ni
El Arenal	6	267	342	2180	924	60
La Salve	10	298	372	2430	782	58
Deusto	10	275	244	1600	157	34
Olabeaga	0.2	62	85	424	23	20
Canal de Deusto	54	876	562	5260	340	62
Elorrieta	8	184	171	978	166	29
Zorroza	4	143	117	678	14	28
Desemboc Asua	142	1600	1070	2380	100	33
Desemboc Kadagua	29	683	450	1850	128	41
Erandio	13	371	300	1480	179	31
Simondrogas	10	211	328	1330	145	35
Axpe	114	1800	1990	5160	315	52
Lamiako	73	1060	916	3140	194	29
La Benedicta	64	885	667	3370	211	29
Portugalete	19	1390	501	1850	166	20
Puerto Santurtzi	40	538	467	2660	152	25
Punta Lucero	78	1808	614	2293	198	–
FGR	2	42	38	60	75	60

Source: Cearreta (1992).

back into the stream. They finally choose the second option as it represented a better adaptive and more integrated solution to the needs of the water cycle, socio-economic development, and human and natural life.

In 1979 the consortium launched the *Plan for the Integral Sanitation and Clean-up of the Estuary*. This plan was approved by the consortium's General Assembly with a view to implementing it from 1984 to 2001. It would need five more years to be implemented. Barreiro and Aguirre (2005, p. 27) point out that: "The objective of the plan was to guarantee the aquatic life of the river system, to recover the coastline and beaches for bathing and recreation, to achieve acceptable esthetic conditions for water, and to drastically reduce toxic spills from industrial activities. They established the aim of achieving a minimum content of 6 mg/L dissolved oxygen at any point and time in the estuary. To achieve this, the Master Plan included a forecast investment of 25.28 billion *pesetas* in 1979 and a long-term implementation period of 18 years."

During the first years of the plan's implementation, the consortium faced numerous difficulties, though the environmental requirements helped it choose the second alternative as the backbone of the project. Essentially, the treatment policy was to build "a set of sewers, collectors and interceptors that collect all wastewater (both domestic and industrial) and a part of the rainwater, to be taken to sewage plants where they undergo complete treatment before its discharge in the estuary" (Eizaguirre, 1996, p. 28).

10.2.5 *Building new water sanitation integrated infrastructures*

The *Plan for the Integral Sanitation and Clean-up of the Estuary* aimed to build and develop two main systems of infrastructures across the estuary: the first was to build a waste water treatment plant, and the second to build a system of pipelines, pumping stations, spillways, and wastewater treatment stations across the estuary to clean and channel the water cycle.

A main infrastructure was built the Waste Water Treatment Plant (EDAR) in Galindo was built to eliminate discharges from Bilbao and the estuary's left bank.[xiv] The Galindo Waste Water Treatment Plant was built between 1985 and 1990 in the town of Sestao, along the left bank of the estuary. Wastewater from human and industrial activity is collected through approximately 250 kilometers of collectors and interceptors. This water is transported from the right bank through two subfluvial tunnels to Galindo. Since the plant was completed in 1990, approximately 350,000 cubic meters of wastewater have been treated per day, with a maximum throughput of 12,150 liters per second.[xv] The consortium[xvi] thus summarizes the treatment process as follows:

> In this treatment plant, two different processes take place: the purification of the waters and the treatment of the sludge derived from the process. Water treatment was initiated by eliminating larger solid waste that is retained by means of different screens. The water then passes into a tank where the sands settle to the bottom and the fats are raised to the surface by means of blown air. These fats are removed by scrapers, and the sands are extracted by pumping. In the primary decantation tanks, the organic matter that is suspended in the water settles at the bottom, forming sludge. The water passes to a distribution channel to continue the purification process.
>
> The next destination is biological treatment. The water reaches vats in which an oxygenation system facilitates the development of bacteria. These organisms feed on organic matter so they make most of the substances dissolved in the water, while those retained in the secondary decantation disappear. Regarding the sludge extracted in the different water decantation processes, after treating them to thicken them, they are dried and transported to furnaces where they are incinerated at 850 degrees Celsius. The steam that is extracted from this process is used in a cogeneration plant that produces enough energy to supply the needs of all the purifier's installations. (Bilbao Bizkaia Ur Partzuergoa-Consorcio de Aguas de Bilbao Bizkaia: "Depuración de Aguas Residuales".[xvii]

One of the main objectives of the plan was to recuperate water oxygen levels (not less than 60%). Between 1996 and 1999, the Galindo facility was responsible for biological purification, while sewage from the right bank of the estuary was channeled. In the following years, it continued to integrate additional discharges into the purification system, eventually eliminating wastewater in the area.[xviii] Part of the waste could not be cleaned because the existing contaminants had already become inert. Azti, the research center which, together with the University of the Basque Country, analyzed the estuary's water quality indicates that these levels

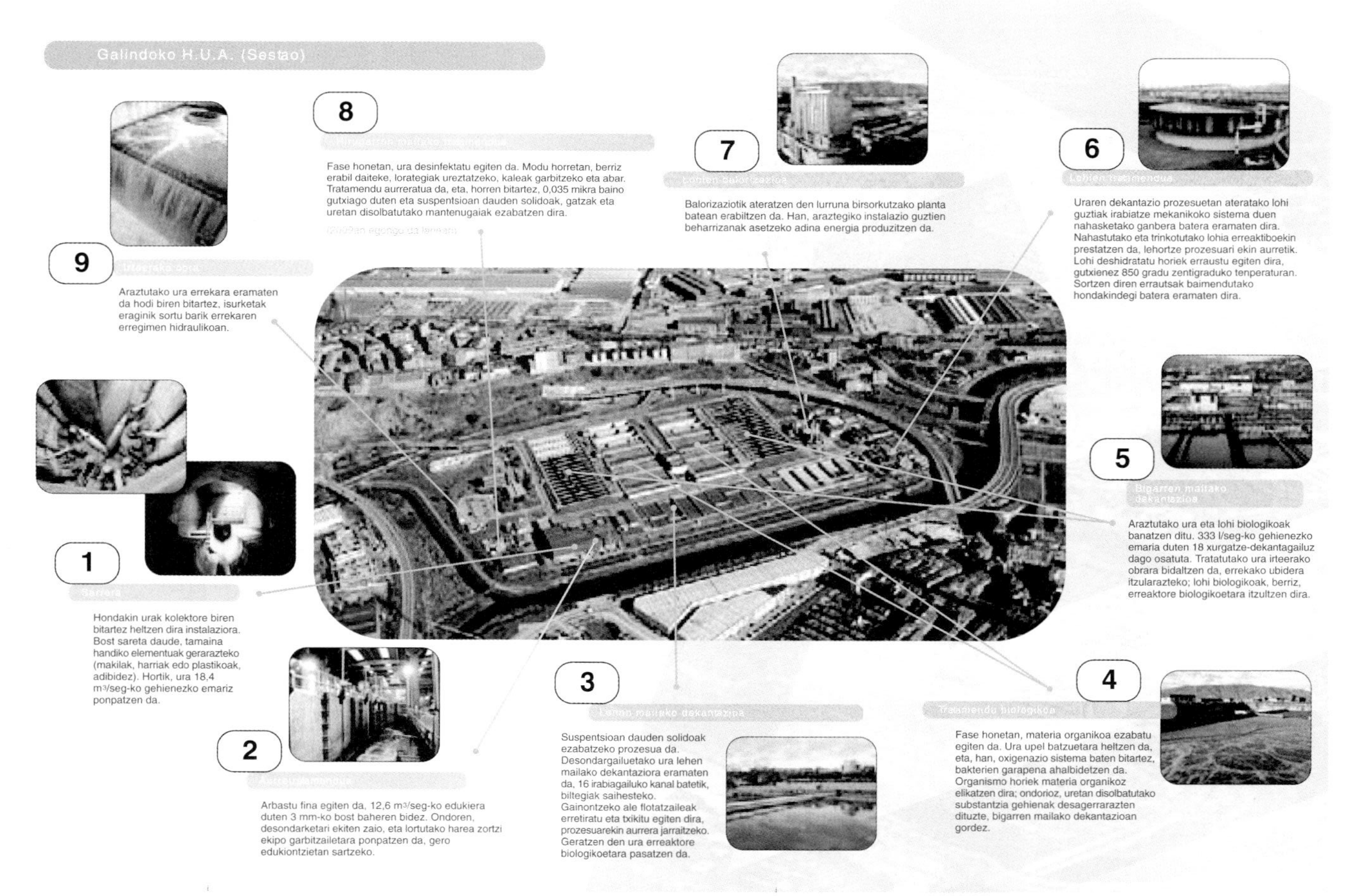

Figure 10.5 Waste water treatment plant (EDR) in Galindo and the estuary water purification process. *Source*: Bilbao Bizkaia Ur Partzuergoa-Consorcio de Aguas de Bilbao Bizkaia: "Depuración de Aguas Residuales". https://www.consorciodeaguas.com/web/CicloAgua/ciclodelagua.aspx?id=depuracion, viewed on 15 March 2017.

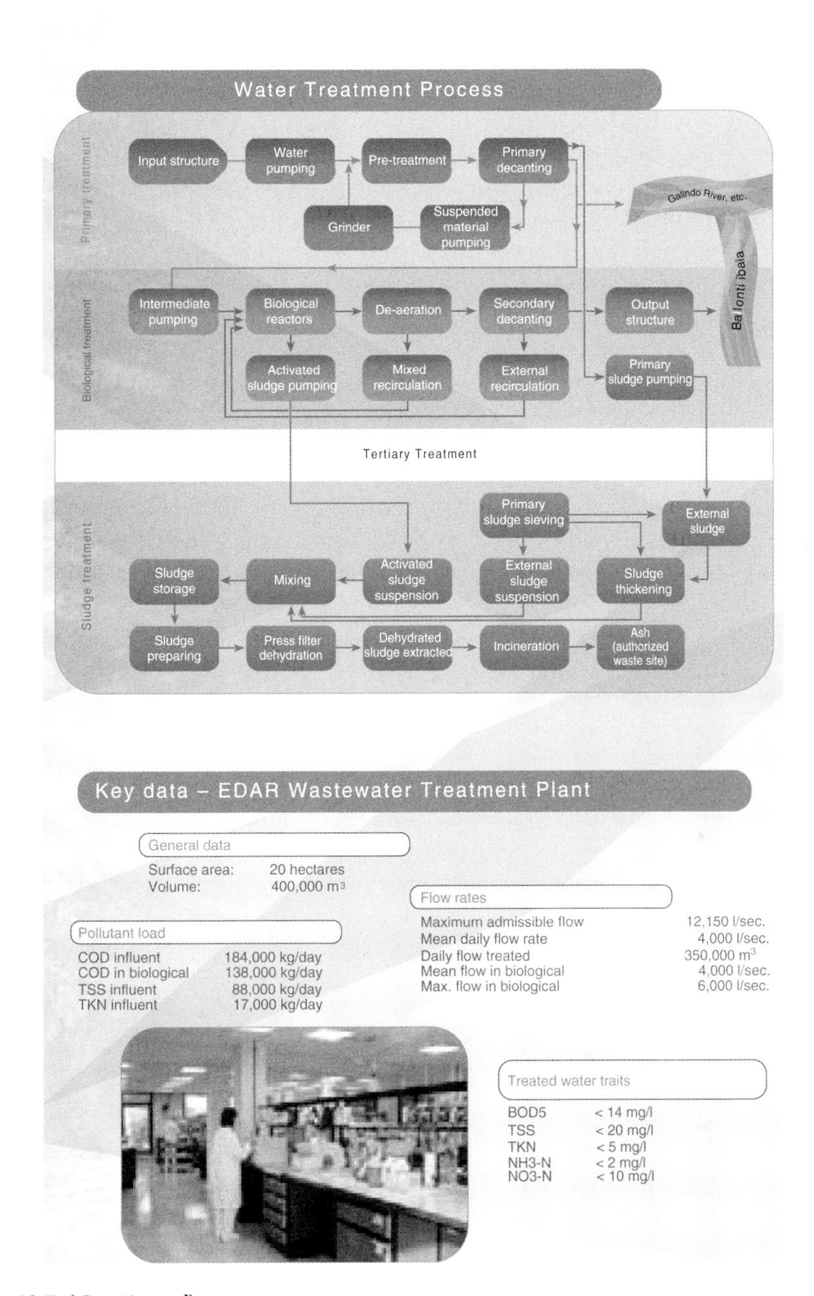

Figure 10.5 (*Continued*)

have gone from 40% in the 1990s to values close to 90% by 2015. [xix] Sludge has been turned to sediment, and pollution reduced. As a result, more than 60 species of fish and some crustaceans have been detected living in the estuary's waters as well as serving as habitat for several birds.

Secondly, the rest of the infrastructure to sanitize the water cycle was built. Currently, there are a total of 51 main pipelines, 42 pumping stations, 84 spillways, and 27 wastewater treatment stations that treat more than 130 million cubic meters per year. These modern infrastructures complement the more than 300 kilometers of collectors and interceptors built and the 100 kilometers of refurbished sewage lines.[xx] As a result of the visual impact and the bad odors generated by the sludge during its treatment, the consortium has also undertaken various projects to improve the surrounding landscape, planting 2,000 trees, creating fences with meshes and constructing a slope that prevents seeing inside the facility from the nearby subway station in Urbinaga.

The plan's implementation was finalized in 2006, 5 years after the initial target date. In 35 years (1979–2015)[xxi] public authorities invested over 1 billion euros to build the collective infrastructure. 25% was financed by the Spanish Government, 17% by the Basque Government, 23% by the Provincial Government of Bizkaia, and 35% by users supplied by the Bilbao-Biscay Water Consortium (Barreiro & Aguirre, 2005). Industry also participated in the investment required, contributing over 118 million euros[xxii] (eliminating discharges and installing wastewater pre-treatment plants to be able to channel their wastewater into the consortium's facilities).

10.3 Inquiring adaptive water management and Knowledge Management approach

Although the Bilbao-Biscay Water Consortium originally adopted a control water management approach as a centralized top-down approach, with time, it transformed its understanding towards a more adaptive and systemic approach, studying the different dimensions of water cycles and hydric resources of the region, especially the estuary and including the long-term use of water units and ecosystems (Pahl-Wostl *et al.*, 2010). In this section we analyze the main dimensions of the adaptive water management approach adopted by Bilbao-Biscay Water Consortium.

10.3.1 Bilbao-Biscay Water Consortium: From control to adaptive water management

Since the launch of Bilbao-Biscay Water Consortium, the technical and managerial team and different members aimed to understand the region's water

Figure 10.6 Panoramic view of Bilbao *ria* as it passed through the University of Deusto at the beginning of the 21st century. *Source*: © University of Deusto.

resource needs, enclosing an advance Knowledge Management focus. This is a main dimension of adaptive water management. The consortium studied fresh water reserves, gathering data from the evolving socio-ecosystems, adopting a Knowledge Management and learning approach based in the analysis of data contrasting different solutions and finally, choosing balanced projects based on different water channelization and treatment systems for industrial and household use and waste. In the process of designing and implementing the water channelization system, the members of the consortium knew that they needed to apply the emergent knowledge generated by environmental and water management scientists, including new approaches to water cycle analysis and complexity of resources. They also studied the needs of multiple users and the mutual dependences among them. Hence, although the Bilbao-Biscay Water Consortium did not call it "adaptive water management", in the 1990s the technical and managerial team furthered their analysis and understanding of the interconnections and complexities across different water cycle components (water units and ecosystems) and different users to maintain and improve the state of water resources in the estuary. This included water flow environmental quality and water available for the ecosystem, the aim being to preserve the quality of the water flow and maintain the integrity of river ecosystems and preserve their ability to provide valuable services to humans.

Secondly, an important dimension of adaptive water management is based on a systematic process to apply continuously improving policies and practices, including learning-by-doing strategies and providing flexibility and adaptability to these boundary conditions (Pahl-Wostl *et al.*, 2007). The analysis of different policies promoted by the Bilbao-Biscay Water Consortium shows that they adopted a "learning-by-doing" approach with a flexible and adaptive process, learning from previous successful experiences and failures. A key challenge for adaptive water management is how to collectively decide on and solve "the allocation of water for multiple uses taking into account different requirements for water quality and the spatio-temporal variability of both supply and demand. The consortium promoted the study of the main vulnerabilities and risks of the different components, in a learning by doing process. So, they learn from different studies and data applying different principles such as social equity, economic efficiency and environmental sustainability all taken into account (Pahl-Wostl, 2006, p. 54).

Third, another main dimension of adaptive water management also includes the analysis of how water ecosystems are affected by fast-changing socio-economic boundary conditions (Pahl-Wostl, 2007). The activities promoted by the Bilbao-Biscay Water Consortium to clean the estuary were developed in parallel to the economic, industrial, and social regeneration of the metropolitan Bilbao region. Water systems and adaptive water management include different "human, physical, biological, and biogeochemical components of water and hydric systems and their interactions" (Pahl-Wostl, 2007, p. 50). In the case of Bilbao-Biscay Water Consortium the water management includes water ecosystem complexity at different scales. It implies facing economic, environmental, and social adversity, as well as radical changes ranging from technical management to the true integration of natural, social, and human dimensions. It also requires making water management more adaptive and flexible, finding operational solutions in fast-changing socio-economic boundary conditions and the impact of climate change on water resources (Pahl-Wostl, 2007).

Fourth, the analysis of water cycles in adaptive water management also has to be analyzed as a common resource shared by multiple users, organizations, and communities affected by mutual interdependences across water geographies (Pahl-Wostl, 2007). The Bilbao-Biscay Water Consortium promoted the idea that the causes of the estuary's extreme pollution and water cycle degradation could be explained by the lack of industrial and environmental regulation, public regulation and the free-rider problem. Different users, cities and towns, individual people, families, companies, and organizations polluted the river basin without understanding their interconnection, extracting resources to secure short-term gains out of self-interest, without regard for the long-term consequences for the entire set of users. The consortium constantly issued messages to try to change this self-interest and promote the principle of collective action and mutual interdependencies.

So, Bilbao-Biscay Water Consortium consistently promoted the growing awareness about the need to co-manage water resources across municipalities and regions (Pahl-Wost *et al.*, 2010), making water management collective and interconnected. Currently, water management is a major challenge for local communities, cities, regions, companies, and industrial parks across the estuary. There is also an emerging consciousness of solidarity between communities and stakeholders across the estuary based on long-term use and sustainability, next generation survival, and social equity.

Fifth, linked to the previous dimensions, the co-management of new commons has led to the emergence of new institutional forms for collective action based on new co-governance structures between public authorities, private organizations, and civil society networks and groups. These structures are based on collaboration for the long-term use of natural resources and their sustainability, peer-production, and mass collaboration (Hess, 2008). This collaboration and cooperation are deployed between different types of stakeholders, with different sizes at the local and broader levels (e.g., global commons). We can see how this approach is clearly promoted by Bilbao-Biscay Water Consortium promoting new collective governance structure among its members. The Bilbao-Biscay Water Consortium has incorporated this participatory process into its governance structure (Pahl-Wostl, 2007): participatory assessment (scenario planning, uncertainties, and perceptions) and participatory implementation and monitoring (adaptation options, water allocation schemes, ecosystem structure, and cooperative governance).

10.3.2 Bilbao-Biscay Water Consortium: Analysis of innovative adaptive water management case

Therefore, the evolution of Bilbao-Biscay Water Consortium represents an innovative case in adaptive water management and Knowledge Management. Our finding show four main dimensions including the analysis of how Bilbao-Biscay Water Consortium evolved:

1. Collective action based on the collaboration of several municipalities and other public authorities to jointly solve problems related to the supply of water and wastewater treatment.
2. The integration of complexity and multiple stakeholders to address water management challenges.
3. The importance of collective governance and the management model created by the Bilbao-Biscay Water Consortium.
4. A system approach to solve water, environmental, social, and economic needs, using the water cycle as a key factor of analysis, including advanced Knowledge Management.

The main findings based on the analysis of water adaptive management and learning management are shown in Table 10.2.

Table 10.2 Main findings including water adaptive management, learning management and systemic approaches

Dimensions	Main activities/ institution	Water adaptive management	Learning management
Collective action	Plan for the integral sanitation and clean-up of the estuary	Collaboration of several municipalities and public authorities to jointly solve the estuary and water cycle challenges and ensure long-term sustainability of a common resource, the estuary	Continuous improving policies and practices Preserving the quality of water flow and the integrity of the hydric ecosystems data analysis and learning processes
Complexity Common resources	Analysis of the water cycle across the region: the estuary	Understanding the estuary ecosystems as common resources, sharing knowledge, and promoting and enhancing collective action between users, cities and industry	Gathering data from water units and ecosystems and different users Studying the causes of polluted ecosystems. Studying the impact of industry and households
Collective Governance	Bilbao-Biscay Water Consortium	Collective water management institution to jointly solve problems related to the supply of water and wastewater treatment Collective governance and participatory processes (multiple municipalities and other stakeholders)	Knowledge generation, dissemination and sharing Co-learning processes between multiple stakeholders to address water management challenges: municipalities and industry
System approach	A system approach to solve water cycle impacts connected to water and environmental ecosystems, environmental, economic development and social needs	Understanding the interconnection and complexities across different water cycle components (water units and ecosystems) Promoting a systemic approach to cleaning the estuary, including biological, but economic and social variables	Use of new systemic water cycle knowledge Towards more holistic and system-based knowledge approach to water cycles and hydric resources, including long-term use

Collaboration and collective action

The Bilbao-Biscay Water Consortium was launched based on the will and endeavor of different municipalities and public authorities to solve collective challenges and ensure the long-term sustainability of a common resource ecosystem, the estuary. All of them assumed that the efforts to supply water and clean the estuary basin could not be solved separately. Originally, local municipalities joined together, engaging the new democratic Basque Government and the Provincial Government of Biscay. Created in 1967 during the last years of the Spanish Dictatorship, people, political parties, and societal organizations in the Basque Country were part of a convulsive period in which citizens demonstrated in favor of democracy and the recovery of self-government, against the economic and industrial crisis and the closure of companies. Citizens also engaged in environmental campaigns such as the fight against the nuclear energy plant in Lemoiz and, later on, even against the construction of the Guggenheim Museum in the 1990s. However, the cleanliness of the estuary did not move citizens like other causes. The natural environment was still not appreciated by the population, contrary to what happens today. According to several interviewees,[xxiii] two key hypotheses explain this lack of popular mobilization: the first is that Bilbao and other cities on the river bank really lived with their backs to the *ria*, seeing it only as a sewer. It is important to note that only few houses located on the banks actually faced the estuary. Secondly, it was an important resource for industry, and, in the 1960s, people in the region didn't question the powerful role of industry in the growth of Bilbao, Biscay, and the entire Basque Country. Consequently, it was also difficult in the early 1970s to ask local and international companies to be environmentally responsible and accountable. Third, the local environmental movement and awareness of the water cycle emerged globally in the 1980s, after the United Nation's Conference on the Environment and Development held in Rio de Janeiro in 1982. In fact, Saiz *et al.* (1996) note that evidence of awareness regarding the issue appeared publically later on, citing different technical reports, including one from Greenpeace. All of these reports appeared in 1984, five years after the clean-up plan's approval.

In parallel, the former Managing Director of the consortium pointed out that the Plan for the Integral Sanitation and Clean-up of the Estuary was presented to its General Assembly once it had been highly discussed and accepted within the community and different municipalities (Eizaguirre, 1996). It was also important to generate a long-term discussion among all the municipalities and political parties to find the best solution for all the affected cities and towns. The environmental consciousness to develop the estuary as an environmental reserve and a valuable ecosystem emerged consistently at the start of the new millennium. This helped a new adaptive water management model emerge. The Bilbao-Biscay Water Consortium made the water cycle and environmental ecosystems key concerns for its new water management focus. In 2002, the Network of Basque Municipalities for Sustainability (Udalsarea) was created. It was based on participatory processes, enabling the population to define their own municipal sustainability strategies.[xxiv] Also, in collaboration with the area's municipalities and the Basque Government, the Agenda 21 school initiative for school-age children was created in 2001, with the *ria* receiving great attention.[xxv]

The Bilbao-Biscay Water Consortium also defined its own environmental policy.[xxvi] Its main principles and goals were as follows: (1) comply with current regional, national, and international environmental legislation and soft regulations linked to environmental challenges (e.g., climate change); (2) apply the environmental obligations for all the municipalities, towns, industries, and agents; (3) minimize environmental impact and prevent pollution in all its activities, using the best clean technologies and practices available for this; (4) optimize the consumption of natural resources, energy, and raw materials needed for its processes; (5) constantly review and improve the environmental performance and environmental management systems adopted; (6) promote the incorporation of sustainability principles in all the processes promoted by the consortium and at all the levels of the organization's activities, facilitating the integration of environmental quality, risk prevention, and labor market efforts, integrating all of these dimensions in an integrated management system that includes environmental values as key variables; (7) provide adequate environmental training to all employees and managers at all level of activity in order to increase their preparation and motivation with respect to the natural environment and their knowledge of advanced environmental science; and (8) disseminate the consortium's environmental commitments and its environmental performance results to all levels of the organization, among its members and throughout the entire value chain (e.g., customers, suppliers, public administrations, industry, and citizens).

Water management complexity and multiple stakeholders

Barreiro and Aguirre (2005) describe the entire process of designing and developing the Plan for the Integral Sanitation and Clean-up of the Estuary as one that had to solve key complexities: "what we find obvious today was achieved after the long-term titanic effort to overcome complex problems. We joined different political and social wills, contributed ideas and coordinated human teams among public bodies with responsibilities in water and sought out economic resources and funding systems."

The consortium was launched based on collective action and successfully engaged multiple organizations to face a common challenge. Different public administrations and associated entities found collective solutions to complex problems (Eizaguirre, 1996). Many different stakeholders participated in these collective actions. In addition to the consortium, other public-private based partnerships were created with this same intent. There are several legal forms to collaborate with public administrations, and Bilbao has turned several times to these solutions in its urban regeneration process. For example, Bilbao

Metropolitan 30[xxvii] was established as an association and Bilbao Ría 2000 as a public corporation.

On the other hand, companies have also been directly involved in cleaning the *ria*. They have built new waste treatment plants before their wastewater reaches the consortium's sewage treatment plants. According to Eizaguirre (1996), more than 2,500 industrial activities were inventoried in the area when preparing the clean-up plan, concluding that 65 of them represented 90% of the pollutant discharges. And, as proof of corporate commitment, Azti[xxviii] carried out inspection on a thousand companies in 2014, only finding a 2.1% degree of non-compliance. Not in vain, 412 business projects have been created since the plan's launch to eliminate discharges and build pretreatment plants.

In the 1970s and 1980s, the area's population, though subject to problems associated to water shortages and the estuary's contamination, was more concerned with high unemployment rates and reindustrialization. However, at a time when social networks did not exist, the consortium actively sought the complicity of citizens through appearances in the media to explain the situation in the region and inform them about the plan's progress.

Local and regional governments and industry understood that they needed to change the way they managed the water supply and use of the estuary, improving the distribution of water and cleaning the estuary in order to improve the water's quality.

Collective governance mechanisms

The Bilbao-Biscay Water Consortium is a regional, collective action institution, in which all the participating city councils take part in the decision-making process through its General Assembly. Currently, and in keeping with its Regulations on Organization and Governance,[xxix] the consortium's main activities and goals are[xxx]: (1) ensure the supply of water for all types of users in the primary network (collection, storage in reservoirs, transportation, treatment, conduction and delivery to large cities and end users or to the warehouses at the heads of the distribution network); (2) treat water in the primary network (general collectors and interceptors which connect wastewater to sewage treatment plants with all the associated waste treatment and disposal facilities, and emitters which return purified water to the receiving aquatic environment); (3) control industrial spills (authorization, monitoring, control and inspection of wastewater discharges in both the primary sanitation network and the sewerage network, as well as applying sanctions for infringements and abusive spills, and terminating water supply contracts); and (4) maintain relations with all the municipalities (control of the measuring equipment, service contracting, consumption levels, billing, and collection management).

Likewise, the consortium consists of five governance offices, with hierarchies and delegations similar to those found in incorporated companies: (1) a General Assembly; (2) a Steering Committee; (3) a President; (4) a Vice President; and (5) a Managing Director.

The General Assembly is formed by all the consortium's members, fundamentally, the municipalities, but also other public administrations such as the Provincial Government of Biscay and the Basque Government as well as different organizations linked to the river basin. All the municipalities name 1 to 5 representatives in the Assembly, depending on their number of inhabitants. The Provincial Government of Biscay, the Basque Government and the basin organizations name one representative each. All have a voice, though only the municipal and Provincial Government representatives have the right to vote. Each municipality's votes are also proportional to their number of inhabitants (from 1 to 50 votes, slightly in favor of the smaller towns).

The Steering Committee includes the President, the Vice President and five members elected from among General Assembly representatives. The aim is for all the institutions to be present in this executive body. Representatives of the basin organizations are also able to attend Steering Committee meetings. The five members representing the municipalities are elected by the President. The President and the Vice President are elected by absolute majority within the General Assembly. However, the President and Vice President cannot represent the same municipality.

The consortium's Managing Director is chosen by the Steering Committee after being nominated by the President. He/she has to have the appropriate professional background, qualifications, and experience leading different consortium services.

System approach: Environmental, social, and economic regeneration

Although the Bilbao-Biscay Water Consortium was created prior to the reestablishment of democracy, it is important to note that all the key plans designed to promote the social and economic regeneration of the area drafted after the restitution of powers to local governments have either developed the consortium's own plans further or leveraged them. Thus, in 1988, the Basque Government initiated a strategic discussion over the future of the Basque Country, called *Perspectivas 2005* ("2005 Prospects") (Rodriguez, 2001). A main goal of this public discussion was to analyze Metropolitan Bilbao's economic regeneration. After a decade of multiple stakeholders' collective discussions and reflection (Departamento de Economía y Planificación-Gobierno Vasco, 1989), the regional government launched its *Strategic Plan for the Revitalization of Metropolitan Bilbao* in 1992 (Rodriguez, 2001). Many different partnering initiatives were created during this time to stimulate investment and a reflective discussion on the model and the initiative to regenerate the region. There are two main initiatives. The public-private think-tank, Bilbao-Metropoli 30,[xxxi] was launched in 1991. Its goal was to research, promote, and plan a new project to revitalize the metropolitan area. Among their main activities and goals, the think-tank defined the regeneration and revitalization of the estuary and its conversion into the backbone of the metropolitan Bilbao-Biscay area. The second key initiative was launched in 1992: Bilbao Ría 2000.[xxxii] The latter was a government-owned enterprise constituted by Spanish and Basque public administrations (Bilbao Port Authority, the city councils of Bilbao and Barakaldo, Biscay Provincial Government, the Basque Government, and the Government of Spain). They owned a part of the lands surrounding the estuary. The enterprise's aim was to become an instrument for the revitalization of Bilbao's port and the region's economy. It also planned on carrying out urban development actions in the areas liberated by moving the port of Bilbao to the delta. One of these areas, for example, was used to build the Guggenheim Museum.

In 1989, the Basque Government and the Provincial Government of Biscay commissioned a consulting firm to draft the Strategic Plan that was finally launched in 1992 (Rodriguez, 2001). The *Strategic Plan for the Revitalization of Metropolitan Bilbao* identified eight critical areas in which to concentrate efforts: (1) invest in Human Resources; (2) become a "metropolis of advanced services in a modern industrial region"; (3) guarantee mobility and accessibility; (4) ensure environmental regeneration; (5) promote urban regeneration; (6) make culture a central focus; (7) coordinate management between public administrations and the private sector; and (8) promote social action.[xxxiii] The main transformations in Metropolitan Bilbao and the Bilbao-Biscay region occurred mostly in the 1990s with multiple initiatives and projects (Rodriguez, 2001).

The Plan for the Revitalization of Metropolitan Bilbao defined that the estuary should be the main pillar guiding the development and regeneration of Metropolitan Bilbao. To this end, the plan included the need for an environmental and systemic sanitation approach to the environment. This regeneration was represented by the ongoing Plan for the Integral Sanitation and Clean-up of the Estuary developed by the Bilbao-Biscay Water Consortium. Furthermore, the revitalization plan also proposed moving Bilbao's port to the estuary's mouth in Sestao and other municipalities, the aim being to free up land along the river bank in downtown Bilbao. This land could then be used to promote new urban regeneration projects and other economic development initiatives based on services, culture, and innovation. In keeping with these indications, the Guggenheim Museum was located on one such lot after the waste and pollution had been thoroughly cleaned.

Fernández Pérez (2005) argues that the Plan for the Integral Sanitation and Clean-up of the Estuary played a key catalytic role. It was an advancement in the urban revitalization of Metropolitan Bilbao that would eventually crystallize in the 1990s. In fact, the 1990s were a rich and ambitious decade for socio-economic and urban plans: the Strategic Plan for Revitalization (1990), Bilbao's General Urban Planning Plan (1990), Metropolitan Bilbao's Territorial Partial Plan (1994), the Galindo-Barakaldo Migration Plan (European Urban Initiative), and the Inter-institutional Agreement for the Revitalization of the Left Bank (1997).

If we look at the Plan for the Integral Sanitation and Clean-up of the Estuary as an instrument for system-based action, it was ahead of its time compared to other plans that would later help shape the social and economic development of Biscay-Bilbao. The environmental regeneration of the *ria* foreshadowed the new activities promoting economic, cultural, and social transformation. In terms of funding efforts, this plan is still the most important and largest investment in the process of regenerating the Basque city. The plan's effects also impacted the cultural industry and the Guggenheim Museum that consequently transformed the city's economic and social reality, promoting cultural tourism. However, without the prior sanitation of the estuary and without a modern water supply system, the local population would have probably fled elsewhere, and tourists would never have been attracted to Bilbao.

Environmental regeneration and cultural centrality were two main concerns. The former Mayor of Bilbao, Dr. Iñaki Azkuna, pointed out that the Guggenheim Museum implied an investment of 132.22 million euros (Mayora, 2014). However, the Plan for the Integral Sanitation and Clean-up of the Estuary required almost 10 times more. Elias Mas Serra (2010), the Director of Bilbao City Council's Architecture Office until 2005, pointed out that, when analyzing Bilbao's and the surrounding area's regeneration and revitalization, the Guggenheim Museum could be considered the driver. This was due to culture. The plan integrated culture as a key dimension of the new economic development model. The term, the "Guggenheim effect", has been aptly coined to study and reference numerous urban revitalization projects around the world (Plöger, 2008; Ward, 2002).

The regeneration of Metropolitan Bilbao included different principles and goals. First, it was based on the aim of generating a new economic reindustrialization model with the promotion of a set of public policies based on transforming the Basque Country as a whole and the Bilbao area (including Bilbao and all the cities along the estuary), in particular, into a leading economic region for business innovation and entrepreneurship. Second, this redesign of the Basque economy included two main values as key for the future: innovation and entrepreneurship. The goal was to transform the Bilbao-Biscay region into a leading innovation and entrepreneurship pole in Europe. Third, based on the debate to generate a new economic model, human development was also integrated as a key factor. The previous model leading to the extreme pollution in the estuary was no longer a viable strategy, and public authorities and citizenship were aware of the need to define a long-term sustainable development model. Based on this consciousness, later on, the Basque Government passed its new General Law on the Protection of the Basque Country's Environment (Law 3/1998), approved in 1998 and specifically including sustainable development as a key aim.[xxxiv] Finally, the initiative focused on a new, key goal: Basque culture[xxxv] and the promotion of the population's cultural awareness. In this respect, the Basque Government and Bilbao City Council started an important discussion on the development of new cultural infrastructures to transform the industrial city into a new cosmopolitan and modern metropolis. Culture was a key factor, attracting the attention of the Guggenheim Foundation to the city when the US-based foundation was looking for a European city in which to open a new museum.

This planning process had four primary results. First, public authorities and private and social stakeholders played a key role in redesigning and regenerating an old industrial economy into a new entrepreneurial, innovative, and service-based economy. In this goal, Knowledge Management and learning has had a key role to co-manage all the different stakeholders and goals. The Basque Government together with the Provincial Government of Biscay and local public authorities worked together to promote innovation and research, aiming to transform the old industrial sector (e.g., mining, metallurgy, energy, and automobiles) into advanced and innovative clusters, working together to grow, expand, and create new value through R&D investments and the generation of new added-value jobs. This played an important role in attracting international economic investment. A second key result was the new port. Public investments were also devoted to building vanguard infrastructures for the new port, including modern facilities for international maritime trade and to also promote the international tourism industry, attracting international tourists to the city. Third, the project aimed to clean up the estuary and build new collective water management infrastructures. Currently, the estuary has been cleaned and has recovered the capacity to generate natural life and human-based water use. Fourth, the city of Bilbao, together with the Basque Government and the Provincial Government of Biscay in addition to many private and social institutions managed to establish international connections and stimulate local investment to attract and build a great cultural infrastructure, the Guggenheim Museum, located on the revitalized river bank in new Bilbao where old docks were once located. The museum opened to the public on October 19th, 1997.

10.4 Conclusions

Our case study on the Bilbao *ria*, the evolving process, and analysis of the role played by the Bilbao-Biscay Water Consortium as a collective action institution and the Plan for the Integral Sanitation and Clean-up of the Estuary illustrate an advanced collective water management based on an inclusive Knowledge Management experience. The consortium and the commitment from multiple public and private agents and social stakeholders to clean up an extremely polluted environment and transform it into a new natural ecosystem, successfully turned waste into value. This case shows how all control water management approaches can be transformed into new adaptive water management approach, based in advanced Knowledge Management and a systemic approach to water cycle management. In contrast to traditional water management, Bilbao-Biscay Water Consortium has moved towards an adaptive water management engaging public administrations, managers and users towards the understanding the complexity of the water cycle and socio-ecosystems and the interconnections among water users, units, resources, and ecosystems. Advance Knowledge Management has been based on the integration of information systems and the development of complex Knowledge Management to understand the complex interconnections among different ecosystems of the estuary and its adjacent hydric resources.

Endnotes

i. Resolution 64/292. The human right to water and sanitation, UN General Assembly, 28 July, 2010.
ii. http://www.un.org/es/comun/docs/?symbol=A/RES/64/292&=E, viewed on 15 January, 2017.
iii. http://www.bilbaoport.eus/en/the-port/ viewed on 30 January, 2017.
iv. http://www.eustat.eus/estadisticas/tema_159/opt_0/ti_Poblacion/temas.html viewed on 30 January, 2017.
v. http://www.consorciodeaguas.com/Web/QuienesSomos/resena_historica.aspx viewed on 15 January, 2017.
vi. http://www.eustat.eus/estadisticas/tema_159/opt_0/ti_Poblacion/temas.html viewed on 15 January, 2017.
vii. http://www.bilbaoport.eus/el-puerto/la-historia/ viewed on 15 January, 2017. http://www.museomaritimobilbao.eus /mmrb/Web/noticias/noticiasficha.asp?cod=8861FCDF-E389-4B09-A1D1-0707D4E6E70C&IdNoticia=67E16826-459A-4AFB-BADD-1F7 0C004159A. viewed on 15 January, 2017.
viii. http://www.eitb.eus/es/noticias/sociedad/detalle/1404442/inundaciones-bilbao-1983-30-aniversario-26-agosto-2013/ viewed on 15 January, 2017.
ix. http://www.consorciodeaguas.com/Web/QuienesSomos/resena_historica.aspx viewed on 15 January, 2017.
x. http://www.bilbao.eus/bld/bitstream/handle/123456789/32972/07.pdf?sequence=1.
xi. Bilbao Bizkaia Ur Partzuergoa-Consorcio de Aguas Bilbao Bizkaia: El ciclo del Agua. Accessible via https://www.consorciodeaguas.com/web/CicloAgua/ciclo_agua.aspx (viewed on January, 15th 2017).
xii. Bilbao Bizkaia Ur Partzuergoa-Consorcio de Aguas Bilbao Bizkaia: El ciclo del Agua. Accessible en https://www.consorciodeaguas.com/web/CicloAgua/ciclo_agua.aspx viewed on 15 January, 2017.
xiii. There are two main laws: 1) Basque and Spanish Laws on water Management, 2) Real Decreto 140/2003 sobre Unidad de Vigilancia y Control de las Aguas de Consumo Humano; and the certification: 1) ISO/IEC 17025.
xiv. http://www.bilbao.eus/bld/bitstream/handle/123456789/32972/07.pdf?sequence=1 viewed on 15 January, 2017.
xv. Provincial Government of Biscay website for Agenda 21: http://www n-integral-de-saneamiento-que-han-devuelto-la-vida-a-la-ria-y-transformado-la-ciudad-2/ viewed on 15 January, 2017.
xvi. Bilbao Bizkaia Ur Partzuergoa-Consorcio de Aguas de Bilbao Bizkaia: "La Estación Depuradora de Aguas de Galindo". Accesible via https://www.consorciodeaguas.com/html/pdf/edarGal.pdf.
xvii. Bilbao Bizkaia Ur Partzuergoa-Consorcio de Aguas de Bilbao Bizkaia: "Depuración de Aguas Residuales". Accessible via https://www.consorciodeaguas.com/web/CicloAgua/ciclodelagua.aspx?id=depuracion. viewed on 15 January, 2017.

xviii. https://www.consorciodeaguas.com/web/CicloAgua/ciclodelagua.aspx?id=depuracion. viewed on 15 January, 2017.

xix. .bizkaia21.eus/atalak/TerritorioSostenible/Lugares/datos.asp?id=3&IdPagina=36&idioma=c viewed on 15 January, 2017.

xx. Webpage on metropolitan Bilbao regeneration projects, https://www.consorciodeaguas.com/web/CicloAgua/ciclodelagua.aspx?id=recogida viewed on 15 January, 2017.

xxi. http://www.azti.es/es/exposicion-la-magia-de-la-ria-sobre-los-35-anos-del-pla https://sites.google.com/site/bilbaometropoli/proyectos/proyecto-12 viewed on 15 January, 2017.

xxii. Bilbao-Biscay Water Consortium website https://www.consorciodeaguas.com/html/pdf/edarGal.pdf viewed on 15 January, 2017.

xxiii. Bilbao- Biscay Water Consortium website http://www.consorciodeaguas.com/Web/oficinaPrensa/noticias_detalle.aspx?id=304&pg=0 viewed on 15 January, 2017.

xxiv. Among other people, we interviewed a former president of a building company that had been working on the infrastructures and sanitation of the estuary. We also interviewed a former director of an insurance company. Most of the people interviewed are now retired.

xxv. http://www.udalsarea21.net/paginas/ficha.aspx?IdMenu=17d05e9d-8396-438d-9530-d4a94ea60184™Idioma=en-GB (viewed on January, 15th 2017).

xxvi. http://www.hezkuntza.ejgv.euskadi.test.hezkuntza.net/r43-573/es/contenidos/informacion/did7/es_2058/r01hRedirectCont/contenidos/informacion/ihitza/es_1018/r01hRedirectCont/contenidos/informacion/ihitza7/es_1032/recursos_c.html (viewed on January, 15th 2017).

xxvii. http://www.consorciodeaguas.com/Web/GestionAmbiental/PDF/Politica/Ingurumen_Politika.pdf (viewed on January, 15th 2017).

xxviii. http://www.bm30.es/ (viewed on January, 15th 2017).

xxix. Itsasnet (2015). Webpage of AZTI: 35 years of the Master Plan of Integral Sanitation and Cleaning-up the Estuary. 23 años del Plan Integral de Saneamiento del Bilbao Metropolitano, 09-09-2015.

xxx. https://www.consorciodeaguas.com/web/Normativa/pdf/estatutosconsorcioc.pdf (viewed on January, 15th 2017).

xxxi. http://www.consorciodeaguas.com/Web/QuienesSomos/competencias.aspx (viewed on January, 15th 2017).

xxxii. http://www.bm30.eus/en/.

xxxiii. http://bilbaoria2000.org/ria2000/ing/home/home.aspx.

xxxiv. Bilbao Metrópoli-30: "Plan de Revitalización". http://www.bm30.es/homemov_uk.html (viewed on January 10th 2017).

xxxv. One of the authors participated in the design, elaboration and parliamentary process to pass this Law: General Law of Protection of the Environment of the Basque Country (Law 3/1998).

References

Arranz S. 2012. Contaminación por metales pesados. El caso de la ría de Bilbao. *Itsas Memoria. Revista de Estudios Marítimos del País Vasco*, Untzi Museoa-Museo Naval, Donostia-San Sebastián, pp. 265–282.

Atienza L. 1991. Un proyecto estratégico para la revitalización del Bilbao Metropolitano. In Rodríguez , *et al.* (eds). *Las Grandes Ciudades: debates y propuestas*, Economistas, Madrid.

Barreiro PM, Aguirre JJ. 2005. *25 años del Plan Integral de Saneamiento de la Ría de Bilbao. Revista Dyna*, vol. Enero-Febrero, pp. 25–31, viewed 10 January, 2017, http://www.revistadyna.com/busqueda/25-anyos-del-plan-integral-de-saneamiento-de-ria-de-bilbao.

Biermann F, Betsill MM, Gupta J, Kanie N, Lebel L, Liverman D, Schroeder H, Siebenhüner B. 2009. *Earth system governance. People, places and the Planet.* Earth System Governance project. Report no. 1, Bonn.

Bilbao Metropoli-30 2001. *Bilbao 2010. Reflexión Estratégica. Bring your dreams to Bilbao*, Bilbao.

Cárcamo J. 2012. Estación bombeadora de Elorrieta. Bilbao. In *Patrimonio Industrial en el País Vasco = Euskadiko industrial ondarea*, vol *2*, pp. 743–747.

Cearreta A. 1992. Cambios medioambientales en la Ría de Bilbao durante el Holoceno. *Cuadernos de Sección de Historia de Eusko Ikaskuntza*, num 20, pp. 437–454, Donostia.

Cox K, Arnold G, Villamayor S. 2010. A review of design principles from community-based natural resource management. *Ecology and Society*, *15*(4), 38.

Departamento de Ordenación del Territorio, Vivienda y Medio Ambiente-Gobierno Vasco (DOTVMA) 1998. *Actuaciones del Programa de Demolición de Ruinas Industriales en la Comunidad Autónoma de Euskadi*, Servicio Central de Publicaciones del Gobierno Vasco, Vitoria-Gasteiz.

Departamento de Urbanismo, Vivienda y Medio Ambiente-Gobierno Vasco (DUVMA), 1994. *Plan Territorial Parcial Bilbao Metropolitano*, Gobierno Vasco, Vitoria-Gasteiz.

Departamento de Economía y Planificación-Gobierno Vasco (DEP), 1989. *Bases para la revitalización económica del Bilbao Metropolitan. Ekonomiaz, 15*, Gasteiz.

Dietz T, Ostrom E, Stern P. 2003. The struggle to govern the commons. *Science, 302*, 1907–1912.

Eizaguirre JM. 1994. El Plan Integral de Saneamiento de la Comarca del Gran Bilbao. *Revista Dyna*, April, pp. 6–10, viewed on 10 January, 2017, http://www.revistadyna.com/busqueda/el-plan-integral-de-saneamiento-de-comarca-del-gran-bilbao.

Eizaguirre JM. 1996. El Plan de Saneamiento del Área Metropolitana de Bilbao. *Revista Dyna*, December, pp. 72–76.

Elejabeitia G. 2015. El Bilbao bajo las Aguas. *Diario El Correo*, viewed on 19 July, 2015, Bilbao, http://www.elcorreo.com/bizkaia/sociedad/201507/19/bilbao-bajo-aguas-20150713161451.html.

Fernández Pérez D. 2005. Las Infraestructuras de saneamiento en un área metropolitana. El caso de la aglomeración urbana de Bilbao. *Revista Ingeniería y Territorio, 71*, 56–65, Colegio de Ingenieros de Caminos, Canales y Puertos, Barcelona, Spain.

Folke C, Hahn T, Olsson P, Norberg J. 2005. Adaptive governance of social–ecological systems. *Annual Review Environmental Resources, 30*, 8.1–8.33.

García Merino LV. 1987. *La formación de una ciudad industrial. El despegue urbano de Bilbao.* Instituto Vasco de Administración Pública = Herri-Arduralaritzaren Euskal Erakundea, p. 164, Oñati.

Gell-Mann M. 1994. Complex adaptive systems. In G Cowan, G Pines, D, Meltzer D. (eds.) *Santa Fe Institute Studies in the Sciences of Complexity*, Proceedings Vol. *XIX*. Complexity: Metaphors, models, and reality (pp. 17–45). Addison Wesley, Reading, MA.

Hardin G. 1968. The tragedy of the commons. *Science 162*, 1243–1248.

Hess Ch. 2008. *Mapping new commons.* Working paper presented at the 12th Biennial Conference of the International Association for the Study of the Commons, Cheltenham, UK, 14-18 July, Syracuse University.

Keohane RO, Ostrom E. 1995. Introduction. In *Local Commons and Global Interdependence.* Keohane RO, Ostrom E. (eds). Sage, London, pp. 1–26.

Lakunza E. 1997. Entrevista a José Miguel Eizaguirre. *Revista Dyna, Ingeniería e Industria, 72*, January, pp. 36–37.

Mas Serra E. 2010. Plan estratégico o estrategia para un discurso?: El caso de Bilbao. *Scripta Nova, Revista Electrónica de Geografía y Ciencias Sociales.* Universidad de Barcelona, vol. *14*, 328.

Mayora J. 2014. *El Guggenheim ha sido una inversión y no un gasto. Diario El Correo*, 22 October, 2014, viewed on 10 January, 2017, http://www.elcorreo.com/vizcaya/20081118/vizcaya/guggenheim-sido-inversion-gasto-20081118.html.

Oki T, Kanae S. 2006. Global hydrological cycle and world water resources. *Science, 313*, 1068–1072.

Olson M. 1965. *The Logic of Collective Action: Public Goods and the Theory of Groups (Revised edition).* Harvard University Press, Cambridge, MA.

Ostrom E. 1990. *Governing the Commons. The Evolution for Collective Action.* Cambridge University Press, New York.

Ostrom E. 1998. A behavioral approach to the rational choice theory of collective action: Presidential address, American Political Science Association. *The American Political Science Review*, *92*(1), 1–22.

Ostrom E. 2009. A general framework for analyzing sustainability of social-ecological system. *Science, 325*, 419–422.

Ostrom E, Burger J, Field CB, Norgaard RB, Policansky D. 1999. Revisiting the commons: Local lessons, global challenges. *Science*, *284*, 278–282.

Pahl-Wostl C. 2007. Transitions towards adaptive management of water facing climate and global change. *Water Resource Management*, *21*, 49–62.

Pahl-Wostl C, Craps M, Dewulf A, Mostert E, Tabara D, Taillieu T. 2007. Social learning and water resources management. *Ecology and Society*, *12*(2), 5.

Pahl-Wostl C, Holtz G, Kastens B, Knieper, C. 2010. Analyzing complex water governance regimes: the Management and Transition Framework. *Environmental Science & Policy*, *13*, 571–581.

Plöger J. 2008. *Bilbao City Report.* Centre for Analysis of Social Exclusion. London School of Economics and Political Science. 5 February, 2008, Accessible via http://sticerd.lse.ac.uk/dps/case/cr/CASEreport43.pdf.

Puertos del Estado 2016. *Anuario Estadístico 2015*. Madrid.

Rodriguez A. 2001. Reinventar la ciudad: milagros y espejismos de la revitalización urbana en Bilbao. *Ciudad y Territorio Estudios Territoriales*, *129*.

Saiz JI, Francés G, Imaz X. 1996. *Uso de Bioindicadores en la evaluación de la contaminación de la Ría de Bilbao.* Servicio Editorial de la Universidad del País Vasco-Euskal Herriko Unibertsitateko Argitalpen Zerbitzua, p. 14, Leioa.

Tilmant A, van der Zaag P, Fortemps P 2007. Modeling and analysis of collective management of water resources. *Hydrology and Earth System Sciences*, *11*, 711–720.

Ward D. 2002. *The Guggenheim effect. The Guardian*, 30 October, 2002, viewed on 15 January, 2017, https://www.theguardian.com/culture/culturevultureblog/2006/jul/10/theguggenheim

11 Virtual and inter-organizational processes of knowledge creation and Ba for sustainable management of rivers[1]

Federico Niccolini[1], Chiara Bartolacci[2], Cristina Cristalli[3] and Daniela Isidori[3]

[1]Department of Economics and Management, Università degli Studi di Pisa, Pisa, Italy
[2]Department of Economics and Law, Università degli Studi di Macerata, Macerata, Italy
[3]Department of Research for Innovation, Loccioni Group, Ancona, Italy

Introduction

The Nonaka's SECI (Socialization, Externalization, Combination, Internalization) model of knowledge creation developed by Nonaka at the organizational level provides a categorization of organizational spaces that can host each phase of the knowledge creation process (Nonaka & Konno, 1998).

[1]This chapter builds on and enhances the paper by Bartolacci C., Cristalli C., Isidori D., & Niccolini F. (2016). Ba virtual and inter-organizational evolution: a case study from an EU research project. *Journal of Knowledge Management*, *20*(4), 793–811.

Handbook of Knowledge Management for Sustainable Water Systems, First Edition. Edited by Meir Russ.

Taking the SECI model as the main point of reference, this chapter offers reflections on how the contemporary evolutions of Knowledge Management (KM) approaches can find effective applications, thanks to use of IT tools, for the management of a water system, especially in terms of safety and sustainability.

The article updates the Japanese concept of "Ba": the place for knowledge creation, in which persons – through dialogue and interaction – can create new knowledge and innovative solutions. We study a virtual application of Ba at the inter-organizational level, with specific reference to the monitoring of a river.

In his numerous works, Nonaka elaborated the SECI model of knowledge creation at the level of analysis of a single organization, also providing a categorization of organizational spaces that can host each of the phases of the knowledge creation process (Nonaka & Konno, 1998).

Looking at current scenarios, it appears inevitable, taking the cue from Nonaka himself, that the knowledge spiral can be extended to the inter-organizational epistemological level. To this aim, information technology tools and virtual communities can build the bridge to establish effective interactions for the exchange of knowledge (Panahi, Watson, & Partridge, 2013) and, congruently, they can make Ba change and evolve (Hessman, 2013).

Nevertheless, the main problem with this kind of virtual and inter-organizational evolution of the concept of Ba is linked to the effectiveness of the socialization phase because, according to the existing organizational and Knowledge Management literature, tacit knowledge and contextual knowledge sharing would seem to be possible only through face-to-face interactions (Tee & Karney, 2010; Sáenz, Aramburu, & Blanco, 2012). Moreover, even inter-organizational knowledge transfer *per se* seems to be problematic (Tuomi, 1999).

These points of view derive from the fact that the literature has yet to investigate these issues from the organizational perspective, identifying a new model that contextualizes the Ba in virtual and inter-organizational environments. This chapter attempts to fill this gap and conceptualize this topic, offering a novel approach and model.

To this purpose, the present work is based on the case study of a Knowledge Management virtual platform that finds application in the framework of a project named Flumen. This project was developed through collaboration between public organizations and a medium-sized Italian industrial organization, the Loccioni Group (hereinafter also LG). The aim of the project was the sustainable management of a two-kilometer segment of a river (Esino River, Marche Region, Italy). The public and private partnership that took place under the Flumen project has led to important structural works and the putting in place of considerable investments in order to re-establish the river's natural banks, guarantee periodic cleaning of the river bed, and surrounding vegetation maintenance by reintroducing and taking care of endemic plant species and building recreational structures (such as bike paths).

Following some dramatic flooding events, the safety and resilience of the river was considered a fundamental prerequisite for the development of the project (Khatri, 2013). Among the river management activities, flood monitoring was considered crucial, particularly in the area where the LG headquarters are located.

Within this context, the Loccioni Group also developed a parallel project called BLESS+ (Bed Level Seeking System) designed to improve the level of resilience and safety of the river by using a system for real-time monitoring of the existing bridge pier for scour risk during floods, to alert the population, and to activate an emergency plan. BLESS+ stems from the systems engineering of the Bless patent owned by the Polytechnic University of Milan; it was implemented using a virtual and inter-organizational knowledge platform developed within the framework of a European research project called "BIVEE: Business Innovation in Virtual Enterprise Environments", in which the Loccioni Group was also involved. In March 2015, BIVEE was rated "Excellent" by the European Union (http://cordis.europa.eu/project/rcn/100275_en.html).

The project gave rise to a virtual platform which implemented an Open Innovation paradigm (Chesbrough, 2006) and built up a cyber-physical system. The virtual Ba experience within the project can plausibly be used for other knowledge creating processes and can be suitably applied also to other sustainability-oriented solutions for river and water system management.

Adopting a Participatory Action Research approach (Ragsdell, 2009; Coughlan & Coghlan, 2002; Brydon-Miller, Greenwood, & Maguire, 2003; Gummesson, 2000), this chapter contextualizes the SECI model within a Web platform for Open Innovation, in order to inquire whether and how a knowledge creation circle process can take place entirely within a virtual environment that links several subjects from different public and non-profit organizations (industries, universities, etc). In doing so, it contextualizes the SECI model within overlapping innovation phases called waves. In particular, it breaks the innovation process into the main waves that are present in any innovation processes, from idea generation to product engineering, and it applies the SECI model to each of them. Thus, it illustrates an evolution of the SECI model for knowledge creation at the inter-organizational level. Moreover, through a learning history, it describes how all the phases of the SECI process, including socialization, can take place or be supported in virtual spaces.

The chapter is organized as follows. The first section presents a theoretical and multidisciplinary framework. The following sections then deal with methodological and approach issues, presenting the BIVEE project and contextualizing the SECI spiral model within the virtual and inter-organizational platform design.

Subsequently, a real learning history is presented to show and explain the functioning of the platform, structuring it accordingly to SECI model phases.

The final aim of the chapter is to provide some insight as to how to make organizational practice in Knowledge Management more effective, particularly in the context of sustainable management. At the same time, the work aims to produce a theoretical generalization to advance the SECI knowledge creation theoretical model so that it can be successfully applied to solve complex problems in the field of sustainable management of water systems.

11.1 Theoretical framework

Looking at the original concept put forward by Japanese philosopher Nishida (Nonaka & Toyama, 2005, p. 428), Ba is to be considered a shared space representing a foundation for knowledge creation, and thus a platform for advancing individual and organizational knowledge (Nonaka & Konno, 1998) through interaction. Then, what represent the essence of Ba are the contexts and the meanings created and shared through interactions happening at a specific time and space, rather than the space itself. As a result, managing organizational knowledge means managing the context and conditions by which knowledge can be created, shared, and implemented (Wei Choo & Correa Drummond de Alvarenga Neto, 2010). In considering the ontological and the epistemological dimensions of knowledge creation processes, Nonaka & Konno (1998) identified four different stages (Socialization–Externalization–Combination–Internalization) to build up a spiral model. Also, they defined a coherent set of Ba with different characteristics suitable for hosting and better supporting the processes and dynamics of knowledge creation that take place during the different phases.

Originating Ba is a dimension where individuals share emotions and experiences sympathizing and empathizing with others, removing any psychological barriers. That is the primary Ba that kicks off the knowledge-creation process with the Socialization phase.

Interacting Ba is an environment that is constructed more consciously by picking people with specific knowledge and capabilities with whom to integrate. Here, during the Externalization process, tacit knowledge is made explicit through dialogue and metaphorical language (Nonaka & Nishiguchi, 2001, p. 20).

Cyber Ba is a virtual world to interact in and combine explicit pieces of knowledge. The Combination phase taking place in this space is enhanced by information technology that allows the use of on-line inter-organizational group-ware, documentations, and databases.

Exercising Ba supports the conversion of explicit to tacit knowledge. This phase is called Internalization and consists in the continuous implementation of explicit knowledge in real life or simulated applications.

This Ba categorization appears to be functional and necessary to successfully support knowledge creation in its different phases. However, it is important to be aware that an organization's Ba does not consist in the mere accumulation of information or documents; rather, it must be interpreted as a continuous dynamic cycle of converting tacit into explicit knowledge and back.

The choice to focus the literature review on Nonaka's original papers on Ba is an intentional one. Indeed, despite the possible limitations of considering just one author's work, the choice was dictated by the decision to rely on the original sources.

Table 11.1 Categorization of Ba within the SECI model. Personal elaboration by Bartolacci (2014)

Ba	Phase	Level		Type of knowledge created	Main tools needed
		Epistemological	Ontological		
Originating	Socialization	Tacit – Tacit	Individual/Individual	Empathic	Direct Interaction
Interacting	Externalization	Tacit – Explicit	Individual/Group	Theoretical	Metaphors
Cyber	Combination	Explicit – Explicit	Group/Organizational	Systematic	Information Technology
Exercising	Internalization	Explicit – Tacit	Organization/Individual	Operational	Learning by doing

Now, what stands out in this model overview is the huge importance explicitly accorded to physical proximity and face-to-face interaction. They represent the key to conversion and transfer of tacit knowledge and thus the trigger for the whole knowledge creation process. Indeed, according to this model, new knowledge always begins with one individual directly sharing tacit knowledge with another (Nonaka, 1991). On the contrary, virtual spaces, namely cyber Ba, are mostly limited to the combination phase. Here, explicit knowledge is generated and systematized by merging information throughout the organization. Obviously, information technology is essential for providing the collaborative environments to support this phase of the SECI model (Table 11.1) (Nonaka, 1991). However, virtual spaces and ICT are mainly considered accessory tools to be used only once implicit knowledge has been converted into explicit knowledge, to merge or store it.

Indeed, Nonaka based his model on the Polanyi (1967) conception of knowledge as being inseparable from individuals and deeply embedded in a specific context. From this perspective, knowledge represents the result of relative and subjective elaborations of objective information, and organizational knowledge represents the result of relationships and interactions (Nonaka & Toyama, 2005; Polanyi, 1967).

In this regard, social media or virtual platforms, ICT devices or applications are able to enhance and support social relationships and knowledge sharing or transfer (Panahi *et al.*, 2013; Siebdrat, Hoegl, & Ernst, 2009); consequently, they could be considered essential elements needed to build up a proper space for knowledge creation or innovation. Moreover, in taking into account the evolution of organizational space itself, it seems entirely appropriate to re-discuss and reconsider the spatial aspects of knowledge-related processes. Contemporary organizations merge the physical dimension with the virtual one, work spaces with data flows. They often represent cyber-physical systems, interconnecting Webs of information and production (Hessman, 2013).

In this vein, also drawing inspiration from Castells' works on urban space as a mirror of social organization, it could be meaningful and valuable to take a moment to reflect on organizational spaces. Being open systems, organizations, as well as other open system (e.g. cities), can be conceived as contemporarily made up of flows and places, and of their relationships (Castells, 2005). The interface between electronic communication and physical interaction, or the combination of networks and places, shapes and deeply transforms the knowledge creation processes. These considerations acquire significance and relevance at the inter-organizational epistemological level of knowledge creation, in particular. Interestingly, the SECI model was developed from an intra-organizational point of view, while the inter-organizational level of analysis had been left as a possible and interesting field of future research, without being analyzed in its dynamics.

Moreover, considering the rise of the Open Innovation paradigm (Chesbrough, 2006) and the constitution of networks and communities of people and of organizations in the current scenario, it is evident that knowledge creation processes are strongly influenced and sustained by information technology (see for example Martínez-Torres, 2014; Hüsig, 2011; Tickle, 2011). Tuomi (1999; Šarkiūnaitė & Krikščiūnienė, 2005) outlined how the SECI model was developed from the viewpoint of the inner actions of one single organization. Indeed, one of the main unexplored questions is that of the functioning of the model in a network of many different organizations or members of those organizations, where the cross-culturalism can cause problems due also to different kinds of organization cultures (Kostiainen, 2002). That's because knowledge transfer is possible through an interactive mechanism that is based on shared rules, norms, organizations, and procedures (Fong Boh Nguyen, & Xu, 2013; Siebdrat *et al.*, 2009).

More specifically, the essential precondition for originating Ba to rise is a strong sense of belonging and of commitment in the network members in order for them to perceive tacit knowledge transfer as part of the purpose of the network itself. To this end, the trigger for the Socialization process to occur is that there be a field in which to interact and share experiences so that it is possible to deeply understand and empathize with others' *modus cogitandi* and mental models (Takeuchi & Nonaka, 2004).

To this purpose, high-fidelity communication media (McLuhan, 1964; Panahi *et al.*, 2013) are very useful since they are able to send complete messages, convey an abundance of tacit and contextual components requiring little extra interpretation. However, Socialization in an online environment remains challenging and, considering the importance accorded to this phase, the consensus is that it should include a face-to-face component (Siebdrat *et al.*, 2009; Tee & Karney, 2010; Saenz *et al.*, 2012).

Nevertheless, according to the theory of innovative *milieux*, networks should be spread outside the restricted region one belongs to, in order to gain new information, competencies, and influences (Kostiainen, 2002). Moving on to organizational contexts, the Open Innovation paradigm (Chesbrough, 2006) recognized the importance of grasping ideas *intra* and *extra moenia* in order for organizations to reach markets faster and to make their competencies evolve accordingly. In this regard, ICT solutions can meet the need for weak ties to expose individuals to new ideas that can trigger new knowledge creation (Alavi & Leidner, 2001).

Nowadays, various communities have already taken advantage of the Web to ease communication and information flows inside and outside of the community (see for example Dong, 2014; Tickle, 2011; Battistella, 2013; Balka, 2014; Siebdrat *et al.*, 2009). Moreover, new knowledge creation processes and innovation itself depend not only on the free flow of information in general, but on the recombination of non-obvious knowledge to trigger innovative solutions to complex problems (Newell, Robertson, Scarbrough, & Swan, 2009). In this view, information technology is not irrelevant, but it is insufficient by itself (Hargadon, 1999; 2002; Panahi *et al.*, 2013). Thus, according to Davenport and Prusak, it has already been proven that IT cannot replace human knowledge or create its equivalent (Davenport & Prusak, 1998, p. xi); that is to say, the human factor is the essential one when it comes to knowledge and knowledge creation processes in which IT can also be a useful tool. Indeed, the combination of people and IT is especially useful for building virtual group communities of interest and for providing support for creating concrete outputs, such as information items that can be accessed by the community.

In Nonaka's SECI model, ICT is contextualized within the realm of cyber Ba. This Ba can be considered a place of dialogue, in which new explicit knowledge is combined with existing knowledge. However, considering the flexibility of modern IT, other forms or features of organizational Ba and the corresponding phases of knowledge creation can be enhanced through several kinds of information systems (Alavi & Leidner, 2001; Lopez-Nicolas & Soto-Acosta, 2010), especially at the inter-organizational epistemological level. Nevertheless, it is necessary to always keep in mind that the Web is not merely an interlinked cluster of machines, but rather, a network of humans negotiating linguistic meanings through machines (Shvaiko *et al.*, 2010).

11.2 Methods

This work contextualizes the SECI model within a Web platform for Open Innovation in order to inquire whether and how a knowledge creation spiral process can take place entirely within a virtual environment linking several subjects from different organizations and universities. To this purpose, it utilizes the example of a platform developed during a three-year project financed by the European Union the outcomes of which were rated "Excellent". The R&D project, called "BIVEE: Business Innovation in Virtual Enterprise Environments", was part of the Factories of the Future FP7 project, sub-programme area: FoF-ICT-2011.7.3 – Virtual Factories and Enterprises (http://cordis.europa.eu/project/rcn/100275_en.html). It involved nine different European partners who worked together until December 2014, and the project was devoted to the creation of ICT platform tools for managing innovation.

11.3 Approach

Given that the project was aimed at linking theory and practice through research, for the purpose of development, Participatory Action Research (PAR) was the approach chosen to investigate the phenomena of knowledge creation circle processes.

Although there is scepticism linked to the supposed lack of scientific rigor and discipline in action research along with the difficulty of generalizing results from this kind of study (McKay & Marshall, 2001), this approach nevertheless appears to strongly fit the analyzed case and the studied phenomena. Indeed, by definition, action research aims to make organizational practice more effective while simultaneously building up a body of scientific knowledge in social science, involving the collaboration and co-operation of the action researchers and members of the organizational system.

The aptness of applying this research approach in this work is backed by Lewin's thought that causal inferences about the behavior of human beings are more likely to be valid and "enactable" when the human beings in question participate in building and testing them (see for example: Ragsdell, 2009; Coughlan & Coghlan, 2002; Brydon-Miller *et al.*, 2003; Gummesson, 2000). In fact, in participatory action research the researcher can gain genuine insights into the organisation and, subsequently, can design and provide methods and tools to be adapted as the process progresses.

In particular, considering the phenomena taken into account in this work, Ragsdell (2009) draws attention to the role that PAR could play in overcoming difficulties associated with developing a Knowledge Management culture and implementations. The PAR process supports social networking, the creation of transparency and trust, ownership, and organisational change.

In this study, two of the four researchers were actors and agents of change, actively promoting and participating in the BIVEE European Project (De Guerre, 2002). They designed the platform according to Open Innovation principles and organized innovation processes while keeping the SECI model phases in mind. Consequently, they conducted PAR in real time, while the other two were involved *ex post*. In this way, it was possible to have a different point of view that could help re-work the facts in a traditional case study written in retrospect; this enabled the research team to have a "learning history" that could be used as an intervention to promote reflection and learning (Gummesson, 2000).

11.3.1 The Flumen and BIVEE projects. A safe and sustainable future for a dangerous and neglected river

For this work in particular, the underlying approach of the project is of special interest. It was aimed at putting people, with their creativity and competencies, at the core of the knowledge creating process and providing a nurturing environment where open thinking and free interaction are more important than formal processes and stringent control.

The section is built on a case study of a Knowledge Management virtual platform that found application within the framework of an environmentally oriented CSR project (Elkington, 1997) developed in a real industrial setting. The project, called Flumen, was born from the collaboration between some public organizations and the Loccioni Group, a medium-sized Italian industrial organization specialized in technological solutions for a diverse group of clients and fields of application. Flumen's aim is the sustainable management of a two-kilometer segment of the River Esino that flows close to the Group's headquarters.

The Flumen project has pursued multiple sustainable objectives: social, recreational, and ecological. From a social perspective, one implicit goal of the Flumen project is to make this segment of the river a sort of social "back to the future" space, as it was only a few decades ago. It was a gathering place used for practical activities such as washing clothes, but also for recreational ones such as jogging, biking and picnicking on the banks. Finally, another sustainability goal of the project involves preserving the ecology and landscape heritage ones; as part of this effort, a natural restoration was made along the river, reintroducing numerous indigenous plant species. These include endemic fluvial trees as well as the main ancient and typical agricultural resources in the territory (e.g. olive varieties such as Coroncina, Mignola, Orbetana, Raggia, Rosciola, fruit trees such as Pope's apple, pink apple, gentle pink apple, Brigoncella plum, Visciola cherry, the Verdicchio dei Colli Esini grape, and vegetables such as the local Jesino artichoke, the two-toned "Monachelle" ("Little Nun") bean, eight-row flint corn, and the "da serbo" yellow tomato) (D'Ascoli, 2017) (see Figures 11.1 and 11.2).

Considered an example of good practice for corporate social responsibility (CSR), the Loccioni Group manufacturing company has, for decades, viewed safety as the cornerstone of any CSR project, and the Flumen project was no exception. In this context, all of the company's investments in the sustainable development of the river were preceded by the necessary investments in the preparatory and basic safety aspects. In other words, the safety and resilience of the river were considered fundamental prerequisites for its sustainable development.

History has taught LG how a river can so quickly change from being a precious and unique source of life, resources, energy, and recreational opportunities to a force that can destroy all things man-made. This is especially evident when humans get too close to the river with their buildings and activities, as LG did. In 1990 a dramatic flood seriously damaged LG headquarters (as shown in Figure 11.3).

Within this scenario, the Loccioni Group decided to also take part in a European project called "BIVEE: Business Innovation in Virtual Enterprise Environments" (http://bivee.eu/) which aimed to create a Knowledge Management virtual platform. This virtual platform has also been used as a tool to develop another project (BLESS+); the goal of this one was to provide an innovative

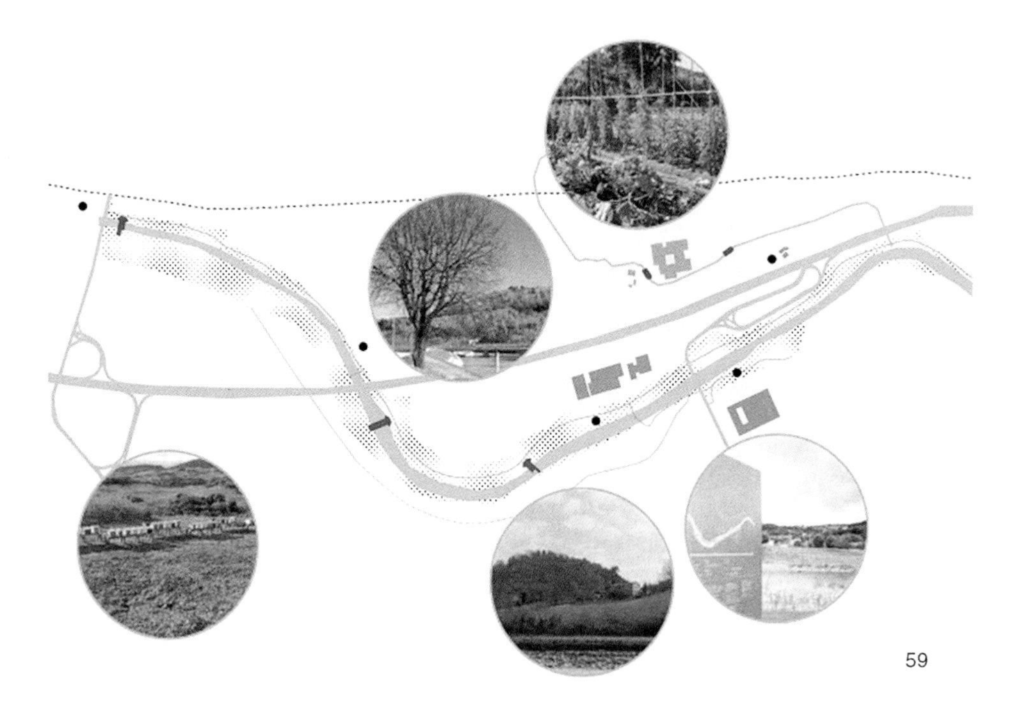

Figure 11.1 The agricultural and heritage restoration aspects of the Flumen Project. *Source*: D'Ascoli (2017).

Figure 11.2 A recreational area along the River Esino. *Source*: D'Ascoli (2017).

Figure 11.3 The effects of the flooding of the River Esino as it affected Loccioni Group Headquarters. *Source*: Loccioni Group Archives.

solution to flooding issues and develop a system for real-time monitoring of the bridge pier for scour risk during floods. This will be explained in greater detail in next paragraph.

The BIVEE project is about the Virtual Enterprise Environment, defined as a temporary alliance of businesses that come together to share skills or core competencies and resources in order to better respond to business opportunities, whose cooperation is supported by computer networks (Chesbrough, 2006). Also, it takes into account and splits two different spatial dimensions: the Value Production Space (VPS) and the Business Innovation Space (BIS). While ideas for improvement are mainly created and elaborated in the VPS, pure or radical innovations emerge from the BIS. This corresponds to the idea of the coexistence of material and abstract components within the organizational innovation process and stresses the difference between improvement and innovation concepts and dynamics (Smith *et al.*, 2013).

In essence, the BIVEE project developed a distributed and collaborative platform of ICT services with two well differentiated scopes: enterprise innovation management and production process improvement of SMEs. Indeed, the BIVEE platform (see Figure 11.4) is based on:

- *A Mission Control Room*, for monitoring Virtual Enterprise value production activities;
- *A Virtual Innovation Factory (VIF)*, for managing the entire cycle of innovation idea development.

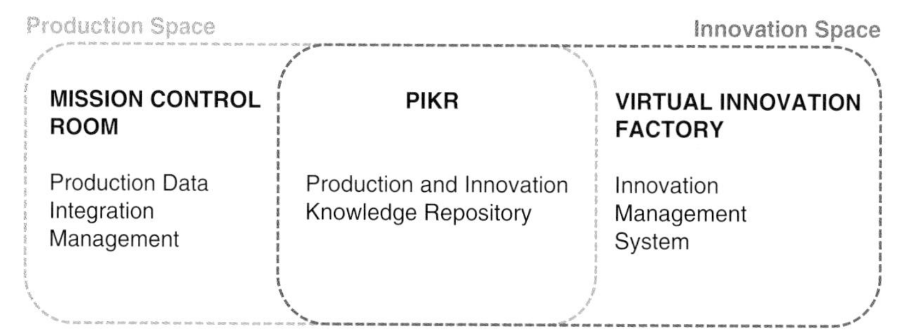

Figure 11.4 Overview of the BIVEE framework. Adaptation from D2.2 "Specification of business innovation reference frameworks (in the context of the VEMF)".

The VIF is a Web application that is accessed by users over a network based on HTML5, CSS3 and Javascript in the search for a better user experience providing the feeling of a desktop application. The framework used is Meteor, a client-database-server, written entirely in JavaScript that uses two types of servers, one for HTTP files and requests and another for DDP (a server customized for Meteor). Meteor includes real-time and highly efficient monitoring to avoid extra operations on the database. The database is MongoDB;

- A *Production and Innovation Knowledge Repository (PIKR)*, for providing a unified access point to heterogeneous knowledge resources (e.g., business processes, documents, technology, business domains, competitors).

The BIVEE Platform heavily relies on the collaboration of different skilled actors to successfully conduct an innovation venture (Figure 11.4). The embracement of an Open-Innovation approach further enforces this aspect, envisioning the participation of different stakeholders also belonging to the surrounding Business Ecosystem or even to the "external world".

Due to the high heterogeneity of networked organizations, the primary issue regards the need for knowledge sharing, efficient access to knowledge resources, and interoperability technologies. Not only does the platform support the social interactions happening through the BIVEE Environment, but also the discovery and categorization of Web contents that could be useful in improvement and innovation activities (Smith *et al.*, 2013), using the semantics-based infrastructure for management of digital documental resources. At the same time, BIVEE can provide a means for monitoring and measuring the success of any process improvement or innovation venture (http://bivee.eu/).

The virtual and inter-organizational knowledge creation process within the project

The knowledge spiral establishes the knowledge exchange interface between the Business Innovation Space and the Value Production Space and also describes

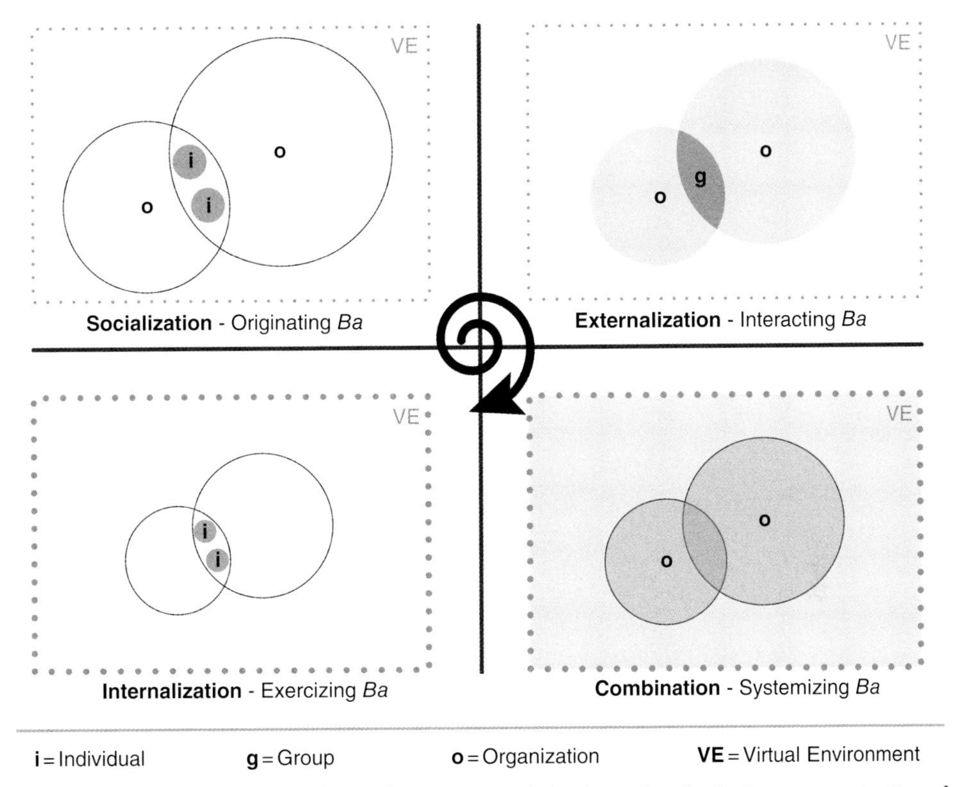

Figure 11.5 The four phases of Nonaka's SECI model, adapted to fit the inter-organizational ontological level within a virtual environment. *Source*: Bartolacci *et al.* (2016).

how knowledge is advanced iteratively (Rossi *et al.*, 2012, p. 14). It is based on Nonaka's SECI model (Figure 11.5) and the process coming from Combination, Internalization, Socialization, and Externalization is carried out in both spaces.

However, implementing the Open Innovation approach means bringing an important change to the SECI spiral model. Indeed, the BIVEE platform takes the inter-organizational dimension into account from the very beginning. So, when considering knowledge creation phases, the BIVEE platform does not consider individuals or groups belonging to the same organization, but rather, it involves individuals, groups, and organizations acting in the same virtual space.

The groups engaged in the "Innovation team" is the final result of a participation process driven by technical and cultural aspects, that in literature is linked to the Communities of Practices (CoPs) concept (Wenger, 1998). In case that we analyze, the presence of shared values and goals (in terms of sustainability, safety and social impact), leads to establish a relationship based on relevant core values as responsibility and trust. This kind of "inter-organizational social capital" enhanced knowledge sharing and transfer.

Nevertheless, while aiming to describe the main dynamics, the knowledge spiral is not meant to be a deterministic model. The BIVEE approach endeavors to find a balance between guidance and freedom. The latter is the natural nurturing ground for innovation, but in absence of the former, it could be possible to encounter endless loops that are very risky in a business context. In order to provide some guidance to the innovation space, the BIVEE project divides and organizes the Innovation Process into four main waves (Knoke, 2012). In fact, within innovation management, the long tradition of analyzing and structuring innovation processes has produced several stages and models, beginning with the innovation process conceived as a linear one, and finally evolving into a model based on feedback loops (see for example Veryzer 1998; Chiesa, Frattini, Lamberti, & Noci, 2009; Frankenberger, Weiblen, Csik, & Gassmann, 2013; Penide, Gourc, Pingaud, & Peillon, 2013; Kotha & Alexy, 2014).

The underlying concept of waves in the BIVEE project strongly characterizes it and differentiates it from linear phase models (Whitehead, 1926; Birkinshaw & Mol, 2006; Thoben, 2007). Indeed, the concept of "wave" is not rigid and can support the variability and the recursiveness of the innovation process. Consequently, this implies that the innovation process within BIVEE is represented by four distinct moments that are not linearly subsequent and consecutive, but which can also overlap (Knoke, 2012, p. 21). In other words, thanks to the wave concept the BIVEE model takes into account the fact that, when prototyping for example, it could be necessary to go back to the feasibility phase or even start the innovation process over (Garcia & Russ, 2017).

Moreover, the main steps considered in these models are generally acknowledged in innovation management literature and, despite different nomenclatures and classifications, they are a constant within any innovation project *latu sensu* (see Figure 11.6).

- *Creativity*: starts with an innovation idea or a problem to be solved, providing a first sketchy idea to be developed. It mainly refers to creative activities and brainstorming.
- *Feasibility*: justifies the actual undertaking and the further development of the original idea in economic and operational terms by collecting and providing the necessary information.
- *Prototyping*: produces a first implementation of the initial idea in the form of a prototype. The idea is drawn into the real world for the first time.
- *Engineering*: consists in testing and overhaul-procedures. The original idea transformed into a prototype is attentively analyzed to generate production and engineering plans (Taglino *et al.*, 2012).

Each of these waves can be considered to have given birth to different but interrelated SECI spiral processes that intertwine information flows with production ones, space of flows with space of places.

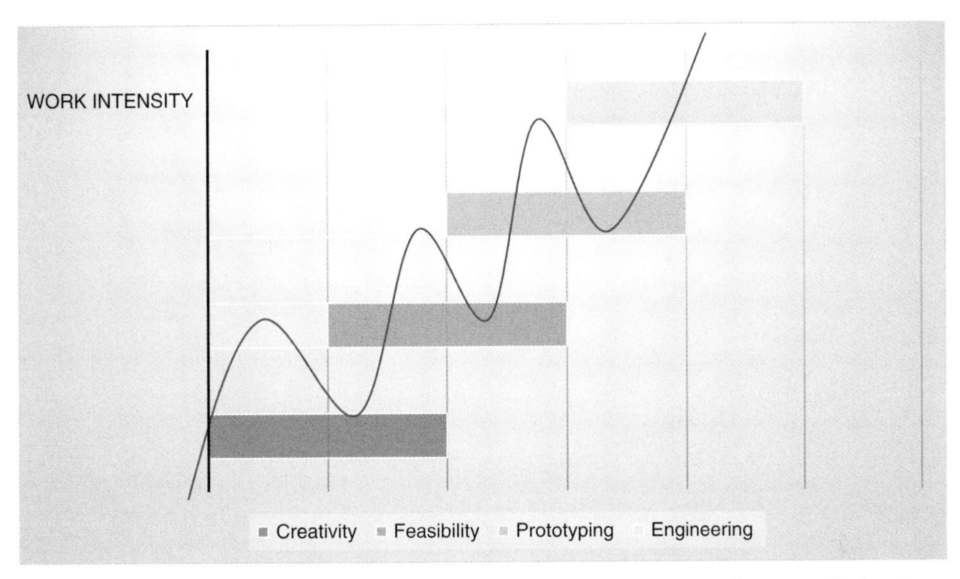

Figure 11.6 The main "waves" in the innovation process. Personal extrapolation from Bartolacci *et al.* (2016).

11.3.2 *The BLESS+ project and the SECI model applied to develop solutions for the safety and the sustainable management of a river*

To understand better how the waves and SECI model can be contextualized in a scenario of safe and sustainable management of a water system we now describe how the BIVEE virtual and inter-organizational platform was used to study and identify effective solutions for monitoring bridge pier scour risk during floods. A "neuralgic" point to obtain this real-time monitoring was the Scisciano Bridge, which connects two of the main buildings of the Loccioni Group. This bridge was considered an optimal point at which to place a sedimeter called BLESS (Bed Level Seeking System), a reliable monitoring system which would be used to control the phenomenon of pier scouring.

The dynamic process of knowledge exchange and decision making are described below, in each phase, and summarized in Table 11.2.

Creativity wave

Socialization. The project started with a specific request by LG top management during an internal meeting with the Research for Innovation team (the LG internal team aimed to develop long-term research projects and innovation). The team was asked to develop a monitoring system able to check for the seismic vulnerability of the nearby bridge.

Table 11.2 The four SECI phases applied to the Flumen innovation project. Actions and decisions made during the four main innovation waves. Personal extrapolation from Bartolacci *et al.* (2016)

	SECI PHASE			
WAVES	**Socialization**	**Externalization**	**Combination**	**Internalization**
Creativity	Intuition	Idea sharing	Identification of technical solutions	Project approval
Feasibility	Technical and economic evaluation	Selection of alternatives	Production of feasibility reports	Prototype approval
Prototyping	First prototype building	Laboratory test and system installation	Production of a prototyping assessment report	Engineering phase approval
Engineering	Evaluation of the production	Sharing of data with public organizations involved	Realization of the system final version	Implementation of the system

One of the engineers working in the Research for Innovation unit had an intuition: the vulnerability of the bridge could be caused by hydro geological risk. She logged onto the BIVEE platform to put forward the idea in the Virtual Innovation Factory and share it with other engineers on the "innovation team".

Externalization. A LG engineer recalled that the Scisciano bridge had partially collapsed during a flood of the River Esino in the 1990s, so she collected information about the causes. In the company's internal database she found the technical report in which pier scouring was cited as one of the problems leading to the collapse. She shared the collected documents and she also selected possible required competencies to test this idea; she then found matching domain specialists among BIVEE users.

Combination. The innovation team uploaded documents and added comments in the idea wizard (a kind of remote desktop where each member can modify, add, or share documents and interact). As a result, they found two different technologies for measuring the pier scour based on sonar and Fiber Bragg Grating (FBG).

Since all the requested fields were completed and several documents uploaded, the idea was considered suitable for turning into an innovation project. At this point, both technical solutions were evaluated in light of existing literature. The conclusion was that the second solution (BLESS system, a solution patented by the Politecnico di Milano) (Cigada, Ballio, & Inzoli, 2009) would be significantly more reliable in case of floods, even if further improvement was required. After having checked out the marketing report and chosen Key Performance Indicators (KPIs) to take into account, the innovation team submitted the Innovation Report to LG management.

Internalization. LG management decided to go ahead with the feasibility wave of the project and the decision was communicated by mail. At this moment, the project entered the feasibility wave.

Feasibility wave

Socialization. Researchers and business people worked in cooperation to understand the technical and economic feasibility of the project. Technical solutions proposed by the innovation team had to be improved and simplified in order to be engineered so as to obtain a high-performing and reliable system.

Externalization. The main project and the sub projects were explained, specifying again who the members of the innovation team would be and what the required competencies were.

First of all, the innovation team chose the components suppliers, looking for the best solution in term of cost and performance. Once found, the solutions were shared through the BIVEE platform and researchers at the University of Milano; the suppliers became partners of the Loccioni Group VE and members of the innovation team. At this point, different components were tested before arriving at a satisfactory solution.

Combination. The Feasibility Study Report was prepared and uploaded onto the platform. The document included the state of the art, the sub-projects, the risk analysis and the cost-benefit analysis, along with some recommendations for the realization phases. Once the last technical aspects were discussed, the innovation team was ready to build the first prototype.

Internalization. Loccioni Group top management had to evaluate again the state of the art of the project through KPIs and documents. They understood that although the risks were high, the project could help to solve an actual and crucial problem which could involve both public and private sectors, and ultimately they decided to go on to the prototyping wave.

Prototyping wave

Socialization. The Innovation Team started to build the first prototype, respecting the tight constraints on the budget while trying to maximize the reliability of the BLESS+ system. Five subprojects were identified: BLESS system, civil engineering project, electrical project, mechanical project, external heating circuit.

Externalization. The first tests conducted in the laboratory proved that the solution was really robust, flexible, and reliable. All the subproject reports were completed in order to move ahead with the installation. Then, when the system was installed on the Esino river bed, close to a bridge pier, technical problems occurred and one of the two devices installed was broken. Fortunately, the two devices were installed for redundancy.

As a result, although there was a positive deviation from planned cost, the results were considered positive overall because the risk level of flood was very high.

Combination. The BIVEE Prototyping Assessment Report was compiled; it presented considerations and feedback to take into account during the engineering wave.

Internalization. After a long and detailed measuring campaign, the innovation team established that the improvements suggested for the system made it ready for the engineering wave.

Engineering wave

Socialization. Loccioni Group top management required a careful analysis of all the components to guarantee cost-effective selection of components for a profitable production of several units of the system. The results obtained from the prototyping wave showed that the cost for the components of this first prototype was too high. However, taking into account and analysing also the strategic factors and market opportunities, the cost was considered acceptable in relation to the innovative potential of the system. Therefore, Loccioni Group management decided to go forward with the engineered version of the BLESS+ system.

Externalization. At the moment the project is still in the Engineering wave, waiting for the final version of the BLESS system to be produced for a new customer. It is worthy of note that the data produced by this system will be used to manage hydro geological risk and will be shared with public administrations for flood damage evaluation and public alert management.

The overall process has shown how the Flumen innovation project followed all the four SECI phases.

11.4 Conclusion

The management of water systems in a sustainability-oriented perspective represents a challenging problem that often requires the involvement of several different professional areas of expertise: not only hydrogeological but also political, ecological, socio-economic, technological, and financial. These skills are normally represented in different organizations. In this scenario, the virtual and inter-organizational actualization of Nonaka's SECI model proposed in this chapter can offer a useful perspective for the topic of this book: *Knowledge Management for Sustainable Water Systems.* This approach can also facilitate a realistic interaction among all of the competences needed for the effective and sustainable management of a hydrological system, such as a river.

Moreover, our proposed updating of Nonaka's model fills a gap in the Knowledge Management literature. While it is well known that the SECI model focused on individuals and their tacit and implicit knowledge, the contributions by Nonaka stop at an intra-organizational level of analysis. They do not carry the analysis to the inter-organizational epistemological level nor do they take into account the modalities of the implied interactions.

Consequently, there is ample room for interesting reflection on the evolution of the spaces dedicated to knowledge creation, namely Ba, taking into account the Open Innovation paradigm and considering organizations as cyber-physical systems. This chapter was built around the BIVEE project and its outcomes. The European project was aimed at building a platform for sustaining and managing innovation by linking different organizations in a virtual environment.

According to the PAR (Participatory Action Research) approach, the design of the structure of the platform itself was influenced *ex ante* by the willingness to concretely apply the SECI spiral model for creating new knowledge and the Open Innovation perspective. That was necessary in order to implement a practical solution that could lead to a better management of Open Innovation and to show a progression in the theoretical model for knowledge creation.

What came out of this research project is a structured and functioning platform involving a network of different organizations. The platform was designed to connect different organizations to allow them to mutually share and obtain knowledge efficiently. In fact, the BIVEE platform adopts a document centric approach which focuses on the documents exchanged during the improvement and innovation activities that take place within a virtual enterprise (Taglino *et al.*, 2012). The framework is built around two basic sets: a standard library, which allows companies to set up environments for collaboration and process alignment, and a set of standardized semantically-defined metrics, which makes it possible to quantify and benchmark processes from a strategic point of view (Rossi *et al.*, 2012).

Moreover, the BIVEE platform was designed to manage Open Innovation processes following a wave partition that reflects the main steps that compose any theoretical model for innovation processes. Thus, a virtual platform that could encompass and host all the knowledge creation phases at an inter-organizational level was born. Indeed, the learning history showed how all the phases of the spiral knowledge could take place in or be supported by a virtual dimension and result in a physical or production dimension.

A significant aspect that emerged regarded the Socialization phase on the platform: it did not strictly require physical interaction, but was possible thanks to the ICT tools provided by the platform. In this case, face-to-face interaction was effectively replaced by virtual interaction provided by rich media (McLuhan, 1964; Panahi *et al.*, 2013).

Moreover, the virtual environment made it possible to overcome the single organization barriers from the very beginning. In fact, the interaction that first took place among individuals and then among groups was not limited to one single company. On the contrary, in implementing the Open Innovation approach, individuals looked for useful knowledge from outside the company and for individuals, groups, or organizations that owned this knowledge. This resulted in a re-framing of the SECI model itself.

Therefore, if the SECI model describes an incremental and gradual process, from the epistemological point of view, that starts with individuals, then spreads to groups within the same organization, and eventually involves the whole organization (Nonaka, 1994), here things were different because they involved inter-organizational interactions right from the very beginning. Individuals and groups did not necessary belong to the same organization but were interacting thanks to the BIVEE tools in the same virtual environment.

Moreover, the active participation of the different actors was sustained by a strong commitment and an explicit purpose, from the start. Being involved in the same research project or in the development of an innovative idea was the key to building common ground where interaction could take place.

Of course, the fact that subjects often had a common technical and technological background supported the creation of the network and the building of the needed trust (Takeuchi & Nonaka, 2004). Indeed, a platform also clusters innovation ideas around different domains in order to bring similar competencies and backgrounds together and ease the evaluation process of technical solutions thanks to the presence of a domain expert. Nonetheless, heterogeneity of the network should be sought, including both private companies and universities from different European countries and integrating different fields of knowledge and expertise, all of which becomes a necessary condition for developing a European project.

Obviously, strong links among individuals or organizations can be a prerequisite for the success of Socialization, that is to say, of tacit knowledge sharing in a virtual environment. However, interactions in virtual environments can also be used to establish weak links (Alavi & Leidner, 2001) or to strengthen existing ones.

Indeed, ICT tools like the BIVEE platform are useful not just to store and retrieve documents or information, separating it from the owners, but also to bring individuals and social interaction back to the center of knowledge sharing and transfer, according to Nonaka's conception of Knowledge Management. They represent the next step in ICT implementation, moving from information management to proper Knowledge Management.

More specifically, the BIVEE platform consists of an advanced knowledge repository (PIKR) that deals with explicit and digitalized enterprise knowledge to build up a collection of digital document resources. The PIKR can provide, on the one hand, a set of reference structures (i.e., ontologies) for the semantic description of enterprise knowledge resources, and on the other hand, semantics-based services for accessing and reasoning over such descriptions. Consequently, here the term knowledge is mostly used to denote different kinds of information, document content, competencies and skills of people, capabilities of an organization, the expectation being that enterprise knowledge is largely in digital format (Taglino *et al.*, 2012, p. 10). Nevertheless, the BIVEE project was developed with the underlying awareness that there is a part of knowledge, often referred to as tacit knowledge, that remains concealed in the heads of people. For this reason, the BIVEE platform also hosts the Mission Control Room and the Virtual Innovation Factory that attempt to manage people's presence and interactions while involving a certain degree of tacitness of the knowledge that is exchanged and used to find solutions, make decisions, and assess value (Šarkiūnaitė & Krikščiūnienė, 2005).

In this context, the virtual dimension is always integrated with and shows results in the physical dimension, allowing participating members to find effective solutions for the safe and sustainable management of a river. Therefore, the spiral process of knowledge creation is always intertwining innovation and production spaces throughout the innovation waves for the real sustainable management of a water system.

In conclusion, the case studied demonstrated how the ability to create knowledge can become crucial for securing and managing a water system. More theoretically, the project studied evidences of how the knowledge creating process can be effective not only using physical spaces, but also virtual ones and how virtual spaces and ICT tools can be used to effectively enhance and support all the phases of SECI knowledge creation processes at the inter-organizational level. Of course, the limitations of having taken into account a single case study are evident, but analytical generalization is not the goal of this work. Along with its objective to improve the organizational practice of the single project, this chapter aims for theoretical generalization. Moreover, because a theoretical model for innovation management is adopted in a generally widespread way in the practical management not only of a water system but more generally of any innovation project, the model presented through the learning history of this case is easily generalizable. Indeed, it is possible to effectively make theoretical deductions, starting from the single case study.

It can be stated that considering the Open Innovation paradigm and the fact that organizations are now cyber-physical systems, Nonaka's Ba categorization needs to be updated. Indeed, it appears fundamental even, that inter-organizational interactions be included in the model. Moreover, it no longer seems correct to identify a Cyber Ba just to contextualize the combination phase. Accordingly, this work suggests exclusively adopting the nomenclature Systemizing Ba that Nonaka, Toyama and Konno (2000) used in their work to indicate the Ba that hosts the combination phase. Indeed, all the SECI phases can take place totally or partially in virtual spaces; consequently, all kinds of Ba can be both physical and virtual.

Finally, it is well known that sustainability is normally a complex issue to be faced, especially when it applies to water system management (Mounce, Brewster, Ashley, & Hurley, 2010). The resolution of most of the sustainability issues generally requires the involvement and the coordination of different actors (such as policy makers or planners), different stakeholders (that are able to represent the large variety of social, economic, environmental, and cultural aspects involved in every issue related to sustainability management) and different kinds of organizations, not only the private ones, but also the non-profit and public ones that are institutionally oriented to promote or guarantee the sustainability of a water system (Newig *et al.*, 2005). Those organizations are often located far and wide within a territory and their managers may find it difficult to meet in physical spaces. Consequently, the virtual and inter-organizational approach to Ba and to the SECI model that we propose can effectively help to overcome sustainability issues, particularly when they are as complex as the ones referred to water systems management.

References

Alavi, M., & Leidner, D. E. (2001). Review: Knowledge management and Knowledge Management systems: Conceptual foundations and research issues. *MIS Quarterly*, 107–136.

Balka, K., Raasch, C., & Herstatt, C. (2014). The effect of selective openness on value creation in user innovation communities. *Journal of Product Innovation Management*, *31*(2), 392–407.

Di Baldassarre, G., Viglione, A., Carr, G., Kuil, L., Salinas, J. L., & Blöschl, G. (2013). Socio-hydrology: conceptualising human-flood interactions. *Hydrology and Earth System Sciences*, *17*(8), 3295.

Bartolacci, C. (2014). *Knowledge Management and Corporate Storytelling*. Aracne, Roma (IT).

Bartolacci, C., Cristalli, C., Isidori, D., & Niccolini, F. (2016). Ba virtual and inter-organizational evolution: a case study from an EU research project. *Journal of Knowledge Management*, *20*(4), 793–811.

Battistella, C., & Nonino, F. (2013). Exploring the impact of motivations on the attraction of innovation roles in open innovation Web-based platforms. *Production Planning and Control*, *24*(2–3), 226–245.

Birkinshaw, M. & Mol, J. (2006). How management innovation happens. *MIT Sloan Management Review*, *47*(4), 81.

Brettel, M., Friederichsen, N., Keller, M., & Rosenberg, M. (2014). How virtualization, decentralization and network building change the manufacturing landscape: An Industry 4.0 Perspective. *International Journal of Mechanical, Industrial Science and Engineering*, *8*(1), 37–44.

Bruntland, G. H. (1987). *Our common future*. World Commission on Environment and Development (WCED), 1987.

Brydon-Miller, M., Greenwood, D., & Maguire, P. (2003). Why action research? *Action Research*, *1*(1), 9–28.

Camarinha-Matos, L. M., & Afsarmanesh, H. (2005). Collaborative networks: a new scientific discipline. *Journal of Intelligent Manufacturing*, *16*(4–5), 439–452.

Castells, M. (2005). Space of flows, space of places: Materials for a theory of urbanism in the information age. *The City Reader*, 572–582.

Chesbrough, H. W. (2006). *Open innovation: The new imperative for creating and profiting from technology*. Harvard Business Press, Brighton, MA.

Chesbrough, H., Vanhaverbeke, W., & West, J. (2006). *Open innovation: Researching a new paradigm*. Oxford University Press on Demand.

Chiesa, V., Frattini, F., Lamberti, L., & Noci, G. (2009). Exploring management control in radical innovation projects. *European Journal of Innovation Management*, *12*(4), 416–443.

Cigada, A., Ballio, F., & Inzoli, F. (2009). *Hydraulic Monitoring Unit, application for international patent n.* PCT/EP2008/059075, Publication n. WO/2009/013151.

Coughlan, P., & Coghlan, D. (2002). Action research for operations management. *International Journal of Operations and Production Management*, *22*(2), 220–240.

Davenport, T. H., & Prusak, L. (1998). *Working knowledge: How organizations manage what they know*. Harvard Business Press, Brighton, MA.

D'Ascoli, A. (2017). *Flumen. Una Filosofia del Paesaggio/ A Philosophy of Landscape*. Verucchio: Pazzini Editore.

Lausen, H., Ding, Y., Stollberg, M., Fensel, D., Lara Hernández, R., & Han, S. K. (2005). Semantic Web portals: state-of-the-art survey. *Journal of Knowledge Management*, *9*(5), 40–49.

De Guerre, D. W. (2002). Doing action research in one's own organization: An ongoing conversation over time. *Systemic Practice and Action Research*, *15*(4), 331–349.

Dong, A., & Pourmohamadi, M. (2014). Knowledge matching in the technology outsourcing context of online innovation intermediaries. *Technology Analysis and Strategic Management*, *26*(6), 655–668.

Elkington J. (1997), *Cannibals with forks: the triple bottom line of 21st century business.* New Society Publishers, Gabriola Island, BC.

Fong Boh, W., Nguyen, T. T., & Xu, Y. (2013). Knowledge transfer across dissimilar cultures. *Journal of Knowledge Management*, *17*(1), 29–46.

Frankenberger, K., Weiblen, T., Csik, M., & Gassmann, O. (2013). The 4I-framework of business model innovation: A structured view on process phases and challenges. *International Journal of Product Development*, *18*(3–4), 249–273.

Garcia, R., & Russ, M. (2017). *Ideas Generation.* European Network of Design for Resilient Entrepreneurship; Endure, Vol. 2: Knowledge. Available at http://www.endureproject.eu/download/e-book/education/knowledge/6.1.%20Ideas%20Generation.pdf

Gummesson, E. (2000). *Qualitative methods in management research.* Sage, Thousand Oaks, CA.

Hargadon, A. B. (1999). Group cognition and creativity in organizations. *Research on Managing Groups and Teams*, *2*(1), 137–155.

Hargadon, A. B. (2002). Brokering knowledge: Linking learning and innovation. *Research in Organizational Behavior*, *24*, 41–85.

Hüsig, S., & Kohn, S. (2011). "Open CAI 2.0" – Computer Aided Innovation in the era of open innovation and Web 2.0. *Computers in Industry*, *62*(4), 407–413.

Jazdi, N. (2014, May). Cyber physical systems in the context of Industry 4.0. In *Automation, Quality and Testing, Robotics, 2014 IEEE International Conference on* (pp. 1–4). IEEE.

Khatri, K. B. (2013). *Risk and Uncertainty analysis for sustainable urban water systems.* UNESCO-IHE, Institute for Water Education.

Knoke, B. (2012). D2. 1 State of the art about the collection and discussion of requirements for systematic business innovation and definition of KPIs. *Deliverable of BIVEE Business Innovation Virtual Enterprise Environments European Project, available at: http://bivee. eu/download/* (accessed 10 January 2017).

Kostiainen, J. (2002). Learning and the "Ba" in the Development Network of an Urban Region. *European Planning Studies*, *10*(5), 613–631.

Kotha, R., Kim, P. H., & Alexy, O. (2016). Turn your science into a business. *Harvard Business Review*, *92*(11), 21.

Lee, J., Bagheri, B., & Kao, H. A. (2015). A cyber-physical systems architecture for industry 4.0-based manufacturing systems. *Manufacturing Letters*, *3*, 18–23.

Li, W. (2010). Virtual knowledge sharing in a cross-cultural context. *Journal of Knowledge Management*, *14*(1), 38–50.

Lopez-Nicolas, C., & Soto-Acosta, P. (2010). Analyzing ICT adoption and use effects on knowledge creation: An empirical investigation in SMEs. *International Journal of Information Management*, *30*(6), 521–528.

Martínez-Torres, M. R. (2014). Analysis of open innovation communities from the perspective of social network analysis. *Technology Analysis and Strategic Management*, *26*(4), 435–451.

McKay, J., & Marshall, P. (2001). The dual imperatives of action research. *Information Technology & People*, *14*(1), 46–59.

Mounce, S. R., Brewster, C., Ashley, R. M., & Hurley, L. (2010). Knowledge management for more sustainable water systems. *Journal of Information Technology in Construction (ITcon)*, *15*(11), 140–148.

Newell, S., Robertson, M., Scarbrough, H., & Swan, J. (2009). *Managing knowledge work and innovation*. Palgrave Macmillan, London.

Newig, J., Pahl-Wostl, C., & Sigel, K. (2005). The role of public participation in managing uncertainty in the implementation of the Water Framework Directive. *Environmental Policy and Governance*, *15*(6), 333–343.

Nonaka, I., & Konno, N. (1998). The concept of "ba": Building a foundation for knowledge creation. *California Management Review*, *40*(3), 40–54.

Nonaka, I., & Toyama, R. (2005). The theory of the knowledge-creating firm: subjectivity, objectivity and synthesis. *Industrial and Corporate Change*, *14*(3), 419–436.

Nonaka, I. (1991). The knowledge-creating company. *Harvard Business Review*, *69*, 96–104.

Nonaka, I. (1994). A dynamic theory of organizational knowledge creation. *Organization Science*, *5*(1), 14–37.

Nonaka, I., Toyama, R., & Konno, N. (2000). SECI, Ba and leadership: a unified model of dynamic knowledge creation. *Long Range Planning*, *33*(1), 5–34.

Panahi, S., Watson, J., & Partridge, H. (2013). Towards tacit knowledge sharing over social Web tools. *Journal of Knowledge Management*, *17*(3), 379–397.

Penide, T., Gourc, D., Pingaud, H., & Peillon, P. (2013). Innovative process engineering: a generic model of the innovation process. *International Journal of Computer Integrated Manufacturing*, *26*(3), 183–200.

Polanyi, M. (1967). *The tacit dimension*. Routledge & Kegan Paul, London.

Powell, W. W. (1998). Learning from collaboration: Knowledge and networks in the biotechnology and pharmaceutical industries. *California Management Review*, *40*(3), 228–240.

Ragsdell, G. (2009). Participatory action research: a winning strategy for KM. *Journal of Knowledge Management*, *13*(6), 564–576.

Russ, M. (2016). The probable foundations of sustainabilism: Information, energy and entropy based definition of capital, Homo Sustainabiliticus and the need for a "new gold". *Ecological Economics*, *130*, 328–338.

Russ, M. (2017). Knowledge Management. *European Network of Design for Resilient Entrepreneurship; Endure,* Vol. 2: Knowledge. Available at http://www.endureproject.eu/download/e-book/education/knowledge/7.1.%20Knowledge%20management.pdf

Rossi, A., Knoke, B., Efendioglu, N., & Woitsch, R. (2012). D2. 2 Specification of business innovation reference frameworks (in the context of the VEMF). *Deliverable of BIVEE Business Innovation Virtual Enterprise Environments European Project* (accessed Jan. 10 2017).

Sáenz, J., Aramburu, N., & Blanco, C. E. (2012). Knowledge sharing and innovation in Spanish and Colombian high-tech firms. *Journal of Knowledge Management*, *16*(6), 919–933.

Sarkiunaite, I., & Kriksciuniene, D. (2005). Impacts of information technologies to tacit knowledge sharing: Empirical approach. *Informacijos Mokslai*, 69–79.

Shvaiko, P., Oltramari, A., Cuel, R., Pozza, D., & Angelini, G. (2010, May). *Generating innovation with semantically enabled TasLab portal*. In *Extended Semantic Web Conference* (pp. 348–363). Springer Berlin Heidelberg.

Siebdrat, F., Hoegl, M., & Ernst, H. (2009). How to manage virtual teams. *MIT Sloan Management Review, 50,* 63–68.

Smith, F., Taglino, F., Barbagallo, A., & Isaja, M. (2013). Semantics-based social media for the shared production control and collaborative open innovation. *BIVEE European Project, Deliverable*, *5*(4) (accessed Jan. 10, 2017).

Taglino, F., Smith, F., Proietti, M., Assogna, P., et al. (2012). D5.1 Production and Innovation Knowledge Repository. *Deliverable of BIVEE Business Innovation Virtual Enterprise Environments European Project.* Available online: http://bivee.eu/download/ (accessed Jan. 10, 2016).

Tee, M. Y., & Karney, D. (2010). Sharing and cultivating tacit knowledge in an online learning environment. *International Journal of Computer-Supported Collaborative Learning*, *5*(4), 385–413.

Thoben, K. D. (2007). *LABORANOVA: Collaborative Innovation Vision Session 1: Mechanisms for Collaborative Innovation in Living Labs*, ESoCE Net Industrial Forum, Rome, Italy.

Tickle, M., Adebanjo, D., & Michaelides, Z. (2011). Developmental approaches to B2B virtual communities. *Technovation*, *31*(7), 296–308.

Tuomi, I. (1999, January). *Data is more than knowledge: Implications of the reversed knowledge hierarchy for Knowledge Management and organizational memory.* In *Systems Sciences, 1999. HICSS-32. Proceedings of the 32nd Annual Hawaii International Conference on* (12 pp). IEEE.

Veryzer, R. W. (1998). Key factors affecting customer evaluation of discontinuous new products. *Journal of Product Innovation Management*, *15*(2), 136–150.

Wei Choo, C., & Correa Drummond de Alvarenga Neto, R. (2010). Beyond the ba: managing enabling contexts in knowledge organizations. *Journal of Knowledge Management*, *14*(4), 592–610.

Wenger, E. (1998). Communities of practice: learning as a social system. *Systems Thinker*, *9*(5), 2–3.

Whitehead, A. N. (1926). *Religion in the making: Lowell lectures 1926.* Fordham University Press.

12 Water metabolism in the socio-economic system

Delin Fang and Bin Chen

State Key Laboratory of Water Environment Simulation, School of Environment, Beijing Normal University, Beijing 100875, China

12.1 Background

Water scarcity and its socio-economic consequences have been perceived as one of the most important risks globally as reported by the World Economic Forum (Howell, 2013), especially in emerging economies, such as China and India (Mekonnen & Hoekstra, 2016). The direct reason for water scarcity seems to be overexploitation of groundwater resources (Vrba & Renaud, 2016), water pollution due to backward technology (Wan *et al.*, 2016), and irrational water consumption patterns (Laura & Stephan, 2016). Actually, water scarcity is exaggerated by socio-economic activities associated with rapid urbanization and industrialization, thus fundamentally disturbing the natural hydraulic circulation (Veldkamp *et al.*, 2015). The socio-economic network can redistribute the water resources though products transaction, which may mitigate or intensify water scarcity in water stressed regions (Feng *et al.*, 2014; Zhao *et al.*, 2015). Furthermore, economic sectors with various water consumption characteristics are connected through a collaborative production system, which constructs a virtual water network and affects water use efficiency and robustness in an indirect mode (Fang & Chen, 2015; Fang *et al.*, 2014). Therefore, how to assess and predict the performance of water circulation in a socio-economic network has become one of the most essential problems for sustainable water resource management (Flörke *et al.*, 2013). For this reason, it is necessary to adopt a metabolism-based perspective to show how water is extracted, consumed, and cycled within the socio-economic system and finally released or exported to its surrounding environment. The framework of "water metabolism" can facilitate our understanding of the complex water circulation in a socio-economic

Handbook of Knowledge Management for Sustainable Water Systems, First Edition. Edited by Meir Russ.

network, as it can track the water flows through various regions or sectors as well as describe robustness level of water circulation from the overall perspective. This chapter is organized as follows: (2) introduction to water metabolism; (3) assessment and prediction approaches for water metabolism; (4) current situation of water metabolism in China; and (5) conclusions.

12.2 Introduction to water metabolism

Metabolism is a metaphor originating from biology, illustrating how material and energy react and flow within an open system. The human dominated socio-economic system can be seen as a super organism, requiring imports of resources from external environments to support its production and operation, and disposal of waste resources back to the natural system (Chen & Chen, 2012; Lu *et al.*, 2015). Water metabolism research will illustrate the water resource flows within a modern socio-economic network as well as its interaction with natural system (Fang & Chen, 2015; Fang *et al.*, 2014). The water cycle in natural ecosystems has been explored by numerous eco-hydrological models (Piao *et al.*, 2010; Schewe *et al.*, 2014). Meanwhile, ecological research on water circulation within society as a human-dominated ecosystem is still insufficient. For example, the socio-economic water system is often viewed simply as a water consumer, or as a consumption process within the basic natural hydrologic cycle including evaporation, precipitation, runoff, and purification. However, such human dominated water circulation can be seen as the nexus between natural and artificial water cycles, which could be better understood by analyses of function and structure, socio-economic factors and driving mechanisms in the context of water metabolism.

The concept of metabolism depicts physical and chemical processes in organisms or ecosystems including resource supply from the external environment, nutrition and energy exchange within the system, and waste emission. Due to the similarity of natural ecosystems and artificial systems, the metabolic perspective has been widely adopted for the analysis of stock-flow based systems subsequently, with applications at urban, industrial, or household scales. Water is embodied in the production, transformation, consumption, disposal, and decomposition processes involving materials and energy to support the operation of socio-economic systems. Water fluxes among diverse production/consumption sectors, mainly in terms of exchanging embodied water in materials and energy. Figure 12.1 illustrates the basic conceptual framework of water metabolism within a socio-economic network. These metabolic sectors can be classified as follows: (1) Surface water, ground water and hydro projects supply water to the human-dominated system, serving as the water producer of the socio-economic network. (2) The economic sectors of agriculture, energy production and raw material extraction are regarded as the primary consumers of water, which convert water into products. (3) The secondary consumers,

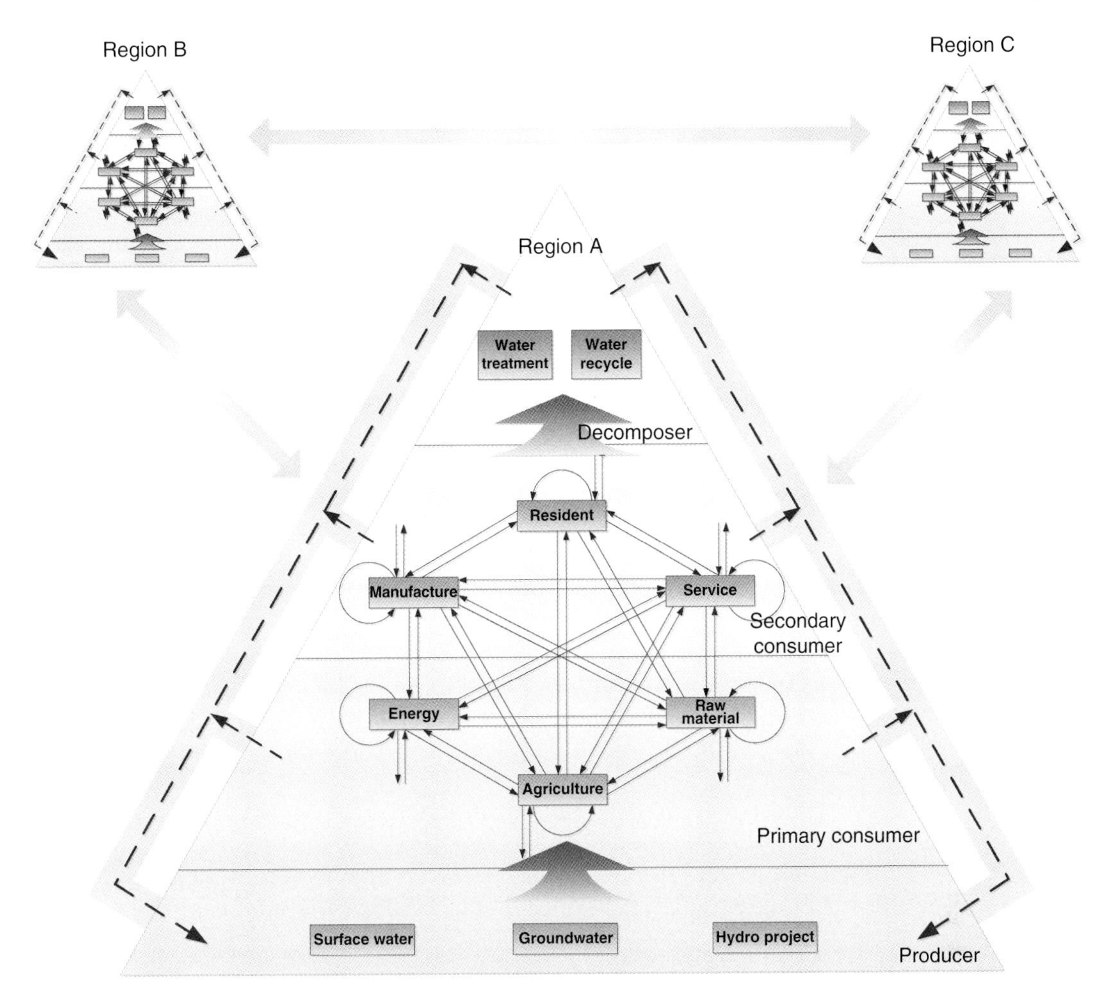

Figure 12.1 The conceptual framework of water metabolism as a socio-economic network.

including manufacture and service sectors as well as residential consumers, import water directly from producer sectors, and import embodied water through the products generated by primary sectors. (4) The sectors of water treatment and water recycling are served as decomposers of the water metabolism system, which treat the polluted water back to the natural water system or to water flow in the socio-economic system. Investigations on metabolic hierarchy shaped by the trophic roles of producer, consumer, and decomposer also greatly contribute to predict system stability in the long-run. Furthermore, on a multi-regional level, the investigation of complex interactions between different regions in terms of water resources also facilitates our cognition of water metabolic processes and properties, and promotes a systematic implementation of water conservation.

12.3 Review of methodologies for water metabolism

The concept of virtual water has been adopted to provide a useful indicator to track the water allocation and circulation in a socio-economic system (Hoekstra & Mekonnen, 2012). Allan (1993) initially introduced the concept of "virtual water" to investigate the total volume of water required during a commodity or service production process, in order to mitigate the water scarce problem in the Middle East (Allan, 1993; Allan, School of Oriental and African Studies, University of London, 1997). A virtual water network (VWN) is shaped by virtual water flows circulating in the socio-economic system, showing the metabolism characteristics of the water system (Fang & Chen, 2015; Fang *et al.*, 2014). The tracking of virtual water fluxes and pathways within the concerned system can facilitate the regulation of virtual water circulation and the adjustment of the sectors' responsibilities (Fang *et al.*, 2014). Moreover, the identification of systematic configuration has also been done to maintain the balance of efficiency and redundancy of the VWN (Fang & Chen, 2015).

There are two kinds of methods for VWN evaluation, i.e., a bottom–up approach based on the detailed information about water consumption (Hoekstra & Mekonnen, 2012; Hoekstra & Wiedmann, 2014; Zeng *et al.*, 2012), and a top–down approach based on input–output tables (Feng *et al.*, 2014; Feng *et al.*, 2012). The bottom–up approach could evaluate the virtual water consumption via water utilization throughout the whole production of goods or services, such as life cycle analysis (LCA). Besides evaluation, the bottom–up approach can detect the major driving factors of water consumption, like the log-mean Divisia index (LMDI) model (Zhao & Chen, 2014; Zhao *et al.*, 2014). The top–down approach is mainly based on an input–output (IO) method, like environment-extended input–output analysis (EIOA) and ecological network analysis (ENA). Table 12.1 shows a detailed comparison between the methods of IO, ENA and LCA for VWN evaluation.

The EIOA can illustrate the supply chain effects on water resource utilization from a comprehensive perspective and promotes evaluation of the drive force of final consumers (Feng *et al.*, 2014; Guan *et al.*, 2014). The multi-regional EIOA can further the investigation of virtual water circulation through sectoral, regional, national and global supply chains, and interpret the responsibility of each region and sector, like importer or exporter. Furthermore, incorporation with the water stress index, which is defined as the ratio of total annual freshwater withdrawals to hydrological availability, ranging from 0 (no stress) to 1 (severe stress), the EIOA can detect the pressure of scarce virtual water flow on regions facing severe water stress (Feng *et al.*, 2014; Pfister *et al.*, 2011; Zhao *et al.*, 2015). Furthermore, the linkage analysis derived from input–output analysis is an effective method to illustrate the function of each sector in the economy, like resource supplier or resource consumer. It can track the water resources through the detailed production line in a socio-economic network, and provide the indicators, such as inside linkages, forward linkage and backward linkage, to specify

Table 12.1 The comparison of methods for water metabolism in socio-economic networks

	IO	ENA	LCA
Concepts	An input–output (IO) table demonstrates a detailed flow of goods and services between producers and consumers and the intermediate linkages (inter-industry analysis) between all producing sectors in a given year. It can describe the structure of an economy, production and consumption patterns and interaction with the environment (Miller & Blair, 2009).	Ecological network analysis (ENA) is a mathematical methodology to study within system interactions for a given system structure (connectance pattern), function (flow regime), and boundary input, which is mostly concerned with interrelations of material, energy and information among system compartments (Fath & Patten, 1999).	Life cycle analysis (LCA) is a technique to assess environmental impacts associated with all the stages of a product's life from cradle to grave (i.e., from raw material extraction through materials processing, manufacture, distribution, use, repair and maintenance, and disposal or recycling) (Pfister *et al.*, 2009).
Notions and utilities	***Leontief inverse matrix*** Showing the total requirements of virtual water resources (both direct and indirect) driving by final demand (Miller and Blair, 2009). ***Multi-regional Input–Output (MRIO)*** In a MRIO framework, different regions are connected through inter-regional trade. It is an accounting framework to assess the regional virtual water flows between different regions (Feng *et al.*, 2012; Sánchez-Chóliz and Duarte, 2003). ***Linkages analysis*** Internal effect (*IE*): resources obtained solely with the processes of the sector block itself;	***Average Mutual Information (AMI)*** It is the measure of average degrees of all flows with determinate directions within the network in the context of ENA, showing the network's capacity to keep its integrity over the long term. It can be used for VWN to evaluate the system efficiency from the structural perspective (Ulanowicz *et al.*, 2009). ***Redundancy (H_c)*** It is the measure of the average degrees of all flows with indeterminate directions within the network in the context of ENA. It can be used to evaluate the redundant degree of VWN, i.e., VWN's capacity to resist the internal and external changes (Ulanowicz *et al.*, 2009).	LCA can help avoid a narrow outlook on environmental concerns by: 1) Compiling an inventory of relevant energy and material inputs and environmental releases; 2) Evaluating the potential impacts associated with identified inputs and releases; 3) Interpreting the results to help make a more informed decision. ***Life cycle inventory (LCI)*** A flow model of the technical system, which is constructed using data on inputs and outputs. The flow model is typically illustrated with a flow chart that includes the activities that are going to be assessed in the relevant supply chain and gives a clear picture of the technical system boundaries.[9]

(continued)

Table 12.1 (*continued*)

IO	ENA	LCA
Mixed effect (*ME*): resources obtained with the participation of other sectors; Net backward linkage (*NBL*): resources used in other sectors, and which is then transferred to the target sector in order to satisfy the final demand, i.e. net imports; Net forward linkage (*NFL*): resources consumed in the target sector, which is used by other sectors as intermediate inputs and which never returns, i.e. net exports (Duarte *et al.*, 2002; Sánchez-Chóliz and Duarte, 2003).	***Robustness (R)*** *R*: It is the integrated measure of both AMI and redundancy (H_c). It can be used to identify the balance between efficiency and redundancy of VWN and measure how stable the VWN system is (Ulanowicz *et al.*, 2009). ***Network control analysis (NCA)*** Quantifying the contribution of each compartment to the other compartment's input and output in the context of ENA. It can be used to evaluate the control/interdependence degree between sectors within the VWN. The control difference matrix (CD) explains the influence of one sector exerted on another within the overall system configuration, which is featured by the integral flow. If the CD value is positive, it stands for the **control intensity**; otherwise, it shows the **dependent intensity** (Fath & Patten, 1999).	***Classical life cycle impact assessment (LCIA)*** consists of the following mandatory elements: (1) selection of impact categories, category indicators, and characterization models; (2) the classification stage, where the inventory parameters are sorted and assigned to specific impact categories; and (3) impact measurement, where the categorized LCI flows are characterized, using one of many possible LCIA methodologies, into common equivalence units that are then summed to provide an overall impact category total (ISO, 2013).

Network utility analysis (NUA)

Quantifying the net gain or loss of utility between pair-wise sectors in the context of ENA. It can be used to investigate the relevant benefit/cost relation between the pair-wise compartments within the VWN, i.e., the net gain of virtual water flows contributes positive utility while the net loss provides negative utility. **Direct utility matrix** reflects the direct utility relationships, and **integral utility matrix** represents both the direct and indirect utility interactions between sectors (Schramski *et al.*, 2006).

the characteristics of each sector and detect the key sector in a particular system (Fang & Chen, 2017; Lopez *et al.*, 2014).

Based on the IO model, ENA can facilitate further understanding of water metabolism in the socio-economic network via function and structural analyses. ENA, introduced by Hannon in 1973, aims to investigate the interdependence of species and functional groups and determine the distribution of both direct and indirect ecological flows in an ecosystem, thus providing a powerful tool for investigating the internal structure of the virtual water flows (Hannon, 1973). Patten *et al.* (1976) and Finn (1976) developed a line of flow-based ecological network analysis, which has been successful in evaluating the direct, indirect and cycling flows of an ecosystem's energy and materials and the mutual relationships between compartments from a whole system perspective. Ulanowicz introduced information theory into ENA (information-ENA) and presented a uniform way to quantify a system's effective performance (efficiency) and reserve capacity (redundancy) with a robustness metric to signify the tradeoff allotment (Ulanowicz, 2009; Ulanowicz, 2011). Currently, ENA has been used to investigate the network's structure and function via water circulation and mutual relationship analyses to show the interdependence and interactions between different sectors, and uncover the indirect flows and influences hidden in the water network from a whole systematic perspective. The information-based ENA and the flow-based ENA are also combined to penetrate into the inner structure and function of VWN. The advantage of using information theory to describe the network structure is that it characterizes water flow between each sector by information metrics. For example, the probabilities of virtual water flows passing through certain pathways are calculated via ascendency to represent the inner organization, and via redundancy to depict the resistance towards external disturbances of the water system. Then, the flow-based ENA is employed to detect the network structure and function via network control analysis (NCA) and network utility analysis (NUA). The knowledge acquitted from the NCA is then adopted to evaluate the dominance of one sector over another via pairwise environs and quantify the influence of one sector exerted on another within the overall water system configuration. NUA is also used to quantify the mutual relationships between different water use sectors in the VWN, showing the mutual benefit between compartments via a matrix of mutualism. So far, ENA has been chosen as a useful tool to examine the structure and function of VWN and the interactions among its economic sectors. Meanwhile, ascendency analysis is introduced to describe the robustness of VWN, showing the balance of efficiency and redundancy of water systems. Then, flow-based ENA including NCA and NUA is performed to investigate the dominant sectors and pathways for virtual water circulation and the mutual relationships between pairwise sectors. The knowledge resulting from this combination of two lines of ENA is very useful for expanding the traditional ENA and updating water metabolism analogy to illustrate the intrinsic characteristics of socio-economic water systems.

12.4 Water metabolism in China and its nexus with other resources

The water resource problem has become a bottleneck for the rapid economic development of China, such as the uneven distribution of water resources, irrational land use (especially agricultural sectors), serious industrial water pollution, extreme drought and flooding caused by climate change and its coupling effects with other resources, like energy, food, minerals and land (Fang & Chen, 2016; Feng *et al.*, 2014; Zhao *et al.*, 2015).

In China, the eastern part near the coast has high precipitation, while the northwestern part located in the hinterland is lacking water. Meanwhile, both the southeastern and southwestern parts of China possess a large number of huge rivers and lakes with abundant rainfall and snowmelt (Feng *et al.*, 2015). Due to the regional disparity of water resource distribution (Figure 12.2), China has implemented over 20 major physical water projects, including the South–North Water Transfer Project, which is the largest in the world, transferring 44.8 Gm^3 water from the Yangtze River Basin to the Huang-Huai-Hai River Basin annually. For example, in the capital region of China, per capita water availability is only 1/8 of national average and 1/24 of world average, with 12.3% of shallow freshwater overexploitation (Beijing Water Authority, 2014; Tianjin Water Authority, 2014). Meanwhile, the Beijing-Tianjin-Hebei region is importing large amounts of real and virtual water from other regions (Zhao *et al.*, 2015); however, some of these have surplus water, whereas others suffer from even more severe water shortages (Feng *et al.*, 2014).

Although the water transfer project might relieve the water scarcity problems in northern China, the irrational land utilization, especially for food production of agricultural sectors, also leads to the declining efficiency and resilience of the water utilization system covering both socio-economic and natural hydrological perspectives, which might exacerbate the overall water scarcity in China (Fang & Chen, 2015; Guo & Shen, 2014; Zhao & Chen, 2014). The rapid industrialization and urbanization have further accelerated the water pollution rate in China, which requires a large quantity of recipient water bodies (grey water) to dilute pollution to a minimum reusable standard. Thus, cumulative water pollution is becoming a key driver to pollution induced water scarcity across China (Guan *et al.*, 2014). Furthermore, climate change has accelerated the frequency of occurrence of extreme weather, such as the severe flooding caused by continuous heavy rainfall in Hubei province in 2016, and the rare drought occurring in Yunnan province in 2010, both of which resulted in huge disaster and economic losses.

Albeit the water scarcity problems involve several different facets, none of them is isolated, as they intertwine with each other connecting an integrated network (Fang *et al.*, 2014; Zhao *et al.*, 2014). Accordingly, the combined scenario simulation of water metabolism in a socio-economic network, covering

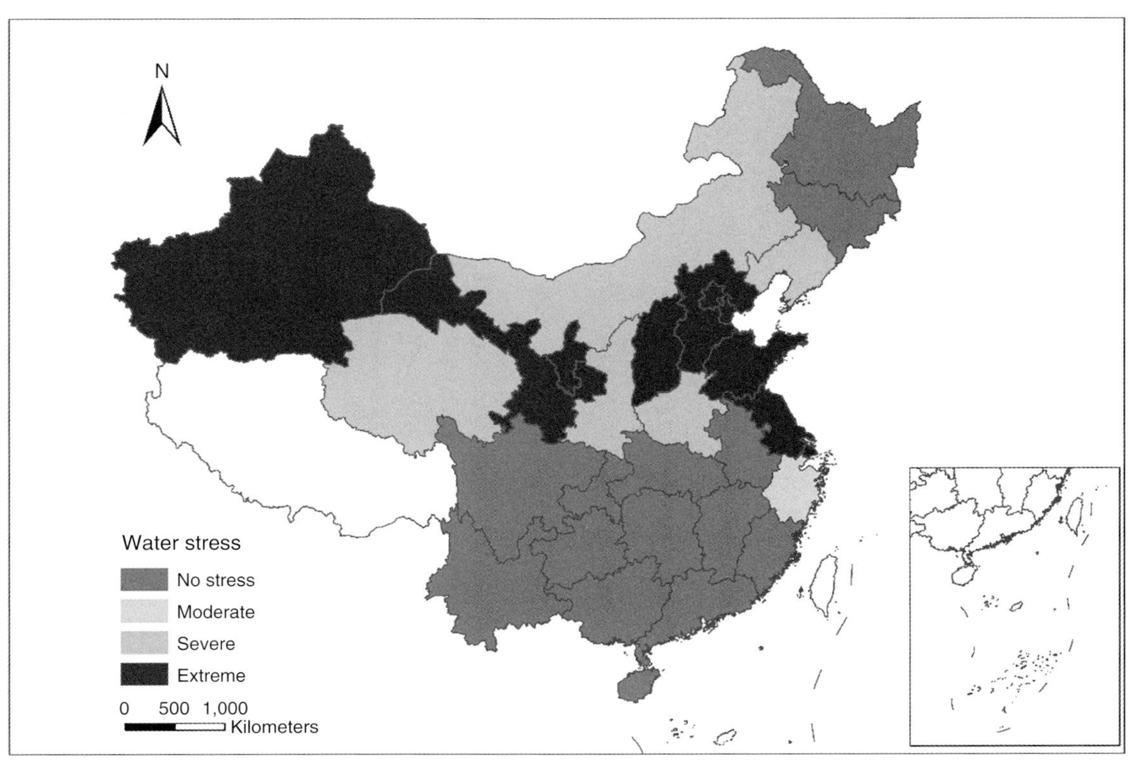

Figure 12.2 Water stress of China's provinces (2007 baseline). Note: The map is based on the water stress index (0.00–0.25: no stress, 0.26–0.50: moderate stress, 0.51–0.75: severe stress, and 0.76–1.00: extreme stress), and the index is calculated based on the research of Feng *et al.*, 2014).

a hydrological model, socio-economic analysis, land utilization evaluation and climate change analysis, is vital for rational and effective water resources management. Water resource governance faces increasing complexity, especially when the inter-dependencies and linkages with energy, food, minerals and land resources are considered. Such coupling effects between water and other resources, also termed as nexus effects, originate from the requirement of one resource as an input for the production of another one. For example, water is a vital input for agricultural, energy production and mineral extraction; modern water systems also utilize energy and minerals for production and distribution. The nexus effects on water resources are more complex when they are related with the supply and demand activities in the socio-economic system. Furthermore, environmental challenges such as climate change and ongoing economic volatilities make the nexus effect on water resources more uncertain and less predictable.

Meanwhile, China is not unique in struggling towards efficient and sustainable water utilization. Therefore, multi-regional coordination of water resource management is critical, which may help share valuable experiences and improve governance efficiency. For example, China proposes to build the Silk Road Economic Belt, which not only brings together Asia and Europe with closer trade and cultural ties, but also facilitates the water resource exchange between multi-countries.

12.5 Conclusions

The investigation of water metabolism in the socio-economic network can facilitate research on water circulation in the human-dominated system. Under the water metabolism framework based on linkage analysis and flow-based ENA, the characteristics of each sector and pathway are revealed in terms of virtual water flows, resulting in knowledge which enables us to judge the water supplier, water consumer, strong water flow pathway or weak water flow pathway. Furthermore, the information-based ENA can investigate both efficiency and robustness of VWN from the perspective of water informatics enhancing our holistic knowledge of the water system. The overall performance of water metabolism, i.e., the balance between water utilization efficiency and system structural resistance towards unexpected disturbances, can thus be shaped by information-based methodology and metrics.

Accordingly, water metabolism in a socio-economic network can be internally analyzed from flow, sector, pathway and structural levels, thus providing a powerful knowledge-based approach for coordinated water resource management, not only from specific production perspective, but also from regional, national or even global levels.

References

Allan, J.A., (1993). Fortunately there are substitutes for water otherwise our hydro-political futures would be impossible. *Priorities for Water Resources Allocation and Management 26.*

Allan, J.A., (1997). "Virtual water": *a long term solution for water short Middle Eastern economies?* School of Oriental and African Studies, University of London.

Beijing Water Authority, Beijing, China. (2014). *Beijing Water Resources Bulletin.*

Chen, S., Chen, B., (2012). Network environ perspective for urban metabolism and carbon emissions: a case study of Vienna, Austria. *Environ. Sci. Technol. 46*, 4498–4506.

Duarte, R., Sanchez-Choliz, J., Bielsa, J., (2002). Water use in the Spanish economy: an input–output approach. *Ecol. Econ. 43*, 71–85.

Fang, D., Chen, B., (2015). Ecological network analysis for a virtual water network. *Environ. Sci. Technol. 49*, 6722–6730.

Fang, D., Chen, B., (2016). Linkage analysis for the water–energy nexus of city. *Appl. Energ.*

Fang, D., Chen, B., (2017). Linkage analysis for the water–energy nexus of city. *Appl. Energ. 189*, 770–779.

Fang, D., Fath, B.D., Chen, B., Scharler, U.M., (2014). Network environ analysis for socio-economic water system. *Ecol. Indic. 47*, 80–88.

Fath, B.D., Patten, B.C., (1999). Review of the foundations of network environ analysis. *Ecosystems 2*, 167–179.

Feng, K., Hubacek, K., Pfister, S., Yu, Y., Sun, L., (2014). Virtual scarce water in China. *Environ. Sci. Technol. 48*, 7704–7713.

Feng, K., Siu, Y.L., Guan, D., Hubacek, K., (2012). Assessing regional virtual water flows and water footprints in the Yellow River Basin, China: A consumption based approach. *Applied Geography 32*, 691–701.

Feng, L., Chen, B., Hayat, T., Alsaedi, A., Ahmad, B., (2015). The driving force of water footprint under the rapid urbanization process: a structural decomposition analysis for Zhangye city in China. *J. Clean. Prod.*

Finn, J.T., (1976). Measures of ecosystem structure and function derived from analysis of flows. *J. Theor. Biol. 56*, 363–380.

Flörke, M., Kynast, E., Bärlund, I., Eisner, S., Wimmer, F., Alcamo, J., (2013). Domestic and industrial water uses of the past 60 years as a mirror of socio-economic development: A global simulation study. *Global Environ. Change 23*, 144–156.

Guan, D., Hubacek, K., Tillotson, M., Zhao, H., Liu, W., Liu, Z., Liang, S., (2014). Lifting China's water spell. *Environ. Sci. Technol. 48*, 11048–11056.

Guo, S., Shen, G.Q., (2014). Multiregional input–output model for China's farm land and water use. *Environ. Sci. Technol. 49*, 403–414.

Hannon, B., (1973). The structure of ecosystems. *J. Theor. Biol. 41*, 535–546.

Hoekstra, A.Y., Mekonnen, M.M., (2012). The water footprint of humanity. *Proc Natl Acad Sci USA 109*, 3232–3237.

Hoekstra, A.Y., Wiedmann, T.O., (2014). Humanity's unsustainable environmental footprint. *Science 344*, 1114–1117.

Howell, L.E., (2013). *Global Risks 2013.* 8th edn. World Economic Forum.

ISO, (2013). *Water footprint – Principles, requirements and guidelines; the International Organization for Standardization.* ISO/DIS 14046.

Laura, S., Stephan, P., (2016). Dealing with uncertainty in water scarcity footprints. *Environ. Res. Lett. 11*, 054008.

Lopez, L.A., Arce, G., Zafrilla, J., (2014). Financial crisis, virtual carbon in global value chains, and the importance of linkage effects. The Spain–China case. *Environ. Sci. Technol. 48*, 36–44.

Lu, Y., Chen, B., Feng, K., Hubacek, K., (2015). Ecological network analysis for carbon metabolism of eco-industrial parks: a case study of a typical eco-industrial park in Beijing. *Environ. Sci. Technol. 49*, 7254–7264.

Mekonnen, M.M., Hoekstra, A.Y., (2016). Four billion people facing severe water scarcity. *Science Advances 2.*

Miller, R.E., Blair, P.D., (2009). *Input–output analysis: foundations and extensions.* Cambridge University Press.

Patten, B.C., Bosserman, R.W., Finn, J.T., Cale, W.G., (1976). Propagation of cause in ecosystems. *Systems Analysis and Simulation in Ecology 4*, 457–579.

Pfister, S., Bayer, P., Koehler, A., Hellweg, S., (2011). Environmental impacts of water use in global crop production: hotspots and trade-offs with land use. *Environ. Sci. Technol. 45*, 5761–5768.

Pfister, S., Koehler, A., Hellweg, S., (2009). Assessing the environmental impacts of freshwater consumption in LCA. *Environ. Sci. Technol. 43*, 4098–4104.

Piao, S., Ciais, P., Huang, Y., Shen, Z., Peng, S., Li, J., Fang, J., (2010). The impacts of climate change on water resources and agriculture in China. *Nature 467*, 43–51.

Sánchez-Chóliz, J., Duarte, R., (2003). Analysing pollution by way of vertically integrated coefficients, with an application to the water sector in Aragon. *Cambridge J. Econ. 27*, 433–448.

Schewe, J., Heinke, J., Gerten, D., Haddeland, I., Arnell, N.W., Clark, D.B., Kabat, P., (2014). Multimodel assessment of water scarcity under climate change. *Proc Natl Acad Sci USA 111*, 3245–3250.

Schramski, J.R., Gattie, D.K., Patten, B.C., Borrett, S.R., Fath, B.D., Thomas, C.R., Whipple, S.J., (2006). Indirect effects and distributed control in ecosystems. *Ecol. Model. 194*, 189–201.

Tianjin Water Authority, Tianjin, China. (2014). *Tianjin Water Resources Bulletin.*

Ulanowicz, R.E., (2009). The dual nature of ecosystem dynamics. *Ecol. Model. 220*, 1886–1892.

Ulanowicz, R., (2011). *Quantitative Methods for Ecological Network Analysis and Its Application to Coastal Ecosystems.* Treatise on Estuarine and Coastal Science. Academic Press, Waltham, pp.35–57.

Ulanowicz, R.E., Goerner, S.J., Lietaer, B., Gomez, R., (2009). Quantifying sustainability: resilience, efficiency and the return of information theory. *Ecol. Complex. 6*, 27–36.

Veldkamp, T.I.E., Wada, Y., de Moel, H., Kummu, M., Eisner, S., Aerts, J.C.J.H., Ward, P.J., (2015). Changing mechanism of global water scarcity events: Impacts of socioeconomic changes and inter-annual hydro-climatic variability. *Global Environ. Change 32*, 18–29.

Vrba, J., Renaud, F.G., (2016). Overview of groundwater for emergency use and human security. *Hydrogeology Journal 24*, 273–276.

Wan, L., Cai, W., Jiang, Y., Wang, C., (2016). Impacts on quality-induced water scarcity: drivers of nitrogen-related water pollution transfer under globalization from 1995 to 2009. *Environ. Res. Lett. 11*, 074017.

Zeng, Z., Liu, J., Koeneman, P.H., Zarate, E., Hoekstra, A.Y., (2012). Assessing water footprint at river basin level: a case study for the Heihe River Basin in northwest China. *Hydrology and Earth System Sciences 16*, 2771–2781.

Zhao, C., Chen, B., (2014). Driving force analysis of the agricultural water footprint in China based on the LMDI method. *Environ. Sci. Technol. 48*, 12723–12731.

Zhao, C., Chen, B., Hayat, T., Alsaedi, A., Ahmad, B., (2014). Driving force analysis of water footprint change based on extended STIRPAT model: Evidence from the Chinese agricultural sector. *Ecol. Indic. 47*, 43–49.

Zhao, X., Liu, J., Liu, Q., Tillotson, M.R., Guan, D., Hubacek, K., (2015). Physical and virtual water transfers for regional water stress alleviation in China. *Proc Natl Acad Sci USA 112*, 1031–1035.

Index

Handbook of Knowledge Management for Sustainable Water Systems, First Edition. Edited by Meir Russ.